H. Sakio, T. Tamura (Eds.)

Ecology of Riparian Forests in Japan

Disturbance, Life History, and Regeneration

H. Sakio, T. Tamura (Eds.)

Ecology of Riparian Forests in Japan

Disturbance, Life History, and Regeneration

Springer

Hitoshi Sakio, Ph.D.
Professor, Sado Station
Field Center for Sustainable Agriculture and Forestry, Faculty of Agriculture
Niigata University
94-2 Koda, Sado, Niigata 952-2206, Japan

Toshikazu Tamura, Ph.D.
Professor, Department of Environment Systems, Faculty of Geo-environmental Science
Rissho University
1700 Magechi, Kumagaya, Saitama 360-0194, Japan

ISBN 978-4-431-76736-7 Springer Tokyo Berlin Heidelberg New York
eISBN 978-4-431-76737-4

Library of Congress Control Number: 2008924622

Springer is a part of Springer Science+Business Media
springer.com
© Springer 2008
Printed in Japan
Typesetting: Camera-ready by the editors and authors
Printing and binding: Kato Bunmeisha, Japan
Printed on acid-free paper

Foreword

The riparian forests in the Asia Monsoon Belt of Japan are subject to a variety of geomorphic and fluvial disturbances that can vary longitudinally at the catchment scale. This is due to a combination of seasonal and extreme floods caused by snowmelt and heavy rainfalls initiated by monsoons and typhoons, as well as the high gradient topography and extensive area of tectonic activity. The Japanese riparian forest (JRF) research group, represented by Dr. Hitoshi Sakio, the editor of this book, has examined various mechanisms for maintaining plant species diversity in riparian zones, and contributed to the development of community and ecosystem ecology.

Many riparian studies conducted in the United States, Canada, and European countries have focused mainly on seasonal or relatively frequent flood disturbances in large rivers, and on the responses of one group of species, the Salicaceae. In contrast, the scope of riparian studies promoted by the JRF research group ranges from headwaters to low-gradient alluvial rivers, and the group has clarified a number of specific or facultative strategies, in various tree species, for coping with the temporal and spatial reliability of safe sites. The group has focused not only on niche partitioning along environmental gradients but also on niche partitioning in life history stages, from reproduction and dispersal to seedling establishment, and the further development of immature trees.

The Japanese River Law, which was revised in 1997, included "conservation and improvement of river environments" as one of the purposes of river management, and clearly stated that riparian areas should be protected. However, reservoirs and erosion control dams built in headwater streams are thought to have had a considerable effect on the structure, species diversity, and regeneration mechanisms of riparian forests by regulating river flows and trapping sediment. Furthermore, most alluvial rivers in Japan have been channelized for land development purposes or through the construction of maintenance facilities. Natural riparian forests are disappearing in all parts of the country at an alarming rate.

With regard to preserving riparian ecosystems that still remain in world streams and rivers, and restoring riparian forests for the future, I believe that the papers included in this book offer invaluable knowledge and open up important perspectives for the future.

Futoshi NAKAMURA
Laboratory of Forest Ecosystem Management
Hokkaido University

Preface

In this book we examine the dynamics, coexistence mechanisms, and species diversity of riparian tree species, focusing on natural disturbances, life-history strategies and the ecophysiology of trees. We reveal that natural disturbances, including flooding, are important to regeneration and coexistence for riparian forests with niche partitioning. This book offers useful information relating to a number of urgent problems such as the conservation of endangered species and the restoration or rehabilitation of riparian ecosystems.

Riparian forests are highly important because of their range of ecological functions and services. They impede increases in water temperature, supply abundant litter and coarse woody debris, and absorb nutrients such as nitrogen. These effects extend to fishes and aquatic insects in well developed environments. Riparian forests are diverse with respect to species, structure, and regeneration processes, and are the most valuable part of many ecosystems. However, most riparian landscapes in Japan were degraded during the 20th century, especially after the Second World War, due to industrial development, agricultural land use and inappropriate concrete construction for river management.

In this collection, more than 20 serious researchers undertook a range of diverse studies from the fields of geography, ecology, and physiology in the context of the remaining Japanese riparian forests. In this book, we discuss riparian forests from the subpolar to warm-temperate zones, and cover headwater streams, braided rivers on alluvial fans, and low-gradient meandering rivers. Topics covered include current problems, such as the coexistence mechanisms of trees, tree demography, tree responses to water stress and the conservation of endangered species.

Some parts of the compilation of these studies were carried out in Project 2 of the Open Research Center Program of the Graduate School of Geo-environmental Science, Rissho University, with financial assistance from the Ministry of Education, Culture, Sports, Science and Technology. The production of this book was supported, in part, from grants from the Ishibashi Tanzan Memorial Foundation of Rissho University and the 8th Minamata Prize for the Environment.

We are deeply grateful to Dr. Yasuyuki Oshima for his encouragement, and for numerous constructive comments on this book. In particular, he went to the many research sites covered in this book, and spoke with authors about riparian research at a workshop on riparian forests that he had organized. Sadly, he died on January 2006. We dedicate this book to Dr. Yasuyuki Oshima. Thanks are also due to the staff of Springer for publication and to Dr. Motohiro Kawanishi, who assisted with editing.

Hitoshi SAKIO
Toshikazu TAMURA

Contents

Foreword V
Preface VII

Part 1: Introduction

1   Features of riparian forests in Japan
    H. SAKIO   3

1.1   Diverse riparian forests in Japan   3
1.2   Climate, geography, and disturbance regimes   8
1.3   Dynamics and coexistence of riparian forests   9
1.4   Research on riparian forests in Japan   10
References   10

**Part 2: Geography, disturbance regime and dynamics of
          sediments in riparian zone**

**2   Occurrence of hillslope processes affecting
       riparian vegetation in upstream watersheds of
       Japan**
    T. TAMURA   15

2.1   Introduction   15
2.2   Segmentation of a valley-side slope   16
2.3   Groundsurface instability on lower sideslope   17
2.4   Function of a valley-head area in a watershed environment   22
2.5   Significance of deep-seated landslides and large-scale landslides
       in an upstream watershed environment   24
2.6   Effects of hillslope processes on riparian vegetation   25
2.7   Concluding remarks   27
References   28

**3   Sediment dynamics and characteristics with
        respect to river disturbance**
        S. YANAI   31

3.1   Introduction: river and riparian geomorphology in Japan   31
3.2   Fluvial geomorphology: landforms and processes   32
3.3   Natural disturbances   35
3.4   Conclusions   42
References   43

Part 3: Riparian community

**4   Vegetation-geographic evaluation of the
        syntaxonomic system of valley-bottom forests
        occurring in the cool-temperate zone of the
        Japanese Archipelago**
        K. OHNO   49

4.1   Introduction   49
4.2   Methods   50
4.3   General views of valley-bottom forests in Japan   51
4.4   Syntaxonomy and vegetation-geographic evaluation of the
        valley-bottom forests   53
4.5   Relationship between the valley-bottom *Pterocarya* forests and
        other azonal vegetation   65
4.6   Vegetation-geographic evaluations of the valley-bottom forests
        in Japan   68
References   69

Part 4: Riparian forests in headwater stream

**5   Coexistence mechanisms of three riparian species
        in the upper basin with respect to their life
        histories, ecophysiology, and disturbance regimes**
        H. SAKIO, M. KUBO, K. SHIMANO and K. OHNO   75

5.1   Introduction   75
5.2   Disturbance regime of the upper basin riparian zone   76
5.3   Population structure   77
5.4   Reproductive strategies of riparian tree species   81
5.5   Responses to light and water stress   84
5.6   Coexistence mechanisms of three riparian species   87

5.7   Conclusions   88
References   88

**6   Population dynamics and key stages in two
         Japanese riparian elements**
         Y. KANEKO and T. TAKADA   91

6.1   Introduction   91
6.2   Riparian habitats and disturbance regimes   91
6.3   Study species and their ecological niches   93
6.4   Spatio-temporal variations in population growth rate under
         typhoon disturbances along a riparian environmental gradient ...   96
6.5   What are the key stages for population dynamics?   99
6.6   Conclusion   102
References   104

**7   Rodent seed hoarding and regeneration of
         Aesculus turbinata: patterns, processes and
         implications**
         K. HOSHIZAKI   107

7.1   Introduction   107
7.2   Methods   108
7.3   Seed dynamics: cache generation, retrieval and consumption   109
7.4   End-points of seed dispersal   111
7.5   Seedling regeneration   114
7.6   Ecological roles of seed dispersal by rodents   116
7.7   Annual variation   118
7.8   Implications for life history in the riparian habitats   119
References   120

**8   Longitudinal variation in disturbance regime and
         community structure of a riparian forest
         established on a small alluvial fan in warm-
         temperate southern Kyushu, Japan**
         H. ITO and S. ITO   123

8.1   Introduction   123
8.2   General description of the case study site   124
8.3   Disturbance regime and site conditions along the stream
         gradient   125
8.4   Habitat segregation and species diversity pattern   127
8.5   Occurrence of infrequent species   132
8.6   Conclusions   133

References   134

Part 5: Riparian forests on wide alluvial fan

**9   Structure and composition of riparian forests with
         reference to geomorphic conditions**
         S. KIKUCHI   139

9.1   Introduction   139
9.2   Overview of the Tokachi River system   140
9.3   Variation in site conditions on the floodplain in comparison to
       the hillslope   141
9.4   Site conditions for dominant tree species   144
9.5   Temporal and spatial variation among tree species   147
9.6   Conservation of riparian forest dynamics   150
References   151

**10   Mosaic structure of riparian forests on the
         riverbed and floodplain of a braided river: A case
         study in the Kamikouchi Valley of the Azusa
         River**
         S. ISHIKAWA   153

10.1   Introduction   153
10.2   Outline of the Kamikouchi Valley   153
10.3   Mosaic structure of riparian vegetation   156
10.4   Geomorphic process and disturbance regime of the floodplain in
       Kamikouchi   157
10.5   Young pioneer scrubs and forests   159
10.6   Seedling growth traits of salicaceous species   160
10.7   Old pioneer and late successional forests   161
10.8   Conclusion   162
References   163

**11   Coexistence of *Salix* species in a seasonally flooded
         habitat**
         K. NIIYAMA   165

11.1   Introduction   165
11.2   Seed dispersal and snowmelt floods   167
11.3   Micro topographic scale distribution of *Salix* species   169
11.4   Soil texture and seedling establishment   169
11.5   Habitat segregation along a river   171

11.6   Conclusion   173
References   173

Part 6:   Riparian forests in lowland regions

12   Process of willow community establishment and topographic change of riverbed in a warm-temperate region of Japan
M. KAMADA   177

12.1   Introduction   177
12.2   Study area and willow communities on the bar   178
12.3   Tolerance of willow seedlings against submerged and drought
       conditions   180
12.4   Actual process of willow bands formation   183
12.5   Stabilization of river-system and its influence on riparian
       ecosystem   186
12.6   Conclusion   188
References   189

13   Growth and nutrient economy of riparian *Salix gracilistyla*
A. SASAKI and T. NAKATSUBO   191

13.1   Introduction   191
13.2   Growth pattern of *Salix gracilistyla*   192
13.3   Biomass and production of *Salix gracilistyla*   194
13.4   Nutrient economy of *Salix gracilistyla*   196
13.5   Nutrient sources   199
13.6   Conclusions   200
References   201

14   The expansion of woody shrub vegetation (*Elaeagnus umbellata*) along a regulated river channel
M. KOHRI   205

14.1   Introduction   205
14.2   Study sites   208
14.3   Spatial and temporal distribution of the population   209
14.4   Seed germination and survival of the seedlings   213
14.5   Life-history strategies of *E. umbellata* in relation to the river's
       disturbance regime   216

14.6   Managing *E. umbellata* populations 218
14.7   Conclusions 219
References 220

Part 7: Riparian forests in wetland

**15   Distribution pattern and regeneration of swamp
       forest species with respect to site conditions**
       H. FUJITA and Y. FUJIMURA 225

15.1   Introduction 225
15.2   Topographical features of swamp forests 226
15.3   Site conditions of swamp forests 227
15.4   Features of swamp forest tree species in cool temperate zone in
       Japan 230
15.5   Conclusions 233
References 234

16   Flooding adaptations of wetland trees
       F. IWANAGA and F. YAMAMOTO 237

16.1   Introduction 237
16.2   Tree growth in flooded areas 238
16.3   Stem morphology and photosynthetic activity 240
16.4   Seasonal changes in morphology 242
16.5   Plant growth regulators in relation to stem morphology 243
References 245

Part 8: Species diversity of riparian forests

**17   Diversity of tree species in mountain riparian
       forest in relation to disturbance-mediated
       microtopography**
       T. MASAKI, K. OSUMI, K. HOSHIZAKI, D. HOSHINO, K.
       TAKAHASHI, K. MATSUNE and W. SUZUKI 251

17.1   Introduction 251
17.2   Methods 254
17.3   Diversity on a larger scale 256
17.4   Diversity at the local stand level 261
17.5   Conservation of diversity at different scales 264
References 265

18   Diversity of forest floor vegetation with landform type
M. KAWANISHI, H. SAKIO and K. OHNO 267

18.1   Introduction 267
18.2   Landform type 268
18.3   Species richness of riparian forest floor plants 269
18.4   Effect of landform on forest floor vegetation 271
18.5   Conclusions 276
References 276

Part 9: Endangered species and its conservation

19   Ecology and conservation of an endangered willow, *Salix hukaoana*
W. SUZUKI and S. KIKUCHI 281

19.1   Introduction 281
19.2   Taxonomy and morphology 282
19.3   Distribution 283
19.4   Community structures 287
19.5   Ecological succession and species diversity 291
19.6   Life-history strategy of *S. hukaoana* 292
19.7   Impacts of river engineering on species conservation 293
19.8   Conservation strategy of *Salix hukuoana* 295
References 296

20   Strategy for the reallocation of plantations to semi-natural forest for the conservation of endangered riparian tree species
S. ITO, Y. MITSUDA, G. P. BUCKLEY and M. TAKAGI 299

20.1   Needs and problems for the conservation of rare riparian trees ... 299
20.2   Case study 1: Estimation of potential habitat for an endangered riparian species 300
20.3   Case study 2: Reallocation of plantation to semi-natural forests based on expected tree density 303
20.4   Conclusion 308
References 309

Part 10:Conclusion

**21   General conclusions concerning riparian forest
        ecology and conservation**
        H. SAKIO   313

21.1   Riparian forest research in Japan   313
21.2   Disturbance regime, life history, and dynamics   313
21.3   Conservation of species diversity   323
21.4   Riparian forest management   324
21.5   Future research   326
References   326

Subject Index   331

Part 1

Introduction

1 Features of riparian forests in Japan

Hitoshi SAKIO

Saitama Prefecture Agriculture & Forestry Research Center, 784 Sugahiro, Kumagaya, Saitama 360-0102, Japan (*Present address*: Sado Station, Field Center for Sustainable Agriculture and Forestry, Faculty of Agriculture, Niigata University, 94-2 Koda, Sado, Niigata 952-2206, Japan)

1.1 Diverse riparian forests in Japan

The landscapes of riparian ecosystems are so beautiful that they are often the subject of photographs, picture postcards, or paintings. These prints illustrate the harmony between forests and water. Riparian forests are of great importance because of their range of ecological functions, diversity, and services.

Riparian forests are situated in the transition zones from terrestrial to aquatic ecosystems, including mountain torrents, rivers, lakes, or swamps. There are many types of riparian forests throughout Japan (Figs. 1-8); most are natural and valuable riparian forests that exhibit extensive biodiversity and ecological functions. Gregory et al. (1991) reported a high degree of structural and compositional diversity in the riparian forests of Oregon. However, most riparian landscapes in Japan were lost during the twentieth century, particularly after World War II, due to increased industrial or agricultural land use and the construction of concrete embankments for river management.

Today, the rehabilitation and restoration of riparian vegetation is a major subject of ecosystem management. Many nature restoration projects are ongoing throughout Japan; however, most of them are not effective for riparian ecosystems due to a lack of knowledge about these systems. In this book, we summarize the current understanding of typical riparian vegetation in Japan as well as disturbance regimes, forest structure, dynamics, and coexistence strategies, together with the life history and ecophysiology of trees and the diversity of vegetation.

Sakio, Tamura (eds) Ecology of Riparian Forests in Japan : Disturbance, Life History, and Regeneration
© Springer 2008

Fig. 1. The erosion zone upstream of volcanic Mount Tokachi in Hokkaido. The eruption of this volcano in 1926 caused a massive mudflow (see Chapter 3)

Fig. 2. Kushiro Mire in Hokkaido, a registered Ramsar site (Ramsar Convention on Wetlands of International Importance). The dominant species are *Alnus japonica, Fraxinus mandshurica* var. *japonica*, and *Salix* species

Fig. 3. Wetland forests of Kuromatsunai in southern Hokkaido. The dominant species are *Alnus japonica* and *Fraxinus mandshurica* var. *japonica*

Fig. 4. Vegetation mosaic of the riparian area along the Yubiso River, Gunma Prefecture, central Japan. The willow shrubland, willow stand including *Salix hukaoana*, and the riparian old-growth stand, which is composed mainly of *Aesculus turbinata* and *Fagus crenata*, developed on the active channel, lower floodplain, and higher floodplain, respectively (see Chapter 19). Photo: Wajiro Suzuki

Fig. 5. Ooyamazawa riparian forest of the cool-temperate zone in the upper basin of the Chichibu Mountains, central Japan. The dominant canopy tree species are *Fraxinus platypoda*, *Pterocarya rhoifolia*, and *Cercidiphyllum japonicum* (see Chapter 5)

Fig. 6. Kamikouchi riparian forest in Matsumoto, central Japan. *Salix* species are dominant in braided rivers and on alluvial fans. *Salix aubutifolia* (formerly *Chosenia aubutifolia*) is found only at Kamikouchi and in Hokkaido (see Chapter 10)

Fig. 7. Riparian *Salix* forests in a middle reach of the Ohtagawa River. The dominant species are *S. gracilistyla*, *S. chaenomeloides* Kimura, and *S. triandra* L. (*S. subfragilis* Anders.). River flooding and rapid stream flow create sandy to gravelly bars, especially in the middle reaches of the rivers (see Chapter 13). Photo: Akiko Sasaki

Fig. 8. Riparian forest in the warm-temperate zone on an alluvial fan in a volcanic caldera in Koike Lake in the Kirishima Mountains, southern Japan (see Chapter 8)

1.2 Climate, geography, and disturbance regimes

1.2.1 Geography and features of rivers

Japan is situated to the east of the Eurasian continent and is composed of four main islands: Hokkaido, Honshu, Shikoku, and Kyushu (Fig. 9). The Japanese archipelago contains many volcanic and alpine mountains over 2,000 m in height that range from the northeast to the southwest of the country. The mountainous areas are characterized by complex topography with steep slopes and a network of streams. Approximately two-thirds of the land area in Japan is covered by forest.

Japan's rivers are very short, their flow is generally quite fast, and river features change from the upper streams to the lower reaches. Nakamura and Swanson (2003) classified river landscapes into three morphologies: headwater streams, braided rivers on alluvial fans, and low-gradient meandering rivers. Toward the lower reaches, the river width increases and the longitudinal slope becomes more gradual. The riverbed material generally consists of large rocks in the upper stream and sand or silt in the delta area.

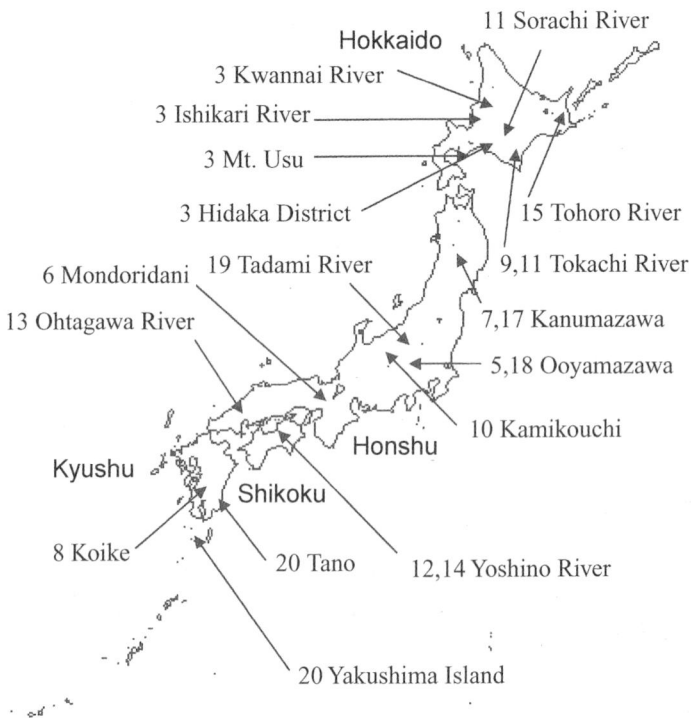

Fig. 9. Research sites of riparian forests in Japan. Numbers indicate chapter numbers

1.2.2 Climate in Japan

The Japanese climate ranges from subarctic to subtropical, and most of the archipelago experiences a temperate climate. The climate is divided into two types: the Pacific Ocean type is characterized by little precipitation in winter and heavy rainfall during the summer months, when the daily rainfall can often exceed 300 mm in western Japan. In addition, large typhoons during the summer and autumn often cause natural disturbances that result from heavy rain and strong winds. Conversely, the northern area of the Japan Sea side experiences heavy snowfall during the winter.

1.2.3 Disturbance regimes in riparian ecosystems

Disturbance regimes in riparian zones vary in type, frequency, and size (Ito & Nakamura 1994; Sakio 1997). Natural disturbances result mainly from heavy rain and strong winds caused by typhoons, as well as volcanic eruptions and snow meltwater. Damage to riparian zones results largely from flooding by typhoons and meltwater. In Japan, riparian zones are often damaged as a result of the variable climate and complex topography. Periods of flooding occur in northern Japan during the spring following heavy winter snowfalls, whereas in western Japan summer typhoons can be detrimental to riparian zones.

Disturbance regimes differ between riparian features. At the headwaters of a stream, landslides and debris flows are dominant. On alluvial fans, frequent channel shifts result in braided rivers such as the Azusa River of Kamikouchi (Fig. 6). Flooding is the dominant disturbance in low-gradient meandering rivers. In addition, disturbances in the headwaters result in continuous destruction and regeneration of the riparian habitat. Landslides and debris flows destroy the mature forest and produce new sites in which seedlings can establish. Flooding in the lower reaches affects the physiology of riparian trees.

1.3 Dynamics and coexistence of riparian forests

Many ecologists support the hypothesis that tree communities are both niche- and chance-determined (e.g., Brokaw & Busing 2000), and recent studies also suggest that both factors play a role in the development of riparian forests (Duncan 1993, Sakai et al. 1999, Sakio et al. 2002, Suzuki et al. 2002). Nakashizuka (2001) stated that it is important to investigate the entire life history of coexisting tree species within a community, and Nakashizuka and Matsumoto (2002) studied the forest ecosystem in the Ogawa Forest Reserve.

The aim of our work was to determine the causes of the dynamics and coexistence mechanisms of riparian tree species, focusing on natural disturbances, life-history strategies, and the ecophysiology of trees. Given this information, we hope to develop a course of action for the conservation of riparian forests. More than 20 authors contributed to this book, addressing many types of riparian forests from Hokkaido to Kyushu, using a variety of methods and analyses.

1.4 Research on riparian forests in Japan

To create this book, we assembled as many scientific experts on riparian forest ecology in Japan as possible, thereby covering research sites from Hokkaido to Kyushu (Fig. 9). Disturbance regimes and geomorphic processes peculiar to riparian ecosystems are covered in Part 2. Part 3 discusses a phytosociological study, and the regeneration and coexistence mechanisms of headwater streams are examined in Part 4. Part 5 addresses braided rivers on alluvial fans and low-gradient meandering rivers with respect to natural disturbance, and the life-history strategy of trees is addressed in Part 6. Part 7 analyzes the vegetation of wetlands and discusses soil conditions and ecophysiology. This book also includes information regarding species diversity (Part 8), conservation of endangered species (Part 9). Finally, Part 10 discusses regeneration recommendations and future conservation of riparian forests.

References

Brokaw N, Busing RT (2000) Niche versus chance and tree diversity in forest gaps. Trend Ecol Evol 15(5):183-188

Duncan RP (1993) Flood disturbance and the coexistence of species in a lowland podocarp forest, south Westland, New Zealand. J Ecol 81:403-416

Gregpry SV, Swanson FJ, Mckee WA, Cummins KW (1991) An ecosystem perspective of riparian zones: focus on links between land and water. BioScience 41(8):540-551

Ito S, Nakamura F (1994) Forest disturbance and regeneration in relation to earth surface movement (in Japanese with English summary). Jpn J For Environ 36(2):31-40

Nakamura F, Swanson FJ (2003) Dynamics of wood in rivers in the context of ecological disturbance. In"the ecology and management of Wood in World Rivers (American Fisheries Society Symposium)" 37:279-297

Nakashizuka T (2001) Species coexistence in temperate, mixed deciduous forests. Trend Ecol Evol 16(4):205-210

Nakashizuka T, Matsumoto Y (eds) (2002) Diversity and interaction in a temperate forest community –Ogawa forest reserve of Japan –. Springer-Verlag, Tokyo

Sakai T, Tanaka H, Shibata M, Suzuki W, Nomiya H, Kanazashi T, Iida S, Nakashizuka T (1999) Riparian disturbance and community structure of a *Quercus-Ulmus* forest in central Japan. Plant Ecol 140:99-109

Sakio H (1997): Effects of natural disturbance on the regeneration of riparian forests in a Chichibu Mountains, central Japan. Plant Ecol 132:181-195

Sakio H, Kubo M, Shimano K, Ohno K (2002) Coexistence of three canopy tree species in a riparian forest in the Chichibu Mountains, central Japan. Folia Geobot 37:45-61

Suzuki W, Osumi K, Masaki T, Takahashi K, Daimaru H, Hoshizaki K (2002) Disturbance
    regimes and community structures of a riparian and an adjacent terrace stand in the
    Kanumazawa Riparian Research Forest, northern Japan. For Ecol Manage 157:285-
    301

Part 2

Geography, disturbance regime and dynamics of sediments in riparian zone

2 Occurrence of hillslope processes affecting riparian vegetation in upstream watersheds of Japan

Toshikazu TAMURA

Department of Environment Systems, Faculty of Geo-environmental Science, Rissho University, 1700 Magechi, Kumagaya, Saitama 360-0194, Japan

2.1 Introduction

Any riparian forest stands on landforms which are formed, maintained and altered by fluvial processes. Activity of fluvial processes is therefore the matter of principal concern in the study of riparian vegetation. Fluvial processes in the broad sense (Leopold et al. 1964) include both stream action, that is fluvial processes in the narrow sense, and mass-movements, that is customarily classified as hillslope processes (Carson & Kirkby 1972). Although relatively frequent events of flood provide fluvial sediments, mass-movements which occur less frequently bring about alteration of landforms and substrata of not only valley sides but also valley bottoms in upstream watersheds.

Montane river basins in the Japanese Islands are well known as the high rate of erosion which amounts to about 7000 $m^3/km^2/y$ in an extreme case (Ohmori 1983). It is actually the result of frequent mass-movements, particularly large-scale ones associated with powerful floods (Machida 1984), which propel thorough alteration of groundsurface including soil parent material (Tamura & Yoshinaga 2007) particularly in mountains of high relief ratio. They are evaluated as destructive processes for riparian vegetation. On the other hand, small or weak mass-movements which occur rather frequently in hillslopes also affect vegetation on valley sides and parts of valley bottoms. The effects of mass-movements are not only restricted in destruction but also extended to increasing diversity of vegetation (Geertsema & Pojar 2007). In turn, unstable geomorphic condition of hillslopes, particularly of their lower segments, is generally prepared by both downward and lateral erosion of streams.

Magnitude, frequency, and spatial extent of various types of hillslope processes act thus as basic factors in the formation and maintenance of habitats in and along

Sakio, Tamura (eds) Ecology of Riparian Forests in Japan : Disturbance, Life History, and Regeneration
© Springer 2008

stream channels in hilly and mountainous areas. Taking the interaction between stream and hillslope processes into consideration, this paper intends to overview the trend of occurrence of mass-movements in respective segments of valleyside slopes in upstream watersheds of Japan. Particular attention is paid to spatial and temporal frequency and magnitude of shallow landslide which is sometimes referred to as surface failure, surface slide or regolith slide.

2.2 Segmentation of a valley-side slope

Fig. 1 illustrates an ordinary arrangement of micro-landforms in a cross-section of an upstream valley. A channelway and a bottomland, which are the direct products of erosion and partly deposition of a stream, and in some places fluvial terraces which are former bottomlands, are sided by lower sideslopes. Showing almost straight or slightly concave cross-sectional form, a lower sideslope frequently exceeds 35 degrees and sometimes attains 60 degrees in its maximum inclination. It should be remarked that such very steep hillslopes also have soil profiles under forest vegetation in many Japanese mountains and hills (Tamura & Yoshinaga

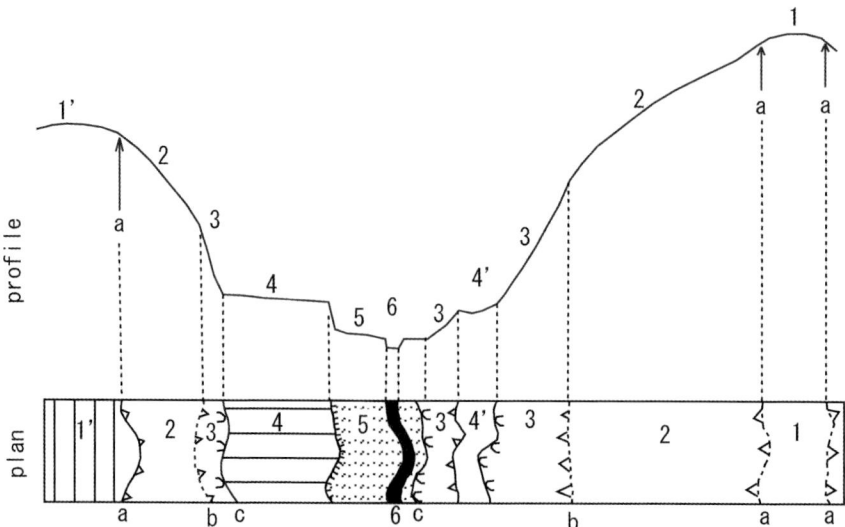

Fig. 1. A schematic cross-section of an upstream watershed
   1 Crest slope (1' crest flat), 2 upper sideslope, 3 lower sideslope, 4 (low) terrace (4' footslope), 5 bottomland, 6 channelway. a Upper convex break of slope, b lower convex break of slope, c concave break of slope.
   Contrast between the upper sideslope and the lower sideslope should be remarked. Proportion of the lower sideslope tends to increase in accordance with the increase of relief energy of the watershed

2007). Geomorphic processes dominant in lower sideslopes are shallow landslides (Tamura 1974, 1981, 1987, 2001). Although a shallow landslide zone is mostly located within the extent of a lower sideslope except its tale which frequently reaches a valley-bottom, its head invades into an upper sideslope in some cases marked by an extremely long runway.

The upper sideslope is an independent hillslope segment separated from the lower sideslope with a convex break designated as the lower convex break of slope in Fig. 1. Most frequently occurring geomorphic process in upper sideslopes is soil creep (Tamura 1974, 1981, 1987, 2001), which usually means slow and less cognizable mass-movements collectively. Continued observation of soil creep in upper sideslopes in the hills around Sendai, northeastern Japan, revealed that the rate of soil creep which is not affected with freeze and thaw amounts to a few centimeter per year at several spots (Matsubayashi & Tamura 2005). It is about ten times of the formerly reported highest rates of soil creep. Subdivided with a few breaks depending on its length, any upper sideslope changes upslope to a crestslope which, including a divide, occupies the uppermost portion of a valley-side cross-section and provides relatively stable habitats unless its cross-sectional extent is too limited.

Taking notice of vegetation disturbance which was considered as chiefly a result of geomorphic activeness, Kikuchi and Miura (1991) divided total hillslopes ranging from bottoms to crests into the lower hillslope and the upper hillslopes, which are separated from each other by the lower convex break of slope.

2.3 Groundsurface instability on lower sideslope

The lower sideslope is characterized by shallow landslide in both its configuration

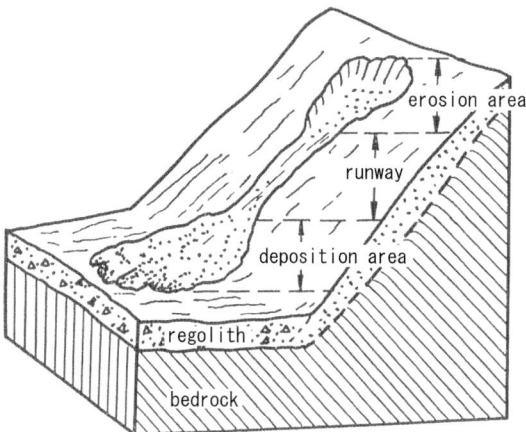

Fig. 2. A typical shallow land-slide zone.

A deposition area is in most cases separated from an erosion area by a runway where no significant erosion is made

and its environmental function. As Fig. 2 illustrates, a typical shallow landslide zone consists of an erosion area where surface material is stripped from, a runway where debris stripped from the erosion area passes through, and a deposition area where debris is accumulated over the former groundsurface. An erosion area occupies rather small portion of a slide zone and the depth of shallow slide movement is usually less than a few meters within the depth of regolith. Thus net volume of debris which is produced in an individual shallow slide rarely exceeds the order of $10^3 m^3$. The length of a slide zone varies from $10^0 m$ to $10^2 m$, depending directly on the length of a lower sideslope or total sideslopes including an upper sideslope. Apparently longer shallow slide zone tends to have a longer runway. In some cases, an erosion area located at a valley-head is followed by very long reaches where debris flows down with abundant water in a narrow valley-bottom.

The shallowness in erosion depth at an ordinary erosion area keeps some trees

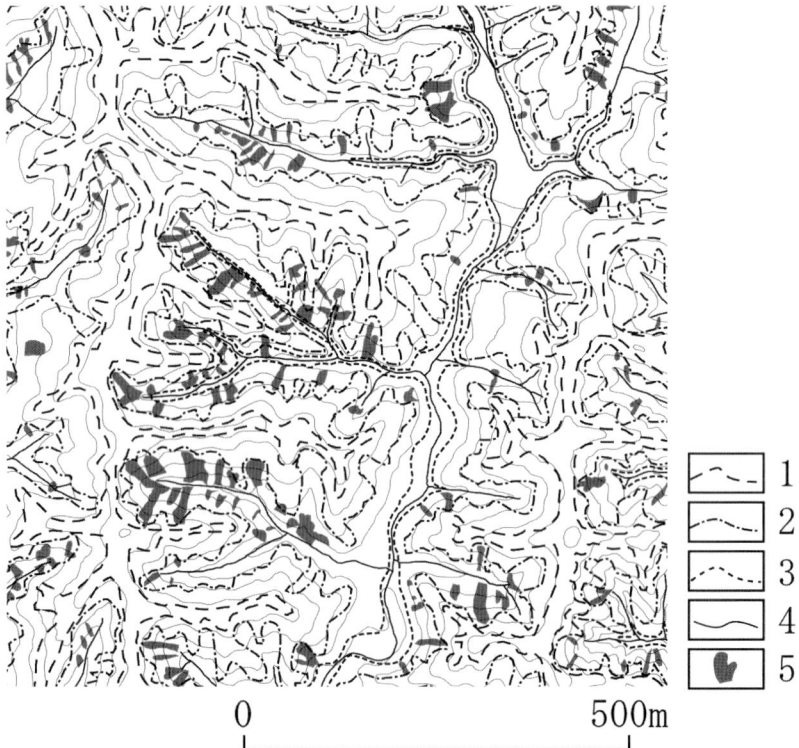

Fig. 3. Example of a shallow slide swarm: The case of rain-induced slides in the Tomiya Hills, 1986 (Tamura et al. 2002a).
1 Upper convex break of slope, 2 lower convex break of slope, 3 concave break of slope, 4 channelway, 5 shallow slide zone (excluding deposition area). Contor interval is 10m

of deep roots up there, and the narrowness in a runway width sometimes enables debris to pass through spaces among trunks. Although shallowly rooted trees, shrubs and herbs are removed with this type slide, they are in some cases alive after translocation to deposition areas unless broken or involved in debris mass, while floor plants which lived in deposition areas are damaged in proportion to the depth of accumulation and/or impacts of collision of debris. The depth of deposits ranges from 10^{-1}m to 10^{0}m and generally decreases after the deposition as seeping out of water from deposits. The length and width of an individual deposition area is, depending on the volume of debris and form of the site, ordinary in the order of a few meters to several tens of meters.

Although a shallow landslide does not produce so massive debris, many shallow slides tend to occur concentratedly in a restricted area as called as a shallow landslide swarm, in which joined debris amounts to considerable volume. In the case of heavy rain on 12-13 July 1972 in the Mikawa Highlands composed chiefly of weathered granitic rocks, near Nagoya, central Japan, about 1600 shallow slides occurred in several watersheds extending about 50km^2 (Kawada et al. 1973), and the heavy rain on 5 August 1986 induced about 2000 shallow slides in the area of 10km^2 in the Tomiya Hills composed of semiconsolidated sandstone-mudstone of Miocene age, near Sendai, northeastern Japan (Fig. 3)

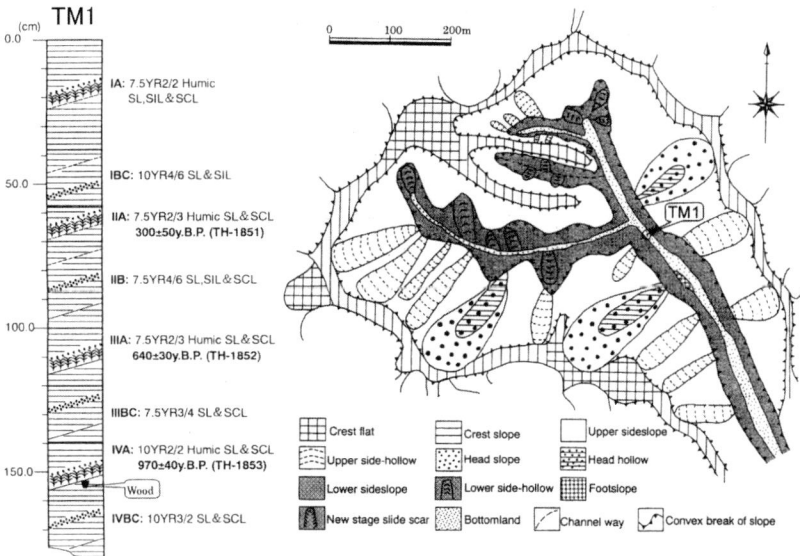

Fig. 4. Regolith profile indicating a recurrence of debris flow which followed concentrated occurrence of shallow slides in a watershed of the Tomiya Hills (Li and Tamura, 1999).
Remark the radiocarbon dates of buried A horizons, printed in thick letters in the regolith profile at TM1. They indicate the occurrences of sudden burying of ground-surface where stable conditions for humus accumulation continued.
Micro-landforms are more precisely classified than those shown in Figs. 1 and 7

(Tamura et al. 2002a). In both cases total volume of debris which flowed down a first- or second-order valley must have exceeded 10^3 m^3. The debris flows affected or destroyed the vegetation on the bottomlands and some lowest parts of the lower sideslopes.

The point at issue is the temporal frequency of shallow slide occurrence on a lower sideslope. In the bottomland of a second-order valley adjacent to the shallow slide concentration area in the Tomiya Hills referred above, Li and Tamura (1999) found several former A horizons in soil profiles composed of fine colluvio-fluvial deposits (Fig. 4). Buried A horizons and overlying colluvio-fluvial deposits in a narrow valley bottom indicate continued stable groundsurface conditions and episodic occurrences of debris flows, respectively. Debris flow in such a small valley is considered to have followed to the occurrence of shallow landslides in adjacent hillslopes. Indeed, several shallow slide scars were recognized on lower sideslopes upstream in the same watershed. Based on radiocarbon dates of former A horizons, it is concluded that three shallow slide swarms occurred at 300 to 400 year intervals on the lower sideslopes of the second-order watershed in the Tomiya Hills (Fig. 4) (Li & Tamura 1999; Tamura et al. 2002a), where normal annual rainfall is about 1200mm.

Several radiocarbon-dated humic topsoil layers intercalated in talus deposits in the northern part of the Kitakami Mountains, northeastern Japan, revealed that about 1000 years or more is the recurrence interval of shallow slides on the lower sidelsope behind the talus (Yoshinaga & Saijo 1989). While, based on dendro-chronologic evidence, about 80 years is reported as the shortest time required for the recovery of soil in the slid site to be suffered from next shallow slide on steep sideslopes of a pyroclastic plateau in Kagoshima, southern Japan (Shimokawa et al. 1989). The former area is underlain by resistant Permian rocks and receives about 1000mm of normal annual rainfall, while the latter plateau is composed of unconsolidated late-Pleistocene pyroclastic flow deposits and its normal annual rainfall exceeds 2000mm. Although it depends on both the frequency of heavy rain and the geologic/geomorphic conditions, the interval of significant shallow slide recurrence on lower sidelopes in most Japanese hilly and mountainous areas seems to be in the range between 100 and 1000 years. It is distinctly shorter than the recurrence interval of shallow slides on upper sideslopes as that will be shown in the following section.

The above figures, 100 to 1000 years, provide a basis for the consideration of disturbance of forest vegetation on lower sideslopes. In addition, it is usually observed that smaller-scale slippage of topsoil, properly a few decimeter deep and less than 1 m^3 of volume, occurs more frequently, for example almost every year, on certain parts of lower sideslopes, and destroys herbs of shorter roots there. Falling of trees and associated uprooting which also disturb soil and floor plants tend to occur frequently on lower sideslopes as well as some particular segments of upper sideslopes.

The lower sideslopes are in most cases characterized by rather sparse vegetation containing few trees (e.g., Tamura 1990; Nagamatsu & Miura 1997; Matsubayashi 2005). In some cases, scarceness of trees as a consequence of total groundsurface instability serves the growth of floor plants, while in other cases

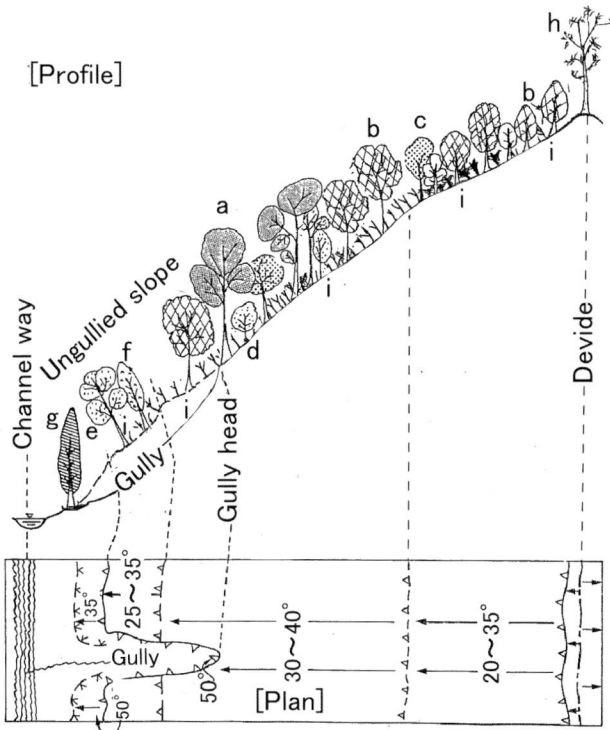

Fig. 5. Contrastive vegetation landscape on the lower and upper sideslope in the hills.
The case of Mitakesan, about 60km north of Sendai, northeastern Japan (compiled from Tamura 1990 and Sasaki 1990). Groundsurface instability in lower sideslopes is indicated by the lacking of *Rhododendron metternichii* Sieb. et Zucc. var. *pentamerum* Maxim., which thickly covers most of upper sideslopes in this cross section.
a *Quercus mongolica* Fischer var. *grosserrata* (Blume) Rehd. et Wils., b *Fagus japonica* Maxim. and *Fagus crenata* Blume, c *Hamamelis japonica* Sieb. et Zucc., d *Enkianthus campanulatus* (Miq.) Nichols., e *Prunus verecunda* (Koidz.) Koehne, f *Acer rufinerve* Sieb. et Zucc., g *Cryptomeria japonica* D. Don, h *Pinus densiflora* Sieb. et Zucc., i *Rhododendron metternichii* Sieb. et Zucc. var. *pentamerum* Maxim.

shrubs and herbs also become sparse because of frequent occurrence of soil slippage or flowage under wet condition provided by shallow soil profile on less permeable rock in lower sideslopes (Miyashita & Tamura 2008). Fig. 5 illustrates a case of contrastive vegetation on lower and upper sideslopes. Relatively unstable nature of lower sideslope surface is of course presented by immature or truncated soil profile too (Locs. 5 and 6 in Fig. 6). The poor development of soil profile is, however, not simply considered as a cause of sparse vegetation on lower sideslopes. It is rather proper to consider that both the poor soil-profile development and the sparse vegetation are the result of groundsurface instability on lower sideslopes.

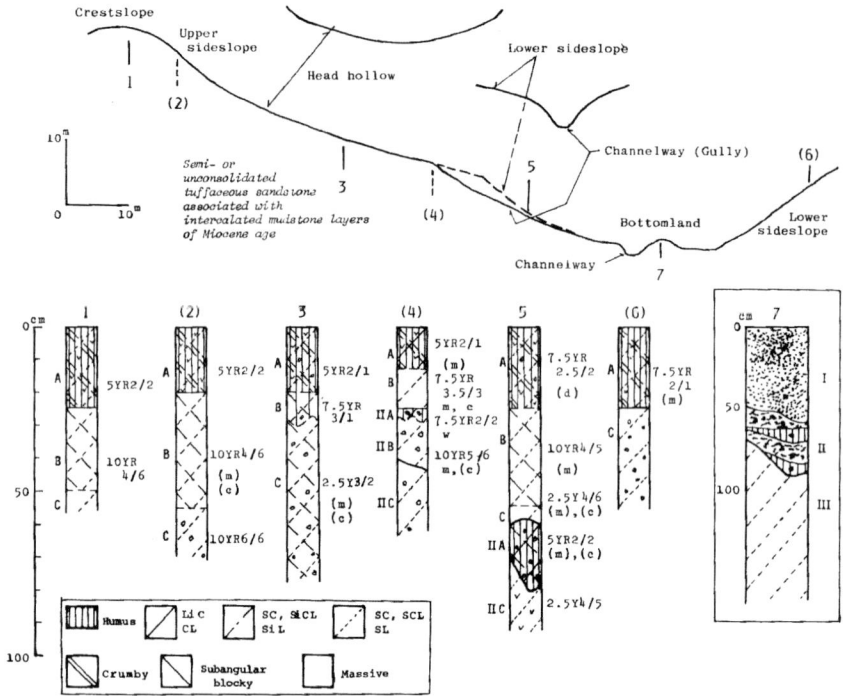

Fig. 6. A soil catena in a cross-section of a second-order valley in the Tomiya Hills, north of Sendai, Northeastern Japan (Tamura, 1987)

2.4 Function of a valley-head area in a watershed environment

A head of a channelway is generally situated at the lower end of a saucer which has no apparent channels. The saucer situated at the uppermost or headmost section of each valley is called a head hollow in the hillslope micro-landform classification by Tamura (1969, 1974) (Fig. 7). It is almost equivalent to the 0-order valley proposed by Tsukamoto (1973). Although a head hollow is in most cases surrounded by rather gentle upper sideslopes, a steep headmost wall as illustrated in Fig. 7 is in some cases discernable. In the head hollow, soil water, which is usually in unsaturated condition and has close connection with hillslope pedologic and geomorphic processes, is concentrated and converted to overland flow and/or streamflow in storm events (Tamura 1987; Tamura et al. 2002b; Furuta et al. 2007). Streamflow is of course the major agents in the landscape formation in and around channelways. Therefore the head hollow is remarked from the viewpoints of both active hydrologic-geomorphic processes and

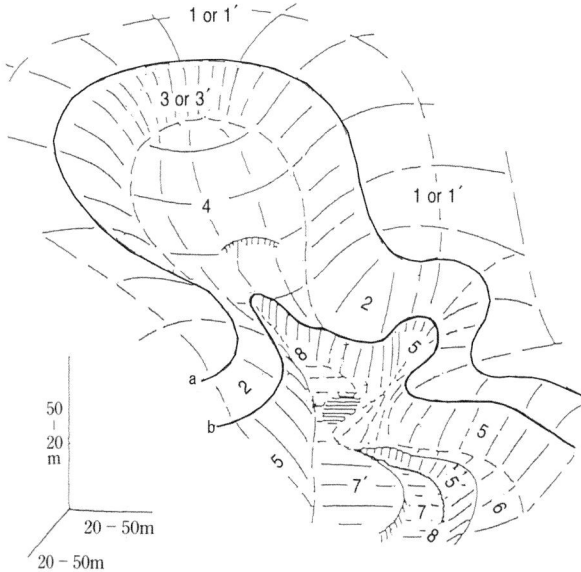

Fig. 7. Valley-head micro landforms (modified from Tamura 1987).
1 Crest slope (1' crest flat), 2 upper sideslope, 3 headmost slope (3' headmost wall), 4 head hollow, 5 lower sideslope, 6 footslope 7 bottomland (7' low terrace), 8 channel-way. a Upper convex break of slope, b lower convex break of slope

geomorphic-pedologic development history in any watershed (Tamura 1974, 1981, 1987; Tamura et al. 2007).

Head hollows are generally considered to provide rather stable groundsurface for soil and vegetation as they are included in the upper hillslope zone according to Kikuchi (2001). It is, however, frequently observed that trees tend to stand more sparsely in a head hollow than in surrounding upper sideslopes and crest slopes. More rapid alteration of some tree species in a head hollow is also reported (Kikuchi & Miura 1991). The cases referred above suggest a little bit unstable character of the head hollows. It is partly due to episodic waterlogging in head-hollow in cases of heavy rain (Tamura et al. 2002b; Furuta et al. 2007). In addition, the following groundsurface disturbances are noticed in head hollows: one is rather continuing removal of surface soil material, e.g. the flow-type soil creep observed in A horizon (Matsubayashi & Tamura 2005), and the other is intermittent colluvial covering of groundsurface, including the slide-type soil creep in B or BC horizon (Matsubayashi & Tamura 2005) or small-scale shallow landslide.

Radiocarbon dates of former A horizons buried in colluvial deposits in a head hollow in the Tomiya Hills, northeastern Japan (Fig. 8), show intermittent debris supply from surrounding upper sideslopes at an interval of about 2500 years. It is significantly longer than 300 or 400 years, which is the recurrence interval of

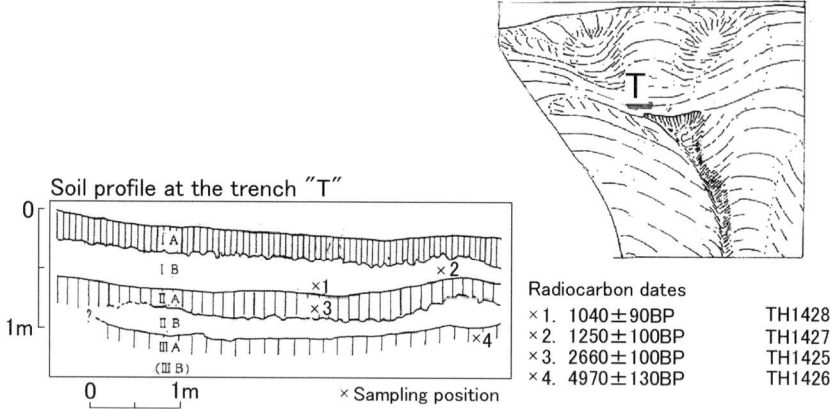

Soil profile at the trench "T"

Radiocarbon dates
×1. 1040±90BP   TH1428
×2. 1250±100BP   TH1427
×3. 2660±100BP   TH1425
×4. 4970±130BP   TH1426

0   1m   × Sampling position

Fig. 8. Landform and a regolith profile in and around a head hollow in the Tomiya Hills

shallow slide in lower sideslopes of a second-order watershed in the same hills (Li & Tamura 1999; Tamura et al. 2002a). Another evidence of radiocarbon-dated humic topsoil intercalated in colluvial and debris-flow deposits in a head hollow in the Takadate Hills, south of Sendai, northeastern Japan (Fig. 9), indicates that significant shallow slides occurred at least twice around the head hollow, perhaps at the headmost wall, in an interval of about 400 years (Li & Tamura 1999; Tamura et al. 2002a). It is considered on the basis of radiocarbon dates of organic material in debris deposited in a narrow bottomland that a little bit deep slide occurred about 100 years ago at the headmost wall of a watershed which, located in a part of the Tomiya Hills, almost lacks a head hollow because of intense headward erosion (Ito & Yoshiki 2004).

As illustrated above, groundsurface of head hollows is less unstable than that of lower sideslopes and bottomlands, while head hollows contain some parts where groundsurface is more unstable than that of ordinary upper sideslopes. It is also demonstrated that some headmost walls are almost similarly unstable to lower sideslopes. The groundsurface instability in and around head hollows inevitably affects riparian vegetation downstream.

2.5 Significance of deep-seated landslides and large-scale landslides in an upstream watershed environment

In contrast to shallow landslides in which the depth of slide movement is ordinarily less than a few meters within regolith as shown in Fig. 2, deep-seated landslides have their slip surfaces in deeper position, which frequently reaches bedrock and, in extreme cases, exceeds 100m deep, and are usually more extensive in surface area also. Therefore the volume of debris which is produced in a single deep-seated landslide can exceeds $10^4 m^3$ in many cases. Because such massive debris is hardly removable in a single slide event, considerable parts of debris remain in erosion areas of many deep-seated slide zones and repeat slide movements. Many, but not all, deep-seated landslide zones are thus characterized by repeated movements, some of which are not so rapid. Although the occurrence of deep-seated landslides is strongly controlled with particular geologic conditions, the areas of such geologic conditions may cover about a quarter of the mountainous and hilly areas of Japan.

Slowly-moving deep-seated slides allow the existence of vegetation on the groundsurface of sliding mass unless it is split, and continuing slow movement affects the growth of trees there (e.g., Higashi 1979). On the other hand, some deep-seated slide zones provide partially unstable groundsurface where not deep-seated but shallow slides occur more frequently. For instance, in a deep-seated slide zone of about 80ha extent in the Chichibu Mountains, central Japan, where the last remarkable action of deep-seated rockslide associated with bouldery debris avalanche is assumed to have occurred about 220 years ago (Sakio 1997; Sakio et al. 2002), many shallow slides seem to have recurred in the main scarp and the sliding masses of the deep-seated slide and consequently small debris flows have been repeatedly induced in downstream valleys. Many well-known deep-seated landslide zones in the Uonuma Hills near Nagaoka, northeastern Japan, responded to severe shock of the Mid-Niigata Prefecture earthquake in 2004 as the occurrence of enormous number of shallow slides, as well as the reactivation of deep-seated slide in other deep-seated slide zones in the same hills (Yagi et al. 2005). Debris produced in those slides flowed downslope and downstream, and partly dammed up valleys. Groundsurfaces disturbed in the quake-induced landslides have continued the unstable condition and have responded to the following heavy snow and rain and snowmelt as the supply of debris which affect riparian vegetation too (Sakurai & Sasaki 2006).

An extraordinary large-scale rockslide, which produces debris of, for instance, the orders of $10^6 m^3$ to $10^9 m^3$, thoroughly alters landscape of not only the slid site but also downstream valleys and plains. Its environmental impact starts with sliding and filling-up of valleys with debris followed by excavation of filled bottoms which provide secondary debris flows and floods. The effects sometimes extend over the watershed and continues several decades or more than a few hundred years in the watersheds and adjacent coasts (Machida 1984; Machida & Miyagi 2001).

2.6 Effects of hillslope processes on riparian vegetation

As Fig.2 illustrates, debris produced in an erosion area of a shallow landslide is transported through a runway to a deposition area of the slide zone. In the case in which the slide zone is situated in a lower sideslope, its deposition area extends to a bottomland and easily contacts with streamflow in a channelway. In the case of slide occurrence in a headmost wall, debris is partly deposited in the head hollow and frequently reaches a channelway, as illustrated in Fig. 9.

Since debris produced in a shallow slide is usually rather fine in its texture and not so big in its total volume, it is directly transported through channelways. In the case when many slides occur in a watershed, it joins to produce a fine-debris flow. Debris flow or debris avalanche sometimes dams up a stream, mostly in combination with driftwoods and coarse rubble. In the case of relatively big individual shallow slide, for example that with debris of the order of $10^3 m^3$, and in the case of deep-seated slide, damming-up occurs directly at the foot of a slide zone. It is followed with fluvial sedimentation upstream and results in raising-up and/or widening of a bottomland.

The effect of damming-up and succeeding geomorphic change to vegetation is not restricted in altered bottomlands and/or former channelways, where a new succession starts after burying, but extended to lower parts of lower sideslopes, where vegetation becomes susceptible to flood according to the extent of aggra-

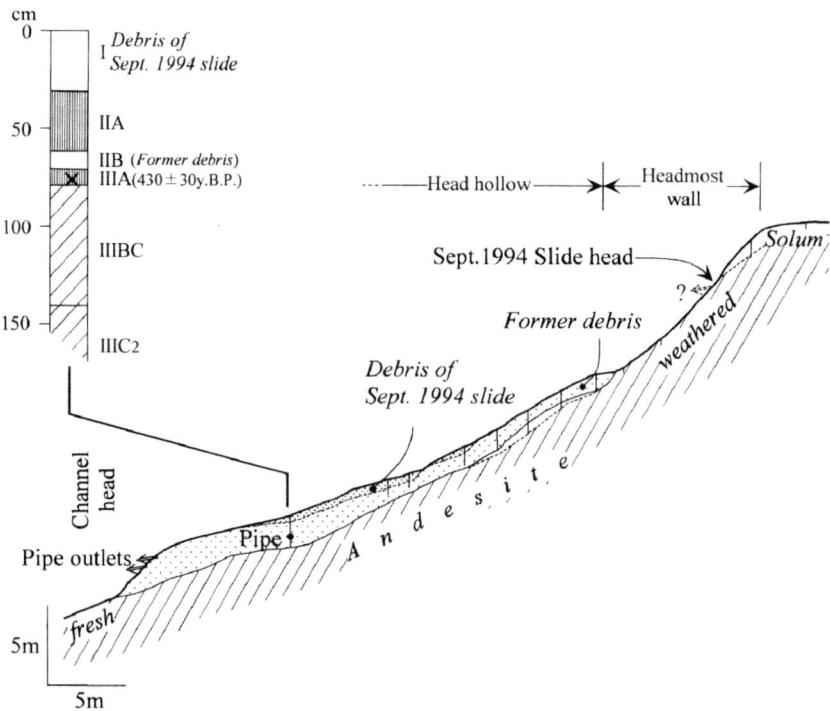

Fig. 9. Landform and a regolith a profile around in a head hollow in the Takadate Hills (modified from Tamura et al. 2002b)

dation as a result of the damming-up. A continued observation of successional vegetation change behind a check-dam which was constructed in an upstream valley-floor presented basic information of the effect of aggradation to vegetation which occurred on not only a bottomland and a channelway but also lower side slopes (e.g., Murakami & Hosoda 2007).

Different from artificial check-dams, natural dams which were formed by debris flows, relatively large shallow slides or deep-seated slides are immediately subjected to erosion of streams which have their temporal heads at the lower sides of the dams. The buried bottomland then undergoes severe downward erosion, which induces the groundsurface instability on adjacent lower sideslopes there. Although the groundsurface instability on lower sideslopes is directly triggered by heavy rain in most cases and severe earthquake shock in some cases, it is potentially prepared by the steepness which is apparently the result of downward erosion in channelways by stream action. The rate of downcutting is a function of channel gradient and streamwater discharge. The rapid progress of downcutting promotes the upstream extension of channelways, which consequently invade into head hollows and induce the groundsurface instability in and around the head hollows.

From the viewpoint of long-term geomorphic development, most Japanese upstream channels are generally considered to have been in the phase of active downcutting since around the beginning of Postglacial times due to the increase in both rainfall amount and frequency of heavy rain (e.g. Kaizuka 1980). The trend must appear in the invasion of channel heads and lower sideslopes to head hollows and upper sideslopes, respectively. The latter process was proved by detailed tephrochronological investigations of hillsolpe development, as reviewed in Chapter 3 of this book. More detailed observations suggest that channel heads at the lower ends of head hollows have shown extension, that is downcutting, and contraction, that is burying, alternately at a time interval of probably several hundreds to thousand years (Furuichi 1995; Tamura et al. 2007).

Actually, the occurrence of downcutting is different by stream segments or reaches as a consequence of local differences in lithology, water collectivity, geomorphic history, and watershed management including various river works. The changing trend of erosion in upstream channels affects directly, and indirectly by the way of changing hillslope processes, to riparian vegetation.

2.7 Concluding remarks

The most popular geomorphic processes which occur in lower sideslopes and headmost walls are shallow landslides which are inferred to recur at almost the same site at an interval of about 100 to 1000 years. Because the interval is considered to depend on both the frequency of heavy rain and the geologic/geomorphic characteristics, it is variable in relation to climate change and tends to have become shorter in the Holocene in Japan. More locally, active incision of streams in some upper reaches induces more frequent occurrence of shallow slides in neighboring lower sideslopes and head hollows. Although an ordinary single shallow slide produces debris of $10^0 m^3$ to 10^3 m^3, many slides tend to occur concentratedly in relatively restricted area and total volume of debris transported to a first- or second-order valley can exceed $10^4 m^3$.

The simple effect of a single shallow slide on riparian vegetation is burying of former groundsurface in the deposition area of each slide zone, which is frequently formed in a bottomland. In many cases, concentratedly occurred plural slides produce debris flow which affects a channelway and a bottomland more extensively. The information as above demonstrates that lower sideslopes as well as headmost walls should be taken notice in the investigation of disturbance of riparian vegetation.

Occupation of plant communities on varied locations in upstream watersheds has been generally studied in strong connection with some disturbances (e.g., Kikuchi 2001; Sakio & Yamamoto 2002; Geertsema & Pojar 2007). In those studies, the occurrence of disturbance has been discussed in its relations to reproduction and growing strategy as well as life history of each plant, and structure of each plant community. Some studies have paid attention to substrata's texture and/or microtopographic characteristics of each stand in order to interpret the type and magnitude of disturbance. In addition, sources of the disturbances, that is erosion areas in the case of shallow landslides, provide important information on the formation of the riparian habitat. Scars of erosion areas in former shallow slides of around thousand years old are usually traceable in detailed geomorphological observation in hills and mountains of Japan. In the case of deep-seated slides and large-scale slides, their scars are more easily traceable after much longer years. The effect of deep-seated slides or rarely occurred large-scale slides and associated big debris avalanches on riparian vegetation appears in many cases to prepare unstable groundsurfaces for the more frequent occurrence of shallow slides.

Although examples referred in this paper tend to be restricted in shallow landslides, the viewpoint presented here will contribute to expand the scope of riparian vegetation study to surrounding hillslopes which have close relation in its formation and maintenance with bottomlands and channelways as habitats of riparian forests.

References

Carson MA, Kirkby MJ (1972) Hillslope form and process. Cambridge Univ Press, 475p

Furuichi T (1995) Swale and soft deposits suggesting scouring and infilling of hollows in the Tomiya Hills, Sendai. Trans Jpn Geomorph Union 16:300 (abstract in Japanese)

Furuta T, Goto K, Tamura T (2007) Infiltration-runoff processes through soil horizons in a forest-covered valley-head. Ann Tohoku Geogr Ass 59:123-139 (in Japanese with English abstract)

Geertsema M, Pojar JJ (2007) Influence of landslides on biophysical diversity: A perspective from British Columbia. Geomorphology 89:55-69

Higashi S (1979) Recognition of changing groundsurface indicated by characteristic vegetation (*Chihyo hendo-ron*). Hokkaido Univ Press, 280p (in Japanese)

Ito A, Yoshiki T (2004) History of the regolith slides in a small valley of the Tomiya Hills, northeastern Japan. Trans Jpn Geomorph Union 25:359-369 (in Japanese with English abstract)

Kaizuka S (1980) Late Cenozoic palaeogeography of Japan. GeoJournal 4:101-109

Kawada G, Kataoka J, Amimoto K (1973) Disasters occurred in the mountainous areas in Aichi and Gifu prefectures induced by heavy rain of July 1972. In: Yano K (ed) Study on heavy-rain induced disasters in July 1972, Report of grant-in-aid for scientific research of natural disasters, pp 108-112 (in Japanese)

Kikuchi T (2001) Vegetation and landforms (*Chikei shokusei-shi*). Univ Tokyo Press, 220p (in Japanese)

Kikuchi T, Miura O (1991) Vegetation patterns in relation to micro-scale landforms with special reference to the lower sideslope. Ecol Rev 22:61-70

Leopold LB, Wolman MG, Miller JP (1964) Fluvial processes in geomorphology. Freeman, 522p

Li Y, Tamura T (1999) Regolith slide recurrence history indicated by radiocarbon ages in the hills around Sendai. Sci Rep Tohoku Univ Ser 7 (Geogr) 49:55-67

Machida H (1984) Large-scale rockslides, avalanches and related phenomena: A short review. Trans Jpn Geomorph Union 6:155-178 (in Japanese with English abstract)

Machida H, Miyagi T (2001) Mass-movements. In: Yonekura N, Kaizuka S, Nogami M, Chinzei K (eds) Regional geomorphology of the Japanese Islands (*Nihon no chikei*), Vol. 1: Introduction to Japanese geomorphology, Univ Tokyo Press, pp 169-177 (in Japanese)

Matsubayashi T (2005) Comparison of vegetation structures along hillslopes at several hills around Sendai. Ann Tohoku Geogr Ass 57:34-35 (Abstract in Japanese)

Matsubayashi T, Tamura T (2005) Mode and rate of soil creep by soil horizons: Device of a monitoring method and its application to forest-covered hillslopes in humid temperate climate. J Geogr 114:751-766 (in Japanese with English abstract)

Miyashita K, Tamura T (2008) Difference in management and structure of secondary forests on hillslopes in the Iwadono Hills, Saitama Prefecture. Ann Tohoku Geogr Ass 60 (in press) (Abstract in Japanese)

Murakami W, Hosoda I (2007) Changes in the topography and vegetation on the upstream side after check dam construction. Ann Tohoku Geogr Ass 59:99-110 (in Japanese with English abstract)

Nagamatsu D, Miura O (1997) Soil disturbance regime in relation to micro-scale landforms and its effects on vegetation structure in a hilly area in Japan. Plant Ecol 133:191-200

Ohmori H (1983) Erosion rates and their relation to vegetation from the viewpoint of world-wide distribution. Bull Dep Geogr Univ Tokyo 65:77-91

Sakio H (1997) Effects of natural distribution on the regression of riparian forests in the Chichibu Mopuntains, central Japan. Plant Ecol 132:181-195

Sakio H, Yamamoto F (2002) Ecology of riparian forests. Univ Tokyo Press (in Japanese)

Sakio H, Kubo M, Shimano K, Ohno K (2002) Coexistence of three canopy tree species in a riparian forest in the Chichibu Mountains, central Japan. Folia Geobot 37:45-61

Sakurai M, Sasaki Y (2006) Vegetation and slope disaster. In: Slope Engineering Sub-committee, Japan Society of Civil Engineers (ed) Studies on monitoring of complex slope hazards induced by the Mid-Niigata Prefecture earthquake, pp 83-90 (in Japanese)

Sasaki H (1990) Plants and vegetation of the Mitakesan Prefectural Nature Conservation Area. Research report of the Mitakesan Prefectural Nature Conservation Area, Miyagi Prefectural Government, pp 15-60 (in Japanese)

Shimokawa E, Jitousono T, Takano S (1989) Periodicity of shallow landslide on Shirasu

(Ito Pyroclastic Flow Deposits) steep slopes and prediction of potential landslide sites. Trans Jpn Geomorph Union 10:267-284 (in Japanese with English abstract)

Tamura T (1969) A series of micro-landform units composing valley-heads in the hills near Sendai. Sci Rep Tohoku Univ Ser 7 (Geogr) 19:111-127

Tamura T (1974) Micro-landform units composing a valley-head area and their geomorphic significance. Ann Tohoku Geogr Ass 26:189-199 (in Japanese with English abstract)

Tamura T (1981) Multiscale landform classification system in the hills of Japan, Part II: Application of the multiscale landform classification system to pure geomorphological studies of the hills of Japan. Sci Rep Tohoku Univ Ser 7 (Geogr) 31:85-154

Tamura T (1987) Landform-soil features in the humid temperate hills. Pedologist 31:135-146 (in Japanese)

Tamura T (1990) Geomorphology and geology of the Mitakesan Prefectural Nature Conservation Area. Research Report of the Mitakesan Prefectural Nature Conservation Area, Miyagi Prefectural Government pp 4-14 (in Japanese)

Tamura T (1990) Lower sideslopes in steeply sloping profiles: Its cognition and function. Ann Tohoku Geogr Ass 42:192 (Abstract in Japanse)

Tamura T (2001) Hill landforms. In: Yonekura N, Kaizuka S, Nogami M, Chinzei K (eds) Regional geomorphology of the Japanese Islands (*Nihon no chikei*) Vol. 1: Introduction to Japanese geomorphology, Univ Tokyo Press, pp 210-222 (in Japansese)

Tamura T, Yoshinaga S (2007) Geomorphic factors. In: Japanese Soc of Pedology (ed) Japanese soils (*Dojo o aishi dojo o mamoru -Nihon no dojo-*), Hakuyusha, pp 21-24 (in Japanese)

Tamura T, Li Y, Chatterjee D, Yoshiki T, Matsubayashi T (2002a) Differential occurrence of rapid and slow mass-movements on segmented hillslopes and its implication in late Quaternary paleohydrology in Northeastern Japan. Catena 48:89-105

Tamura T, Kato H, Matsubayashi T, Furuta T, Chattergee D, Li Y (2002b) Stepwise change in pipe-flow and occurrence of surface-slide with increase in rainfall: Observations in the hills around Sendai, northeastern Japan. Trans Jpn Geomorph Union 23:675-694 (in Japanese with English abstract)

Tamura T, Furuta T, Furuichi T, Li Y, Miyashita K (2007) Head hollows with a nest-of-box structure: Ttheir formation and hydrologeomorphic functions. Abstracts, Jpn Geoscience Union Mtg 2007 (abstract in Japanese and English, http://www.jpgu.org/publication/cd-rom/2007cd-rom/program/pdf/Z164/Z164-002.pdf http://www.pgu.org/publication/cd-rom/2007cd-rom/program/pdf/Z164/Z164-002_e.p df)

Tsukamoto Y (1973) Study on the grouth of stream channel (I): Relationship between stream channel growth and landslides occurring during heavy rain. Shin-Sabo 87:4-13

Yagi H, Yamasaki T, Iwamori T, Atsumi K (2005) Ladslides triggered by the Mid-Niigata Prefecture earthquake in 2004 and their characteristics based on a detail mapping. J Jpn Soc Eng Geol 46:145-152 (in Japanese with English abstract)

Yoshinaga S, Saijo K (1989) Slope development during Holocene reconstructed from alluvial cone and talus cone aggradation process. Trans Jpn Geomorph Union 10:285-301 (in Japanese with English abstract)

3 Sediment dynamics and characteristics with respect to river disturbance

Seiji YANAI

Hokkaido Institute of Technology, 4-1, 7-15 Maeda, Teine, Sapporo, Hokkaido 006-8585, Japan

3.1 Introduction: river and riparian geomorphology in Japan

The archipelago of Japan stretches between Taiwan and the Kamchatka Peninsula along the margin of the Pacific Ocean. Japan is comprised of four main islands: Hokkaido, Honshu, Shikoku, and Kyushu. Two-thirds of the narrow land area is mountainous, with small rivers divided nearly equally into small watersheds. Torrents and rapids transport heavy sediment from mountainous areas, creating alluvial fans and floodplains, before finally flowing into the ocean. More than 260 rivers are at least 20 km long and have watersheds of 150 km^2 and relief greater than 100 m. The largest watersheds of approximately 10,000 km^2 are the Tone, Shinano, and Kitakami of Honshu, and Ishikari of Hokkaido (Takahashi 1990). The slopes of continental rivers are generally gentler than those of rivers in Japan, which tend to have steep slopes and cataracts (Sakaguchi et al. 1995).

Japan has a warm, moist climate, with an annual average precipitation of approximately 2000 mm. Much of this precipitation falls as heavy rains brought by the Baiu front and as typhoons and heavy snowfall from the northwest. As a result, rivers have ample water, and the flow fluctuates by season and year. Maximum flow occurs during spring snowmelt and the Baiu and typhoon seasons of spring through autumn. Local variation is also great. The ratio of maximum to minimum flow in Japanese rivers is much larger than that for rivers in other parts of the world, and flow is highly unstable. The flood discharge per unit of drainage area is one scale larger than that for continental rivers (Sakaguchi et al. 1995). Discharge is also comparatively abundant relative to the small watershed areas. Even in the largest river, heavy rainfall at the headwater reaches the river mouth within 2 days

Sakio, Tamura (eds) Ecology of Riparian Forests in Japan : Disturbance, Life History, and Regeneration
© Springer 2008

and can cause flooding. Usually, the peaks of rainfall and discharge coincide. The coefficient indicating seasonal differences in river discharge is also particularly large in Japan. This value is enhanced by artificial water use such as irrigation and hydroelectric power generation.

Rivers in Japan are also characterized by high erosion rates, sediment transport power, and amounts of sediment (Takahashi 1990). Large amounts of sand, gravel, and silt produced in the upper mountain ranges are transported by flooding during heavy rainfall. The erosion rate (transported sediment per watershed per time period, $m^3/10^6 \cdot year = mm/1000$ years) tends to be larger in the Asian monsoon region than in other regions of the world (Yoshikawa 1974). Tectonic movements have also created a complex geological structure in mountainous regions of Japan (Fujita 1983). The uplift rate is extraordinarily high in central Honshu compared to the worldwide average. This mass movement results in a high potential for sediment transport.

People in Japan have exploited the rich sediments deposited near river mouths for nearly six centuries. Agriculture, including rice cultivation, has been concentrated in these areas, as have industrial and urban centers. Except in Hokkaido, which was settled relatively late and more sparsely, most Japanese rivers have been artificially altered for several hundred years. In the late 19th century, the Meiji Restoration brought increased urbanization and industrialization, as well as an increase in agriculture and rapid decline in virgin forests. Urbanization and industrialization following World War II resulted in further dramatic changes in riverine environments. Water quality and biodiversity have declined and habitat ranges have narrowed. Large-scale civil engineering projects have created continuous levee systems and straightened channels completely in the middle and lower reaches of many rivers. However, since the 1990s, following trends in European countries, more natural river engineering has been promoted and implemented nationwide, particularly since the enactment of the Law for the Promotion of Nature Restoration in 2002. However, large portions of rivers remain artificially channeled. Restoring these rivers and ensuring biological diversity are the main technological and social challenges in Japan today.

3.2 Fluvial geomorphology: landforms and processes

3.2.1 Classifications of fluvial landforms

Geomorphologists and civil engineers have classified various fluvial landforms associated with the upper to lower reaches of river systems. Fluvial geomorphology can be divided by basic processes characteristic of the mountainous upstream areas, alluvial fans, alluvial plains, and deltas (Suzuki 1997). Mountain topography can be subdivided into valleys and fluvial terraces. Yamamoto (2004) also classified rivers into four segments depending on slope break points: the alluvial fan, natural levee, delta, and mountain segment (Fig. 1).

In upstream areas, deep valleys form steep "V" shapes. Riverbeds consist of gravels, cobbles, boulders, and bedrock. Step and pool sequences develop in normal flow conditions. During flooding, the flow becomes rapid, increasing downward degradation. In upstream areas, slopes are generally steeper than a 1/50 gradient. Deep valleys form along fluvial terraces with downward and lateral erosion affecting the bed materials of gravel and bedrock. At alluvial fans, channels widen and currents slow, promoting gravel deposition. Slope angles range from 1/60 to 1/400, and streams often develop braided patterns and point bars.

In the mid-channel regions, water flow sometimes disappears before flowing out again at the lower fan edge. Rivers with natural levees develop a meandering pattern and have sandy bed materials and gentle slope gradients ranging from 1/400 to 1/5000. The channel becomes narrower and the cross section becomes deeper. Swales called back swamps develop between the natural levees and hill slopes; these areas collect fine silt and clay deposited during flooding. Back swamps are generally moist and poorly drained; in Honshu, these wetland areas have been used for rice paddies and have historically suffered from flooding damage. Artificial banks were built on the natural levees to protect communities and agricultural lands against flooding. Near river deltas, slopes become gentler (>1/5000), and clayey materials dominate the deposits. However, every river does not have all of these geomorphological sequences. Some rivers such as the Kurobe River and Oi River in central Honshu lack natural levees and/or deltas and flow directly from alluvial fans into the ocean.

3.2.2 Factors controlling the amount and size of sediment along channels

Riverbed material, slope angles, and the mean maximum annual discharge control the channel characteristics (Yamamoto 2004). Bedrock weathering processes and the amount of sediment production affect the riverbed material. Bedrock is weathered by both physical and chemical processes, and bedrock properties such as strength and permeability determine the resulting landforms (Suzuki 2002). Sequential weathering generally degrades bedrock into boulder, gravel, sand, and finally, clay; however, non-sequential processes such as bedrock to gravel or bedrock to sand frequently occur in granitic areas (Koide 1973). In volcanic rock regions, fine clay particles are not easily produced; large alluvial fans and debris deposits develop at the foot of volcanoes and are composed of large boulders and gravel. In comparison, unconsolidated sedimentary rock tends to be directly decomposed from bedrock to clay. Alluvial fans are poorly developed in these areas. Uplift also controls the grain-size distribution (Ikeda 2000). Even under the same geological conditions, mountainous areas with large uplift ratios produce larger amounts of gravel as a result of landslides than do stable mountainous areas with small uplift ratios. Thus, riverbed quantity and quality are closely related to the underlying geological and tectonic conditions that affect the fluvial geomorphology of watersheds. Riverbeds flowing through mountainous areas or through steep alluvial fans have large (>20 mm) evenly distributed particles. Rivers flowing in weathered granitic areas have variable grain sizes (1–5 mm). Alluvial plains and/or the lower reaches of rivers have fine grain sizes (<3 mm; Yamamoto 2004).

3.2.3 Geomorphological processes that create riparian landscapes

Climate shifts between the Pleistocene and Holocene eras created marked increases in erosion, affecting river landscapes from the headwaters to the mouths. These changes still affect the modern landscape (Oguchi 1996; Oguchi et al. 2001). River discharge and sediment transport forces fluctuate with climatic variation. Aggradations and degradation determine the longitudinal and cross-sectional river profile. During the period of maximum glaciation, the annual mean temperature dropped by 5–10°C, causing a sea regression of approximately 120 m. The Tsushima Strait between Kyushu and the Korean Peninsula nearly closed, restricting the flow of the warm marine current into the Japan Sea; this in turn led to decreased snowfall and snow cover, accelerating periglacial processes. The snow line dropped to 1200 m in Honshu, and periglacial conditions covered one-

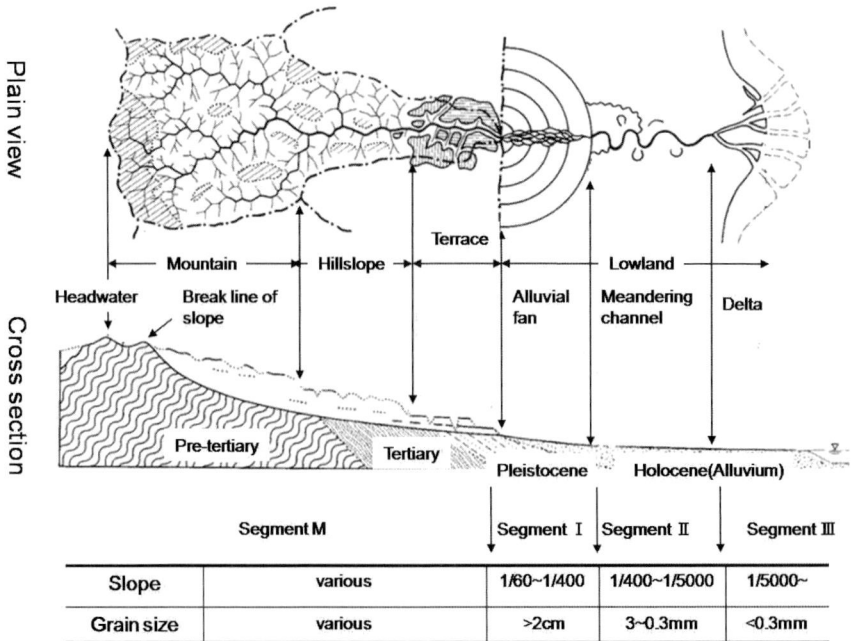

		Segment M	Segment I	Segment II	Segment III
Slope		various	1/60~1/400	1/400~1/5000	1/5000~
Grain size		various	>2cm	3~0.3mm	<0.3mm

Fig. 1. Fluvial geomorphology from mountain to the sea, modified from Suzuki (1997) and Yamamoto (2004)

third of Japan (Yonekura et al. 2001). Periglacial freeze and thaw processes produced massive amounts of debris in the mountains, and buried and braided river channels developed. In contrast, degradation and downward incision were suppressed downstream by the lower sea levels. In the Holocene, the mean temperature rose, and debris production decreased as vegetation recovered. Fluvial processes became dominant. Degradation occurred in the middle and upper reaches, whereas vast floodplains developed in the lower reaches.

Hill slopes can be categorized as smooth or incised. Smooth slopes were formed mainly by freeze–thaw action during the last glacial age, whereas incised slopes were formed by landslides and gully development during the Holocene (Hatano 1979). Slopes in central Hokkaido have been dated using tephro-chronology, which has revealed overlain tephras of several ages from the Pleistocene and Holocene. Valley-side slopes are evenly covered with tephra layers 9000 years old, and incised slopes are covered with only younger tephra deposited approximately 350 years ago (Yanai 1989; Yoshinaga 1990). Representa-tive cross-sectional valley profiles suggest that headwater frames were formed during the last glacial stage and lower valley slopes, separated by break lines, developed by the dissection of older slopes under regimes of increasing fluvial processes. Similar processes have also been observed in northern and central Japan (Yoshinaga & Koiwa 1996; Saijo 1998).

Alluvial fans can also be classified into two types: dissected and undissected. Most dissected fans have Pleistocene surfaces, whereas most undissected fans are composed of Holocene surfaces (Saito 1988). Alluvial fan formation began during the last interglacial stage. At the Tokachi Plain in Hokkaido, where most of the fan topography is still preserved, several periods of expansion have been observed (Ono & Hirakawa 1975). The Toyohira River, a main tributary of the Ishikari River, which flows through the city of Sapporo in Hokkaido, has a large Pleistocene surface and a Holocene surface developing by the dissection of the older surface (Daimaru 1989). The relaxation time of watershed responses to Pleistocene–Holocene climatic change may often exceed 10,000 years; in Japan, most fluvial geomorphological processes are still in the early stage of hill slope incision and limited fan aggradations (Oguchi 1996).

3.3 Natural disturbances

3.3.1 Rapid and slow landslides in headwater areas

Headwater areas (in first- and second-order streams) are sources of water, sedi-ments, and organic matter, as well as important transitional zones from forest to stream ecosystems. Lower steep slopes have shallow soil layers because of periodic soil removal by landslides at intervals of several decades to hundreds of years. Forests beneath slope breaks have a mosaic-like age structure (Sakai 1995). This complex structure is the result of frequent slope failures that disrupt logs and boulders, which may then be transported to the foot of the slope and washed

Fig. 2. Slope failure occurred in headwater area of coastal terraces at Hidaka district, Hokkaido in 1981. The shallow soil layers slipped down with wood from slope to stream

downstream (Fig. 2). This type of mass movement is triggered most frequently by torrential rainfall events or typhoons, and sometimes by snowmelt and earthquakes. Generally, rainfall intensity >50 mm/h is required to induce landslides. The fundamental slope unit for a landslide is called a zero-order valley (Tsukamoto 1973). Landslides are an important process on forested slopes. Landslide frequency is expressed as a function of time; weathering rates decrease with time as the removal of weathered material by rainfall reaches a critical depth. A similar process occurs in bedrock weathering and the production of the soil layer. This pattern ultimately leads to slope regression and an incised valley. The "immune" term refers to the depth of surface layer weathering at which a slope is resistant to failure (Iida 1993).

The slope-failure recurrence interval is determined by bedrock weathering and hardness and the production rate of surface soil layers ranging from 10^2 to 10^3 years. Especially in areas of granite, volcanic pyroclasitic flow, younger and loose tertiary materials, and Quaternary bedrock, slope failure tends to occur at shorter intervals. Shimokawa and Jitozono (1995) estimated the slope-failure occurrence interval using tree ring analyses. In a granitic area of Kyushu in southern Japan, this method revealed various slope ages and a slope-failure recurrence interval of 200–250 years. Shimokawa et al. (1989) applied a similar method in a pyroclastic flow area and found a recurrence interval of 127 years.

Another technique using younger tephra layers was applied to measure landslides in the Hidaka coastal terrace region of southern Hokkaido, where extensive slope failures occurred in 1981. Slope ages were estimated for tephra deposited over 300 years; two-thirds of the slope failures occurred on slopes without tephra. Recurrence intervals were estimated in detail by excavating a valley floor and counting the interbedded tephra layers; recurrence intervals were estimated as 100–150 years (Yanai & Igarashi 1990). A similar method was used to estimate landslide frequency since the early Holocene in the Hidaka range; in this area, a longer recurrence interval (650 years) was observed for the last 8000 years (Shimizu et al. 1995).

Slumps and earthflows also occur in headwater areas. This type of mass movement results in distinct geomorphological features such as steep, horseshoe-shaped cliffs or plateau-shaped gentle debris slopes where ponds and swamps are often formed. Debris outwash forces irregular stream curvature and supplies massive debris, including large boulders that provide favorable stands for the establishment of riparian vegetation in steep valleys. If landslides did not occur, valley slopes would be steep and narrow and considerably limit the area available for riparian forests. Landslides occur throughout Japan, but are especially prevalent in the Kyushu, Shikoku, Hokuriku, Tohoku, and Hokkaido regions, which have Tertiary sedimentary rock, tectonic faulting, and hot springs (Koide 1973; Fujiwara et al. 2004). These types of landslides strongly affect human life. Settlements have long been concentrated in areas of gentle slopes and plains with abundant groundwater and deep soil layers for crops. Landslides create various micro-topographical features that lead to diverse ecosystems (Inagaki et al. 2004). Past landslide movements have created fluvial landform diversity such as wide, braided channels and debris deposits; in these features, debris movement is relatively slow compared to slope failure and lasts longer (i.e., several hundred years). Old landslides occurred during the last glacial period and/or the transition from the last glacial period to the post-glacial period (Fujita 1990). Landslides may reoccur when stream erosion removes the debris edge. Masses of debris build up dams in the stream and sometimes become debris flow when a dam breaks down.

3.3.2 Debris flow and sediment transport processes

Debris flow is usually triggered by torrential rainfall. This flow of material, which can include large materials such as logs and boulders to small materials such as mud particles, moves downstream with the floodwaters. A representative debris flow landform has three components: erosion, transformation, and deposition zones (Koide 1973; Kohashi 1983). The erosion zone occurs in the upstream slopes, where deep gullies and valleys are created. Debris flows create depositional landforms with rounded cross-sectional profiles and spindle-shaped and lobe-like structures at the top of the flow. Large boulders >1 m in diameter are found at the top and margins of these deposits and create a levee structure along the channel. This complex depositional landform creates a favorable generation site for riparian tree species such as *Pterocarya rhoifolia* Sieb. et Zucc. (Sato 1992). The transportation force declines markedly as the depositional area becomes wide, whereas in a mountainous area, the transportation power is maintained by the narrow valley slopes (Shimazu 1994). The deposition of sediments along a channel causes the riverbed to become higher and eventually alters the river channel toward a lower position. Channels change course periodically, and sediments are scattered evenly from the mouth of the valley and plain, ultimately creating an alluvial cone and/or fan (Umitsu 1994). These processes create alluvial fans with similar circular shape and numerous remnants of braided channels on the surface. The spatial scale ranges from hundreds of meters to more than several tens of kilometers. The debris flow frequency varies by region and landscape. Some areas of Honshu have had no debris flows for several centuries, whereas other areas experience frequent occurrences every several years following volcanic eruptions, landslides, or earthquakes.

Volcanoes are characteristic landforms in Japan and are distributed along the Pacific coast, San-in, and central Kyushu Island. Japan has more than 80 active volcanoes, which directly and indirectly affect vegetation. Volcanic eruptions cover hillslopes with fine ash, which decreases their permeability. Rainfall that cannot percolate into the soil layer erodes both the newly fallen ash and former deposits and can ultimately develop into a debris flow. Frequent debris flows diffuse onto the piedmont. Volcanic Sakurajima Mountain in southern Kyushu has active eruptions and frequent debris flows. In the 19-year period between 1976 and 1994, 5–21 debris flows occurred per year in eight streams along the volcano (Osumi River and National Road Office, Ministry of Land Infrastructure and Transportation, unpublished data). Mount Usu in southern Hokkaido is another

Fig. 3. Debris flow caused by volcanic eruption at the piedmont of Mt. Usu, southwestern Hokkaido. Debris flow moved so rapidly that it could pluck boulders size of the cars from the floor of the canyons

active volcano, having eruptions at 20–30-year intervals. The eruption in 1977 caused serious erosion around the summit, which had accumulated dense volcanic pumice. These deposits were eroded by rainfall, creating small ditches that grew to deep gullies; degraded material mixed with water rolled downhill, where it was diffused or deposited on alluvial fans (Kadomura et al. 1983) (Fig. 3). Vegetation recovery decreases the debris flow frequency, although additional volcanic eruptions destroy recovered vegetation. Mount Tokachi in central Hokkaido erupted in 1926 and 1962. The 1926 eruption caused a massive mudflow that devastated surrounding villages. The eruption began in May, when dense snow still covered the area. The hot ejecta first melted the snow cover, producing massive water flows that included wood and rock debris; flooded rivers carried this large debris to settlements 20–30 km away, burying houses and towns. Sixty years later, the reestablished forests along traces of the mudflow suggested a distinct relationship between sediment size and forest type (Yajima et al. 1998). In sandy soil areas, birch (*Betula* spp.) stands were established with dwarf bamboo covering the forest floor; in contrast, spruce (*Picea glenii* Masters) dominated areas of gravelly substrata. The heterogeneous spatial distribution of regenerated stands and their complex structures suggest differences in edaphic conditions, reflecting the various disturbance intensities.

3.3.3 Sediment transport and deposition through channel processes

In upper to midstream areas, initial disturbances involve aggradations and degradations of gravel deposits by periodic flooding. Flooding can erode deposits on the concave curve of a meander and aggregate material on the convex point bars, resulting in terrace-shaped landforms. Araya (1987) examined forests growing on

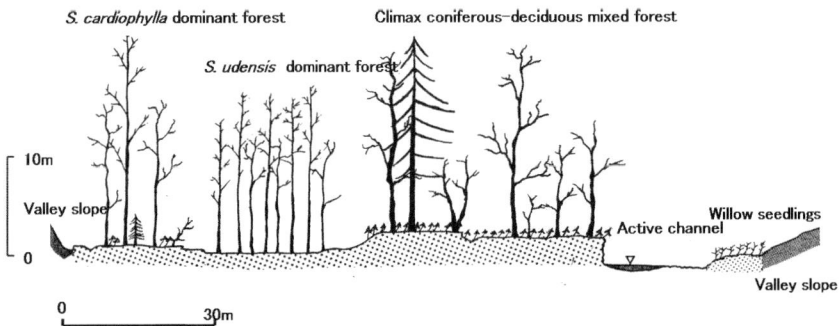

Fig. 4. Representative successional pattern after flooding observed in headwater area of Ishikari River, central Hokkaido. Primary succession is started from young willow seedlings, *S. udensis* dominates 20-30 years later, *S. cardiophylla* var. *urbaniana* become dominant 50 years later and coniferous forest is finally established 200-300 years later

Fig. 5. Invading willow (*Salix cardiophylla* var. *urbaniana*) seedlings on an abandoned channel of upstream of the Ishikari River, central Hokkaido

terrace levels and estimated sedimentation frequency based on tree ring analysis. Younger trees tend to be established near the water margin, whereas older riparian forests occupy higher depositional terraces. Other tree-based analyses can also be used to measure flooding frequency. For instance, unusual sprouting, injury marks indicated in tree rings, and buried root systems are excellent markers of past flooding events. Nakamura and Kikuchi (1996) classified fluvial deposits estimated from tree rings in a riparian forest. The older depositional surfaces tended to be higher and farther from the riverbed and indicated that the predominant disturbances occurred in the unconstrained, wide reaches where floodplains develop. This pattern indicates that eroded areas of floodplain deposits in each age class linearly increased with sediment volume, and that the proportion of the total area eroded decreased exponentially as the age of the sediment increased.

To compare individual forests and disturbance ages, succession patterns of a riparian forest were observed around the well-conserved Ishikari River (Fig. 4). Primary pioneers after flooding were willow (*Salix cardiophylla* Trautv. & Mey. var. *urbaniana* (Seemen) Kudo, formerly *Toisusu urbaniana* (Seeman) Kimura; Fig. 5 and *Salix udensis* Trautv. & Mey., formerly *S. sachalinensis* Fr. Schm.) and alder (*Alnus hirsuta* Turcz.). Species of the same age were well mixed for 20–30 years; after that time, *S. udensis* began to decline. *S. cardiophylla* became dominant in the second-stage forest, which also contained an understory of coniferous seedlings until approximately 100 years. After 200–300 years, coniferous trees such as *Picea jezoensis* (Sieb. et Zucc.) Carr. dominated the canopy, along with elm (*Ulmus laciniata* Mayr; Yanai et al. 1980). However, these forests were destroyed by major flooding, and succession was initiated again. The riparian forest that establishes after disturbance plays an important role in controlling disturbances (Sakatani et al. 1980). Driftwood is easily created by the scouring of tree roots; this washed out wood accumulates on the edge of point bars, trapping sediment and raising the riverbed level. When sediment accumulates to a certain level, the river channel will move to a lower level. Trapped sediment also provides a regeneration site for riparian trees.

These cycles of forest destruction, driftwood production, channel changes, and vegetation regeneration are commonly observed in natural streams. However, artificial dams can disrupt this process. For example, dams decrease flooding and limit the regeneration of endangered willow (*Salix arbutifolia* Pall., formerly *Chosenia arbutifolia* (Pall.) Skvorts.) stands (Takagi & Nakamura 2003). Furthermore, riparian vegetation has commonly been considered obstacle for flood water and has been removed in Japan for the past several decades.

3.3.4 Meandering channels in floodplain and delta regions

Alluvial plains or lowlands develop downstream from alluvial fans. Alluvial plains are characterized by meandering channels, natural levees, and back swamps. During flooding, rivers may overflow their banks; as flume deposits decline, sand and silts contained in the floodwater are deposited. In the lower reaches, such disturbances are generally not strong enough to excavate the ground surface or generate bare ground, as observed in the upper reaches. Constrained vegetation is adapted to longer flooding conditions. Although the scale of the natural levee is different from the scale of the watershed and occurrence interval of flooding, the relative height is usually 2–3 m from the river table. Nakamura et al. (2002) observed the distribution patterns of riparian vegetation in the Kushiro mire of eastern Hokkaido. Willow (*Salix* spp.) generally dominated the natural levees, which were characterized by low water levels, coarse sediments, and high conductivity. *Alnus japonica* Steud., the most common tree species in the mire, favors a high water table and high organic content. This species has adapted to high water levels by developing lenticels with hypertrophied and adventitious roots, multiple sprouting, and vegetative reproduction (Yamamoto et al. 1995).

Most of the moorlands along the seacoast of northern Japan have been created through interactions of fluvial sedimentation and vegetation over several thousand years. However, the mire has changed rapidly because of recent land-use developments. The Kushiro mire presently faces the serious problem of turbid water flooding. The shortening of stream channels associated with agricultural development is a major cause of streambed aggregation. This aggregation reduces the carrying capacity of the channel, resulting in sediment-laden water spilling into the wetlands during flood events. Sedimentation brought by repeated inundation by turbid water has significantly altered the edaphic conditions and thus, the composition and structure of the marsh forests (Nakamura et al. 2002).

Fig. 6. Straighten channel at downstream of Ishikari river, Hokkaido. This river used to be 356km meandering channel has shortened to 268km after work

Likewise, massive gravels deposited at the fluvial plain were extracted for concrete manufacturing during the period of high economic growth (Ogura & Yamamoto 2005). Channel straightening has also dramatically altered the fluvial processes. When the meandering channel was shortened, the slope angle became 1.5–2 times steeper, allowing floodwaters to arrive at the lower reaches much faster. The Ishikari River, which flows through central Hokkaido, used to be one of the most meandering rivers in Japan and had natural levees, swamps, abandoned channels, and cut-off ponds. However, government-funded projects shortened and straightened the channel. As a result, the channel became 100 km shorter and flooding became less frequent (Fig. 6). The riparian forests that remain along the oxbow lakes are isolated and receive no disturbance.

3.4 Conclusions

I reviewed the major dynamic and disturbance regimes characteristic of riparian areas in Japan. In Japan, river headwaters generally form in steep mountainous regions, where sediment production is great. Heavy precipitation produces power-ful stream-flow which transports these materials downstream, with some material overflowing and depositing on alluvial fans or plains. This fluvial morphology is classified into mountain features, alluvial fan, alluvial plain, and delta from upstream to downstream. The grain-size distribution pattern is controlled by geological conditions and the slope angle. In headwater regions, drainage morphology can be classified by glacial and post-glacial dissection regimes, with steep lower slopes and gentle, undissected upper slopes characterized by a deep, weathered soil layer. The dissecting front line is delineated by the slope break; shallow slope failure is induced by heavy rainfall, which is the main factor generating slopes. Forests on steep slopes are strongly affected by slope disturbances at intervals generally of a few hundred years. The accumulated disturbed soils provide sites for vegetation regeneration. Another important disturbance process is deep-sheeting soil slumps, originating from several thousand years ago; this process removes the top portion of the soil mass, creating a stream with an irregular bend and providing massive sediment for riparian regeneration. Debris flows or torrents are major forces for transporting material and creating alluvial fans. Such disturbances are caused by heavy rainfall or massive events such as volcanic and earthquake activity. The disturbance frequency depends on the region and type of disturbance; the most frequent disturbances occur around active volcanoes following eruption. As channel changes take place in upstream to midstream areas, sediment is deposited on terraces; the age of the riparian forest stand can initially be determined by the relative height of these terraces and distance from the water. However, if a younger high terrace is formed by a logjam or other obstruction, sediment can be trapped to cause the channel course change. The close relationship between the forest and stream characterizes the disturbance regime in the upstream to midstream reaches. Periodical flooding is a major disturbance force in the lower reaches, especially in the floodplain. Floodwater transports fine sediments, creating natural levees and changing the meander pattern of the channel, which can lead to the creation of cut-off ponds. However, recent human effects have greatly altered the natural disturbance regime. The relationship between natural and human disturbances and effects must be investigated to better conserve valuable riparian vegetation.

References

Araya T (1987) A morphological and chronological study on the process of sediment movement in Saru River. Res Bull Hokkaido Univ For 44:1217-1239 (in Japanese)

Daimaru H (1989) Holocene evolution of Toyohira River alluvial fan and distal floodplain, Hokkaido, Japan. Geogr Rev 62:589-603 (in Japanese)

Fujita K (1983) A theory of Japanese mountain development—between geology and geomorphology. Soju Shobo, Tokyo (in Japanese)

Fujita T (1990) Landslide—geology of a mountain disaster. Kyoritsu-Shuppan, Tokyo (in Japanese)

Fujiwara O, Yanagida M, Shimizu N, Sanga T, Sasaki T (2004) Regional distribution of large landslide landforms in Japan: implication for geology and land. J Jpn Landslide Soc 41:335-344 (in Japanese)

Hatano S (1979) Mapping of post-glacial dissected hillslopes and its application to landslide prediction. Abstracts, Jpn Erosion-Control Eng Soc 1979:16-17 (in Japanese)

Iida T (1993) A probability model of slope failure and hillslope development. Trans Jpn

Geomorph Union 14:17-31 (in Japanese)

Ikeda H (2000) Gravel transportation and change in grain size from mountain to the sea. Kaiyo Monthly 357:151-155 (in Japanese)

Inagaki H, Kosaka H, Hirata N, Kusaka H, Inada T (2004) Diverse ecosystem of the Mikabu landslide, Shikoku district. J Jpn Landslide Soc 41:245-254 (in Japanese)

Kadomura H, Yamamoto H, Imagawa T (1983) Eruption-induced rapid erosion and mass movements on Usu Volcano, Hokkaido. Zeitschrift für Geomorphologie Neu Floge Supplementary Band 46:123-142

Kohashi S (1983) Landslide, slope failure and debris flow: prediction and countermeasures. Kashima Shuppankai, Tokyo (in Japanese)

Koide H (1973) Land of Japan: nature and development. Univ Tokyo Press, Tokyo (in Japanese)

Nakamura F, Jitsu M, Kameyama S, Mizugaki S (2002) Changes in riparian forests in the Kushiro mire, Japan, associated with stream channelization. Reg Riv Res Appl 18:65-79

Nakamura F, Kikuchi S (1996) Some methodological developments in the analysis of sediment transport processes using age distribution of floodplain deposits. Geomorphology 16:139-145

Oguchi T (1996) Relaxation time of geomorphic responses to Pleistocene–Holocene climatic change. Trans Jpn Geomorph Union 17:309-321

Oguchi T, Saito K, Furumai H, Tockner K (2001) Fluvial geomorphology and paleohydrology in Japan. Geomorphology 39:3-19

Ogura N, Yamamoto K (2005) Natural disturbance, artificial impact and response to fluvial geomorphology as a base of ecosystem. Gihodo Shuppan, Tokyo (in Japanese)

Ono Y, Hirakawa K (1975) Glacial and periglacial morphogenetic environments around the Hidaka Range in the Würm Glacial age. Geogr Rev 48:1-26 (in Japanese)

Saijo K (1998) Slope development since the last interglacial stage in the north and middle parts of the Kitakami Mountains, Northeast Japan. Trans Jpn Geomorph Union 19:209-219 (in Japanese)

Saito K (1988) Alluvial fans in Japan. Kokon Shoin, Tokyo (in Japanese)

Sakaguchi Y, Takahashi Y, Omori H (1995) Rivers in Japan. Iwanami Shoten, Tokyo (in Japanese)

Sakai A (1995) Effect of ground-surface disturbance due to dissection of river valleys on forest vegetation. Jpn J Ecol 45:317-322 (in Japanese)

Sakatani Y, Yanai S, Onodera H (1980) Channel change and structure of riparian forest at Kwannai River (1) Accumulation of large woody debris and channel change. Proc Jpn For Soc Hokkaido branch 29:188-190 (in Japanese)

Sato H (1992) Regeneration traits of saplings of some species composing *Pterocarya rhoifolia* forest. Jpn J Ecol 42:203-214 (in Japanese)

Shimazu H (1994) Segmentation of Japanese mountain rivers and its causes based on gravel transport process. Trans Jpn Geomorph Union 15:111-128

Shimizu O, Nagayama T, Saito M (1995) Distribution and frequency of landslides for the last 8000 years in a small basin, central Hokkaido. Trans Jpn Geomorph Union 16:115-136 (in Japanese)

Shimokawa E, Jitozono T (1995) A study of landslide frequency with vegetation. Kokon Shoin, Tokyo (in Japanese)

Shimokawa E, Jitozono T, Takano S (1989) Periodicity of shallow landslide on Shirasu (Ito pyroclastic flow deposits) steep slopes and prediction of potential landslide sites. Trans Jpn Geomorph Union 10:267-284 (in Japanese)

Suzuki T (1997) Geomorphological basis for map reading. Kokon Shoin, Tokyo (in Japanese)

Suzuki T (2002) Rock control in geomorphological processes: research history in Japan and a perspective. Trans Jpn Geomorph Union 23:161-199

Takagi M, Nakamura F (2003) The downstream effects of water regulation by the dam on the riparian tree species in the Satsunai River. J Jpn For Soc 85:214-221 (in Japanese)

Takahashi Y (1990) River civil engineering. Univ Tokyo Press, Tokyo (in Japanese)

Tsukamoto Y (1973) Study on the growth of stream channels (1) Relationship between stream channel growth and landslides occurring during heavy storm. Shin-Sabo 87:4-13 (in Japanese)

Umitsu M (1994) Late quaternary environment and landform evolution of riverine coastal lowlands. Kokon Shoin, Tokyo (in Japanese)

Yajima T, Nakamura F, Simizu O, Shibuya M (1998) Forest recovery after disturbance by the 1926 mudflow at Mount Tokachi, Hokkaido, Japan. Res Bull Hokkaido Univ For 55:216-228

Yamamoto F, Sakata T, Terazawa K (1995) Growth, morphology, stem anatomy, and ethylene production in flooded *Alnus japonica* seedlings. IAWA J 16:47-59

Yamamoto K (2004) Structural alluvial potamology. Sannkaido, Tokyo (in Japanese)

Yanai S (1989) Age determination of hillslopes with the tephrochronological method in central Hokkaido. Trans Jpn Geomorph Union 10:1-12 (in Japanese)

Yanai S, Igarashi Y (1990) History of the slope failure and paleoenvironment on the marine terrace of Hidaka district, central Hokkaido, Japan. Quat Res 29:319-326 (in Japanese)

Yanai S, Sakatani Y, Onodera H (1980) Channel change and structure of riparian forest (2) Regeneration and destruction of riparian forest. Proc Jpn For Soc Hokkaido branch 29:191-193 (in Japanese)

Yonekura S, Kaizuka S, Nogami M, Chinzei K (2001) Introduction to Japanese geomorphology. Univ Tokyo Press, Tokyo (in Japanese)

Yoshikawa T (1974) Denudation and tectonic movement in contemporary Japan. Bull Dep Geogr Univ Tokyo 6:1-14

Yoshinaga S (1990) Slope development since the last glacial stage in the Tokachi Plain, Northern Japan. Geogr Rev 63A:559-576 (in Japanese)

Yoshinaga S, Koiwa N (1996) Slope development in forested mountains in Japan since the latest Pleistocene to early Holocene. Trans Jpn Geomorph Union 17:285-307 (in Japanese)

Part 3

Riparian community

4   Vegetation-geographic evaluation of the syntaxonomic system of valley-bottom forests occurring in the cool-temperate zone of the Japanese Archipelago

Keiichi OHNO

Graduate School of Environment and Information Sciences, Yokohama National University, 79-7 Tokiwadai Hodogaya-ku, Yokohama 240-8501, Japan

4.1 Introduction

Most of the Japanese Archipelago (Fig. 1), except Hokkaido, is located in an oceanic climate strongly influenced by seasonal rain fronts and typhoons (Yoshino 1978; Murata 1995). The soil is thus greatly eroded by rain wash in various places in Japan, especially in the headwaters and upper reaches of river and in steep river valleys where fluvial terraces are formed. On the other hand, because of high sedimentation during freshets and floods, alluvial fans and deltas are formed in middle and lower reaches of rivers.

Zonal vegetation as a climatic climax is generally formed where climate provides the overriding environmental influence on the vegetation. Azonal vegetation, such as edaphic climax vegetation, may develop where physical destruction by erosion and sedimentation happens frequently (Ellenberg 1978). In the Japanese Archipelago, which is characterized by an unstable natural climatic environment and geological features, certain kinds of riparian forests, develop along many rivers, as an edaphic climax.

The valley bottoms in the upper reaches of rivers are generally unstable habitats and are frequently disturbed by floods and debris flows often occur during typhoons and heavy rainfall (Sakio 2002a). Particular forest types, as an azonal edaphic climaxes, occur and develop on these renewed valley bottoms or fluvial terraces along streams, especially in the montane belts in the *Fagetea crenatae* (Miyawaki et al. 1964) region of the cool-temperate zone. These forests along mountain streams are called brook-channel forest or valley-bottom forest. In this

Sakio, Tamura (eds) Ecology of Riparian Forests in Japan : Disturbance, Life History, and Regeneration
© Springer 2008

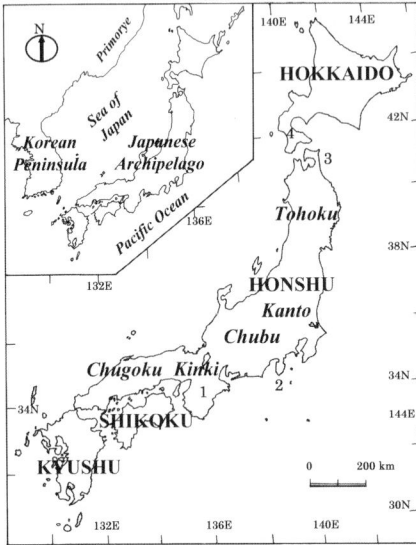

Fig. 1. Japanese Archipelago and its parts. (1 Kii Peninsula, 2 Izu Pen., 3 Shimokita Pen., 4 Oshima Pen.)

chapter, the author mainly discusses these valley-bottom forests among the riparian forests and reviews the plant-sociological system of the forest communities i.e. its syntaxonomic system.

The vegetation units (called syntaxa) of higher ranks in the system of valley-bottom forests are united in the alliance *Pterocaryion rhoifoliae* (Miyawaki et al. 1964). This alliance belongs to the order *Fraxino-Ulmetalia* (Suzuki 1966) within the class *Fagetea crenatae* (Miyawaki et al. 1964) or *Saso-Fagetea crenatae* (Suzuki 1966). Since the 1960s, when these units were recognized much plant-sociological (phytosociological) research has been done on the valley-bottom forests throughout Japan, and many associations have been classified and authorized (Miyawaki 1981-1988). The vegetation-geographic attributes in each unit, however, were neither analyzed nor evaluated in these earlier investigations (Sasaki 1979; Ohno 1983).

In order to review and reconstruct the syntaxonomic system of the valley-bottom *Pterocarya* forests, we compared the species composition of the main forest communities that constitute the main vegetation in the cool-temperate zone of Japan and South Korea. The distributional characteristics of the associations and the syntaxa of higher ranks reviewed in this article are elucidated vegetation-geographically. The ecological characteristics of the valley-bottom forests are defined on the basis of these results.

4.2 Methods

In order to review the syntaxonomic system of the valley-bottom forest communities occurring on the montane belts in the cool-temperate zone of Japan, I made a synthetic table combining relevant vegetation relevés published by the author (Ohno 1982, 1983; Kubo et al. 2001) and by others (Suzuki 1949; Suzuki et al. 1956; Miyawaki 1979, 1981-1988). The nomenclatural type and the systematical legitimacy of each syntaxon were verified based on the synthetic table. In this case, the author also compared against relevés of *Zelkova* forests on steep slopes or ravines (Ohno 1981, 1983) and against wetland *Ulmus* forests (Miyawaki 1985-1988), both of which are close to the valley-bottom *Pterocarya* forests in site conditions and flora. In addition, the species composition of all Japanese relevés was compared with that of syntaxa of the same types from areas of the class *Quercetea mongolicae* (Song 1988) in South Korea.

For analyzing and evaluating the vegetation-geographic attributes of the valley-bottom forests, the author adopted the concept of vegetation class groups advocated by Braun-Blanquet (1959) and Tüxen (1970) and the concept of "outside order" of the class group (called extra-order) that had been used by Suzuki (1966) for systematizing the natural forest communities in Japan. The class group and the extra-order used in this article are not hierarchical syntaxa but rather are geographic vegetation units that represent multi-dimensional vegetation types and also show distributional characteristics (Tüxen & Kawamura 1975; Ohno 1994, 1999).

The class group is defined as a group of classes that, according to Braun-Blanquet (1959), show with high similarity in flora and also in ecological properties and physiognomy. An extra-order is a syntaxon at the order rank that extends over two or more classes or class groups. In addition, the concept of a so-called intra-order is also used. This is a geographic vegetation unit of the order rank that occurs entirely with in a specific class or a class group.

When all syntaxa of the valley-bottom forests were reviewed, the nomenclatural type of each syntaxon identifyied, with literature citation, a lectotype, synonyms and its original source. Then, the character and differential taxa, ecology of habitat and distribution of each syntaxon are described.

4.3 General views of valley-bottom forests in Japan

4.3.1 Structural characteristics of the valley-bottom forests

The summergreen mixed forests dominated by trees *Pterocarya rhoifolia* Sieb. et Zucc., and occasionally *Fraxinus platypoda* Oliv., and attended by other trees such as *Cercidiphyllum japonicum* Sieb. et Zucc*., Aesculus turbinata* Blume and *Ulmus laciniata* Mayr, are called valley-bottom or brook-channel forests.

P. rhoifolia of the Juglandaceae family occurs on fluvial terraces in the montane belts of Honshu, Shikoku, Kyushu and part of Hokkaido, and often forms even-aged forests. In Hokkaido, the Oshima Peninsula (Fig. 1) is its northern limit of distribution, as well as for *Fagus crenata* Blume (Ito 1987). On the other hand, *F. platypoda* occurs mainly on valley bottoms of montane belts to the south of Kanto on the Pacific side of Japan and often forms mixed forests with *P. rhoifolia* in those places. *C. japonicum*, a semi-endemic woody species of Japan (Hotta 1974), is not a dominant tree like the above-mentioned ones. This tree can be seen in valley-bottom forests at various places in Japan, though the number of individuals is usually small. Mixed forests are often formed in the valley-bottom regions of southwestern Japan, where the ranges of the various woody species overlap.

Forests dominated by *P. rhoifolia* generally occur on valley levees in the upper courses of rivers in the cool-temperate Japan. *Euptelea polyandra* Sieb. et Zucc. and *Acer carpinifolium* Sieb. et Zucc., as arborescent (sub-tree) and shrub species, grow dominantly in these forests. These species occur in the forests of various places in Japan except Hokkaido. The understorys of these forests are rich in hygrophilous, and small herbaceous plants such as *Meehania urticifolia* (Miq.) Makino and *Stellaria diversiflora* Maxim., and also such genera as *Chrysosplenium, Elatostema* and *Mitella*. In the herbaceous layer strong growth of luxuriant ferns is a characteristic of these forests.

The major climatic zones of the Japanese Archipelago are generally divided into a Sea of Japan side and a Pacific Ocean side by the central mountain ranges. In Honshu, it is known that plants called "Sea of Japan type" grow only on that side, in areas of heavy winter snowfall, whereas "Pacific-type" plants occur only where the amount of winter snowfall is small (Yamazaki 1959; Hotta 1974; Kira et al. 1976). Some taxa show pairs of vicarious species at the variety level that occur in two regions. For example, *Polystichum retroso-paleaceum* (Kodama) Tagawa is a Sea of Japan species and *Polystichum retroso-paleaceum* var. *ovato-paleaceum* (Kodama) Tagawa is a closely related Pacific species, both growing in the herbaceous layer of valley-bottom forests.

The climatic conditions of each side affect the flora and the species composition of the valley-bottom forests in the same way as in the beech forests, dividing the forest vegetation, in the syntaxonomic system, into a Sea of Japan alliance (called *Saso-Fagion*) and a Pacific alliance (called *Sasamorpho-Fagion*) (Miyawaki et al. 1964).

4.3.2 Environmental characteristics of the valley-bottom forests

The microclimate, microtopography and soil conditions of the valley-bottom forests are described in the literature (Suzuki 1949; Ellenberg 1978). In narrow ravines, the areas where valley-bottom forests develop are generally shaded from sunlight, and the microclimate is cooler and more humid than outside the forests. Fluvial terraces, debris cones and talus deposits in and along mountain streams are the main habitats where these forests are able to develop (Kikuchi 2001; Kubo et al. 2001; Kawanishi et al. 2004, 2005, 2006).

Although there are some differences in sediment texture, the valley-bottom forests are generally moist because of good water supply and may be saturated with water from stream or spring water running down the slopes (Ito 1987; Sakio 2002b). Most habitats of valley-bottom forests are rocky or stony colluviums, poor in fine soil particles but more or less rich in inorganic bases. At the bottoms of ravines, though, organic matter such as fallen leaves and branches is derived from the upper part of the slope and partly from landslides. The biological activity of the soil fauna and bacteria is accelerated by the accumulation of this organic matter on the ground, and as a result, the supply of nutrients to the soil increases (Ellenberg 1978; Nakamura 1995).

4.4 Syntaxonomy and vegetation-geographic evaluation of the valley-bottom forests

The syntaxonomy and ecological characteristics of the geographic distributions of the valley-bottom forests, as azonal edaphic climax vegetation occurring in the *Fagetea crenatae* region in the cool-temperate zone, are as follow.

4.4.1 Class groups as a geographic unit

In the phytosociological system, the climatic climax forests of the cool-temperate zone of the Japanese Archipelago have been united into the class *Fagetea crenatae* (Miyawaki et al. 1964) or *Saso-Fagetea* (Suzuki 1966). According to Suzuki (1966), many of the climatic climax forests in the cool-temperate zone in the northern hemisphere, mainly composed of summergreen tree species of Fagaceae and Aceraceae, are classified into the class group *Acero-Fagetea*. He has included the *Saso-Fagetea* of Japan in this class group.

In the region surrounding the Sea of Japan, Kim (1992) has demoted each class of *Fagetea crenatae* in Japan and of *Quercetea mongolicae* (Song 1988) in Korea and its adjoining regions to the level of sub-class and has united them in a sub-class *Querco monglicae-Fagetea crenatae*. These two classes, however, occur under different climates, oceanic and continental, and are quite different as the climatic climax forests. Moreover, each class can be distinguished by particular character species. These two classes are rather independent syntaxonomically, so it is difficult to group them in one class.

As has been mentioned, the class group is not used as the highest syntaxon (called division) according to Westhoff and Maarel (1973) and Dierschke (1994) in the hierarchical syntaxonomy of phytosociology but is rather a vegetation-geographic unit that shows the vicarious relationships in the flora. There are many plants that are common to both classes, such as *Acer mono* Maxim., *Carpinus cordata* Blume and *Sorbus alnifolia* (Sieb. et Zucc.) C. Koch, because the Japanese Archipelago and the Korean Peninsula (Fig. 1) are located in the same floristic region, that of the so-called "East Asian Flora" (Maekawa 1961; Hotta

1974). The *Fagetea crenatae* can be united in one class group with the *Quercetea mongolicae* based on vegetation-geographic attributes, such as strong relationships in the flora. Therefore, this *Fagus-Quercus* class group is evaluated as a vegetation-geographic unit associated with the same edaphic environment in the areas around the Sea of Japan (Table 1).

4.4.2 Syntaxa of higher ranks and vegetation geographic units of valley-bottom forests

Syntaxa of higher ranks of valley-bottom *Pterocarya* forest communities belonging to the above-mentioned *Fagus-Quercus* class group are classified, and the characteristics of their geographic distributions are also evaluated, as follows.

4.4.2.1 Syntaxon within the class *Fagetea crenatae*

Order: *Fraxino-Ulmetalia* Suz.-Tok. 1966
Lectotype: *Ulmion davidianae* Suz.-Tok. 1954 (from Suzuki 1954; Table 6)

Table 1. Syntaxonomic system of the valley-bottom forests in the cool-temperate zone of Japan and geographic characteristics

Phytosociological system	Main character and differential species (canopy tree*)	Region of distribution	Altitude of distribution (m)	Vegetation-geographic unit
CLASS				
FAGETEA CRENATAE	*Fagus crenata** *Quercus crispula** *Tilia japonica** *Acer palmatum* var. *matsumurae** *Prunus grayana**	Japanese Archipelago		*Fagus-Quercus* class group
ORDER				
FRAXINO-ULMETALIA	*Ulmus laciniata** *Laportea macrostachya* *Meehania urticifolia* *Cardamine leucantha*	Japanese Archipelago, Korean Peninsula, Russian Far East (Primorye)		Intra-order of *Fagus-Quercus* class group
ALLIANCE				
PTEROCARYION RHIFOLIAE	*Pterocarya rhoifolia** *Cercidiphyllum japonicum** *Aesculus turbinata** *Acer carpinifolium* *Euptelea polyandra* *Philadelphus satsumi* *Mitella pauciflora*	Japanese Archipelago		
SUB-ALLIANCE				
FRAXIENION SPAETHIANAE	*Fraxinus platypoda** *Acer shirasawanum* *Pterostyrax hispida* *Deinanthe bifida* *Polygonum debile* *Polystichum retropaleaceum* var. *ovatopaleaceum*	Montane belts in the Pacific Ocean regions of Hohshu, Shikoku and Kyushu	600-1700	Association group of the Pacific Ocean region

Table 1. (continued)

Phytosociological system	Main character and differential species (canopy tree*)	Region of distribution	Altitude of distribution (m)	Vegetation-geographic unit
ASSOCIATION				
1 Hydrangeo sikokiani-Pterocaryetum rhoifoliae	*Hydrangea sikokiana* *Symplocos coreana* *Acer mono* var. *marmoratum* *Lamium humile*	Central Kyushu Mountains Shikoku Mountains and Kii Mountains	700-1550 700-1700	Geographic race of the Sohayaki regions
2 Chrysosplenio-Fraxinetum spaethianae	*Veronica cana* var. *miqueliana* *Isophyrum stoloniferum* *Aster dimorphophllus* *Cirsium effusum*	Montane belts of the Fossa Magna area in the Pacefic Ocean region	900-1450	Geographic race of the Fossa Magna regions
3 Cacalio yatabei-Pterocaryetum rhoifoliae	*Acer argutum* *Cacalia yatabei* *C. farfaraefolia* *Cornopteris crenulatoserrulata* *Scopolia japonica*	Montane belts in southeastern Chubu Montane belts in northern Kanto	600-1600 1100-1600	
4 Dryopterido-Fraxinetum commemoralis	The same diagnostic taxa as the sub-alliance	Montane belts in southern Chubu	800-1350	
SUB-ALLIANCE				
DRYOPTERIDENION MONTICOLAE	*Cephalotaxus harringtonia* var. *nana* *Dryopteris monticola* *Matteuccia struthiopteris* *Polystichum retrosopaleaceum*	Montane belts in the Sea of Japan regions of Hohshu Lower montane belts in Shimokita Peninsula of northernmost Honshu	650-1600 100-500	Association group of the Sea of Japan region
ASSOCIATION				
5 Polysticho-Aesculetum turbinatae	*Trigonotis brevipes* *Cirsium nipponicum* var. *yoshinoi*	Chugoku Mountains Montane belts in northern Kinki	650-1000 650-800	Geographic races of the western Chugoku Mountains and of the eastern Chugoku-northern Kinki
6 Polysticho-Pterocaryetum	*Hydrangea macrophylla* var. *Megacarpa* *Viburnum plicatum* var. *glabrum* *Diphylleia grayi*	Montane belts to the north of Chubu (until to the south of latitude 39°N) Lower montane belts in the northernmost Honshu (to the north of latitude 39°N)	700-1600 100-500	

Suzuki (1966) has unified such edaphic climax forests as the valley-bottom *Pterocarya* forests and the wetland *Ulmus* forests together in the order *Fraxino-Ulmetalia* within the class *Saso-Fagetea* (*Fagetea crenatae* according to Miyawaki et al. 1964). According to Ohno and Song (unpublished), several main character species of the *Fraxino-Ulmetalia*, such as *Ulmus laciniata*, *Laportea macrostachya* (Maxim.) Ohwi, *Meehania urticifolia* and *Cardamine leucantha* (Tausch) O.E.Schulz, are also found in the class *Quercetea mongolicae* (Song 1988) in South Korea. That is, this order belongs to two classes, the *Fagetea* in Japan and the *Quercetea* in South Korea, and it can be interpreted as an intra-order within the *Fagus-Quercus* class group (Table 1). This order also occurs in the Sikhote-Alin Mountains of the Russian Far East (Primorye) (Figs. 1, 2).

4.4.2.2 Syntaxon within the order *Fraxino-Ulmetalia*

Alliance: *Pterocaryion rhoifoliae* Miyawaki, Ohba et Murase 1964
Lectotype: *Chrysosplenio-Fraxinetum spaethianae* Miyawaki, Ohba et Murase 1964 (from Miyawaki et al. 1964; Table 2-12)

The author also compared the species composition of the valley-bottom forests in Japan and those of South Korea. As a result, it was determined that the alliance *Pterocaryion rhoifoliae* was divided by character species peculiar to Japan, such as *Acer carpinifolium, Euptelea polyandra, Philadelphus satsumi* Sieb. and *Mitella pauciflora* Rosend. (Table 1), though some common species were contained. This means the alliance occurs only in Japan. Moreover, we have shown that the area of this alliance is confined to Honshu, Shikoku and Kyushu, with no distribution in Hokkaido (Figs. 1, 2).

4.4.3 Associations and ecological characteristics of valley-bottom forests

The following six associations belonging to the alliance *Pterocaryion rhoifoliae* are reviewed syntaxonomically, and their vegetation-geographic and ecological characteristics are evaluated as follows (Table 1; Fig. 2).

4.4.3.1 Valley-bottom forests in the Sohayaki region

Association: *Hydrangeo sikokiani-Pterocaryetum rhoifoliae* Ya. Sasaki in Miyawaki 1982
Lectotype: Relevé no. SS58 (from Miyawaki 1982; Table 134)
Synonym: *Dryopterideto-Fraxinetum comanthosphacetosum* 1962 (Yamanaka 1962); *Galio paradoxi-Pterocaryetum* (Yamanaka 1962) Ohno 1983 (Ohno 1983); *Chrysosplenio-Farxinetum spaethianae* Miyawaki, Ohba et Murase 1964 (Miyawaki 1984)

The valley-bottom forests dominated by tree species of *Pterocarya* and *Fraxinus*, occasionally mixed with *Cercidiphyllum japonicum*, develop in the valley bottoms in the montane belts of Kyushu, on the climatic Pacific side (Suzuki 1962), and in Shikoku (Yamanaka 1962) and the Kii Peninsula (Miyawaki 1984), which represent a climatic transitional zone between the Sea of Japan and the Pacific Ocean (Fig. 1). The habitats of these valley-bottom forests are unstable because surface sediment often moves intensely in landslides or floods.

The species composition was analyzed based on the relevés (Miyawaki 1981, 1982, 1984) obtained for the investigation of the valley-bottom forests in the above-mentioned regions. As a result, it was determined again that these valley-bottom forest communities were the *Hydrangeo sikokiani-Pterocaryetum rhoifoliae*, of which the character and differential species are *Hydrangea sikokiana* Maxim., *Symplocos coreana* (Lev.) Ohwi, *Acer mono* Maxim. f. *marmoratum* (Nichols.) Hara and *Lamium humile* (Miq.) Maxim. (Table 1).

There are various herbaceous plants and ferns in the understory of this

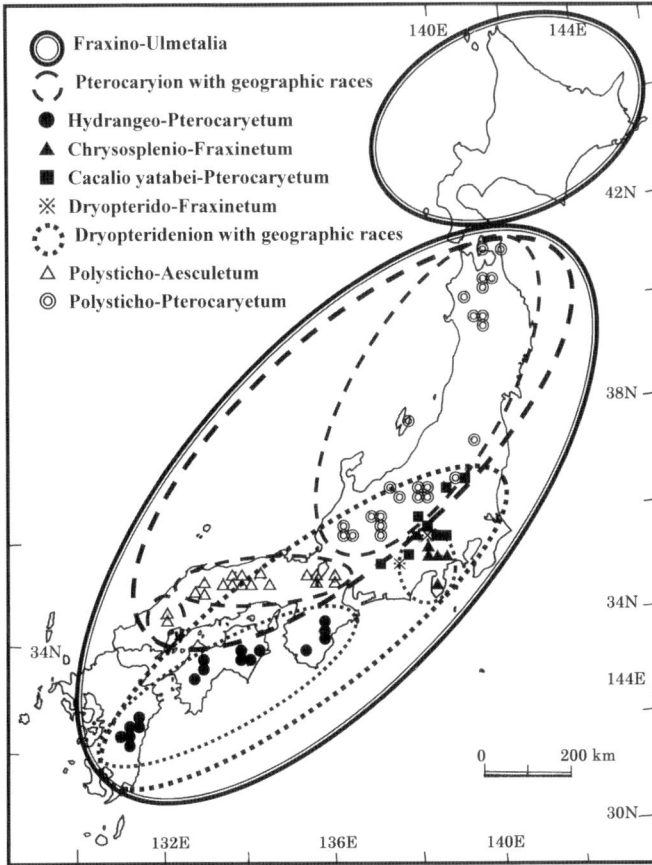

Fig. 2. Locations and areas of the several syntaxa of valley-bottom forests in the Japanese Archipelago

syntaxon. Moreover the list of species in the undergrowth varies with differences in micro-topography, disturbance processes, form of colluvial sediment, and the amount of fine soil between rocks and stones (Kawanishi et al. 2005, 2006).

In the herbaceous layer of this association, there are many species of the so-called Sohayaki element (Maekawa 1961; Hotta 1974), such as *Cacalia tebako-ensis* (Makino) Makino, *Kirengeshoma palmata* Yatabe, *Deinanthe bifida* Maxim. and *Impatiens hypophylla* Makino, which generally occur in the Pacific coastal region of central Kyushu, in Shikoku, and in the Kii Peninsula (Fig. 1). Judging from the distribution and flora of this syntaxon, it is probably to be interpreted as a geographic race occurring in the Sohayaki regions (Table 1; Fig. 2). Species differentiation in the same genera was also observed following the different distribution areas of this association, for example, *Leucosceptrum stellipilum* (Miq.) Kitam. et Murat var. *tosaense* (Makino) Kitam. et Murata is a main differential taxon in Kyushu and Shikoku, and *Leucosceptrum stellipilum* (Miq.) Kitam. et Murat is the same in the Kii Peninsula.

It is likely that the upper limit of the *Fagetea crenatae* region is at 1,800 m above sea level, and the lower limit at about 1,000 m in the mountainous areas of Shikoku and southern Kinki; in the Kii Mountains the upper limit is at 1,700 m (1,550 m in central Kyushu) and the lower limit at 700 m (Miyawaki 1981, 1982, 1984; Table 1). The vertical distribution of this association adjoins the *Camellietea japonicae* (Miyawaki & Ohba 1963) region at lower limit in the Kii Mountains, though it occurs in the *Fagetea* region in most mountains of Kyushu and Shikoku. Therefore, there are comparatively many elements of *Camellietea* in this syntaxon near its lower limit in the Kii Mountains.

It is shown that this association has an azonal distribution pattern as an edaphic climax forest. For that reason, it descends to the *Camellietea* region of the colline to submontane belts. The valley-bottom forest communities in the *Camellietea* region often adjoin the *Zelkova* forests that develop on the steep slopes or ravines of shady valleys (Ohno 1981, 1983).

4.4.3.2 Valley-bottom forests in the Fossa Magna region

Association: *Chrysosplenio-Fraxinetum spaethianae* Miyawaki, Ohba et Murase 1964

Lectotype: Relevé no. 57 (from Miyawaki et al. 1964; Table 2-13)

Synonym: *Isopyro-Fraxinetum spaethianae* Miyawaki et al. 1977 (Miyawaki et al. 1977b); *Isopyro-Fraxinetum spaethianae* Miyawaki, Ohba et Murase 1964 (Sasaki 1979); *Hydrangeo sikokiani-Pterocaryetum rhoifoliae* Ya. Sasaki in Miyawaki 1982 (Miyawaki 1985)

The regions included in the Fuji volcanic zone, such as the Tanzawa Mountains and Hakone Mountains (Kanagawa Prefecture) in southwestern Kanto and the foot of Mt. Fuji (Yamanashi Pref.) and the Izu Peninsula (Shizuoka Pref.) of southeastern Chubu, are designated as "Fossa Magna" in geology (Maekawa 1949, 1961; Uematu 1951; Takahashi 1971). These regions also belong to the Pacific climatic division with humid oceanic conditions (Suzuki 1962). Valley-bottom forests dominated by *Fraxinus* mixed with *Pterocarya* are developed on the debris-flow terraces and alluvial cones of valley bottoms in the headwaters of rivers in the cool-temperate zones of the Fossa Magna regions (Fig. 1).

The species composition of the valley-bottom forest communities was analyzed by using the vegetation materials obtained in the Tanzawa Mountains (Miyawaki et al. 1964; Ohno & Ozeki 1997; Fig. 3), at the foot of Mt. Fuji (Miyawaki et al. 1977b), and in the Amagi Mountains in the Izu Peninsula (Miyawaki 1985). This analysis confirmed the association *Chrysosplenio-Fraxinetum spaethianae*, of which the character and differential species are *Veronica cana* Wall. var. *miqueliana* (Nakai) Ohwi, *Isopyrum stoloniferm* Maxim., *Aster dimorphophyllus* Franch. et Savat. and *Cirsium effusum* (Maxim.) Matsum. (Table 1).

Judging from the distribution and flora of this syntaxon, it is probably to be evaluated as a geographic race occurring in the Fossa Magna regions (Fig. 2). The Sohayaki elements (Maekawa 1961; Hotta 1974), such as *Cacalia tebakoensis* and

Fig. 3. Interior of the *Chrysosplenio-Fraxinetum spaethianae* occurring on talus in the Tanzawa Mountains, western Kanto. The undergrowth of this forest is poor, due to feeding by Japanese deer

Viola shikokiana Makino, also grow in the herbaceous layer of this syntaxon as well as in the above-mentioned *Hydrangeo-Pterocaryetum*, and there are many plants, such as *Leucosceptrum japonicum* (Miq.) Kitam. et Murata f. *barbinerve* Kitam. et Murata of the Fossa Magna element (Maekawa 1949, 1961; Uematu 1951; Takahashi 1971) that differentiate the areas concerned.

It is probable that the vertical limit of the *Fagetea* in the southern parts of the Fossa Magna region is located at 1,700 m to 1,800 m above sea level and lower limit at 700 m to 800 m (Miyawaki 1985, 1986). This association occurs between 900 m and 1,450 m, which means the distribution of this syntaxon is within the *Fagetea* region (Table 1).

4.4.3.3 Valley-bottom forests in montane belts of Chubu and Kanto

Association: *Cacalio yatabei-Pterocaryetum rhoifoliae* (Miyawaki et al. 1979) Ohno in Miyawaki 1985

Lectotype: Serial no. 1 (from Miyawaki 1979; Table 96)

Synonym: *Dryopterido-Fraxinetum spaethianae* Suz.-Tok. 1949 (Miyawaki 1979); *Dryopterido-Fraxinetum commemoralis* Suz.-Tok. 1949 (Miyawaki 1985); *Chrysosplenio-Fraxinetum spaethianae* Miyawaki, Ohba et Murase 1964 (Miyawaki 1985, 1986; Kubo et al. 2001)

The montane belts of inland Chubu and Kanto (Fig. 1) are in the Pacific

Fig. 4. Interior of the *Cacalio yatabei-Pterocaryetum rhoifoliae* occurring on debris flows in Nakatugawa Gorge, western Kanto. The forest floor is rich in shrubs and herbaceous plants

climatic division but with less rainfall (Suzuki 1962). The valley-bottom *Fraxinus-Pterocarya* forest communities, of which the tree layer is dominated by *Pterocarya rhoifolia*, occasionally mixed with *Fraxinus platypoda*, occur on the fluvial terraces and debris cones in the valley bottoms of these regions (Fig. 2).

These forests have been studies syntaxonomically using the relevés obtained in research on Mt. Ena of Gifu Pref. (Miyawaki 1985), in mountainous areas of Shizuoka Pref. (Miyawaki 1985), in Minami-Saku of Nagano Pref. (Miyawaki 1979), in Minami-Koma of Yamanashi Pref. (Miyawaki 1985), along the upper Nakatsu River in Saitama Pref. (Kubo et al. 2001; Fig. 4), in Okutama of Tokyo (Miyawaki 1986), along the upper Tone River in Gunma Pref. (Miyawaki 1986), and at the foot of Mt. Nasu-dake of Tochigi Pref. (Miyawaki 1986). The author determined the association *Cacalio yatabei-Pterocaryetum rhoifoliae*, which is characterized by such character and differential taxa as *Acer argutum* Maxim., *Cacalia farfaraefolia* Sieb. et Zucc., *Cacalia yatabei* Matsum. et Koidz., *Scopolia japonica* Maxim. and *Cornopteris crenulato-serrulata* (Makino) Nakai (Table 1).

This association was once classified into the *Chrysosplenio-Fraxinetum* or the *Dryopterido-Fraxinetum*, based on plants such as *F. platypoda*, *Chrysosplenium macrostemon* Maxim. and *Dryopteris polylepis* (Franch. et Savat.) C. Chr., which are common species and indicate a Pacific type distribution (Hotta 1974). However, this association was distinguished from these two syntaxa by the above-mentioned character and differential species.

It is likely that the upper vertical limit of the *Fagetea* region in Chubu and Kanto is at 1,700 m to 1,800 m above sea level and the lower limit at 500 m to 600 m (Miyawaki 1985, 1986). This association occurs between 500 m (1,100 m in northern Kanto) and 1,600 m in montane belts in southern Chubu (Table 1). Thus the distribution of this association is within the *Fagetea* region.

4.4.3.4 Valley-bottom forests in montane belts of southern Chubu

Association: *Dryopterido-Fraxinetum commemoralis* Suz.-Tok. 1949
Lectotype: Relevé no. M30 (from Suzuki 1949; Table 1)
Synonym: *Dryopteridieto-Fraxinetum commemoralis* Suz.-Tok. 1952 (Suzuki 1952); *Dryopteridi-Fraxinetum commemorlis* Suz.-Tok. 1952 (Suzuki 1966); *Dryopteridieto-Fraxinetum spaethianae* Suz.-Tok. 1952 (Sasaki 1979)

The association *Dryopterido-Fraxinetum commemoralis* is a syntaxon of the valley-bottom forests described for the first time in Japan according to Suzuki. Its range is located in the *Fagetea* region, and it is only reported from two places (Fig. 2), Mt. Shirakura in Shizuoka Pref. (Suzuki 1949) and Mt. Daibosatu-rei in Yamanashi Pref. (Miyawaki et al. 1977b).

The physiognomy of this association dominated by *Fraxinus*, occasionally mixed with *Pterocarya*, is almost the same as in the three syntaxa described above. In addition, it is similar to those three syntaxa in containing many common plants, such as *Stewartia monadelpha* Sieb. et Zucc., *Chrysosplenium macrostemon* and *Dryopteris polylepis*, which show the Pacific distribution type (Table 1). This association, however, is distinguished from the others because it lacks some   peculiar character species. Moreover, it generally has few species, possibly because the habitat usually has large stones or rocks, with little fine soil. On stony or rocky colluvium, the fern *D. polylepis* often grows thickly in the herbaceous layer.

Syntaxonomically, the above-mentioned three syntaxa belong to the sub-alliance *Fraxinenion speathianae* within the alliance Pterocaryion (Table 1). This sub-alliance is probably distinguished by such species as *F. platypoda, Acer shirasawanum* Koidz., *Pterostyrax hispida* Sieb. et Zucc., *Deinanthe bifida, Persicaria debilis* (Meisn.) H. Gross and *Polystichum retroso-paleaceum* var. *ovato-paleaceum*, as floristic elements of the Pacific type. It is also evaluated as an association group of the Pacific region (Table 1; Fig. 2).

4.4.3.5 Valley-bottom forests in the Chugoku Mountains

Association: *Polysticho-Aesculetum turbinatae* Horikawa et Sasaki 1959
Lectotype: Relevé no. 102 (from Horikawa & Sasaki 1959; Table 6)
Synonym: *Polysticho-Pterocaryetum* Suz.-Tok. et al. 1956 (Miyawaki 1983); *Trigonotido brevipedis-Fraxinetum platypodae* Ohno 1982 (Ohno 1982); *Polysticho-Pterocaryetum* Suz.- Tok. et al. 1956 (Miyawaki 1984)

The association *Polysticho-Aesculetum turbinatae* that occurs locally on ravines or valley bottoms in the Sandan-kyo (Hiroshima Pref) and its surrounding areas, in the western Chugoku Mountains, is a syntaxon described by Horikawa &

Fig. 5. Stand from the *Polysticho-Aesculetum turbinatae* occurring on terraces along a brook channel in a valley bottom in the Chugoku Mountains

Sasaki (1959). It is a valley-bottom forest community characterized by a dominant tree, *Pterocarya rhoifolia*, occasionally mixed with *Aesculus turbinata* and *Zelkova serrata* (Thunb.) Makino. Valley-bottom forests with the same physiognomy and substance as this association occur also in the central and eastern Chugoku Mountains and the northern parts of Kinki (Figs 1, 2), where the climatic environment is under the influence of the semi-climatic division of the Sea of Japan (Suzuki 1962).

The species compositions of the valley-bottom forest communities occurring in the Chugoku Mountains and the northern parts of Kinki were compared and analyzed by using the relevés that had been obtained in previous local vegetation surveys (Ohno 1982; Miyawaki 1983, 1984). The results show that the association *Polysticho-Aesculetum turbinatae* is characterized by such character and differential species as *Trigonotis brevipes* (Maxim.) Maxim. and *Cirsium nipponicum* (Maxim.) Makino var. *yoshinoi* (Nakai) Kitam. (Table 1).

The flora of this association is also characterized by floristic elements of both the Sea of Japan type, e.g. *Cephalotaxus harringtonia* (Knight) K. Koch var. *nana* (Nakai) Rehder, *Viola vaginata* Maxim. and *Polystichum retroso-paleaceum* and the Pacific type, e.g. *Pterostyrax hispida* and *Polystichum retroso-paleaceum* var. *ovato-paleaceum*, due to its transitional climatic environment. In addition, this association can be classified into two geographic races (Oberdorfer 1968, 1977; Ohno 1994, 1999) representing the western Chugoku Mountains (Fig. 5) and eastern Chugoku plus northern Kinki (Table 1). The former is distinguished by

Rubus pectinellus Maxim. and *Hosiea japonica* (Makino) Makino, and the latter by *Chrysosplenium fauriei* Franchet and *Rabdosia shikokiana* (Makino) Hara..

4.4.3.6 Valley-bottom forests on the Sea of Japan side of northern Honshu

Association: *Polysticho-Pterocaryetum* Suz.-Tok. et al. 1956
Lectotype: Relevé no. G31 (from Suzuki et al. 1956; Table 3)
Synonym: *Cacalio yatabei-Pterocaryetum rhoifoliae* (Miyawaki 1979) Ohno in Miyawaki 1985, *cacalietosum delphiniifoliae* (Miyawaki 1987)

The association *Polysticho-Pterocaryetum* authorized by Suzuki et al. (1956) is the valley-bottom forest community that develops on the fluvial terraces or debris cones in the valley bottoms of Mt. Gassan in Yamagata Prefecture. *Pterocarya rhoifolia* is a dominant tree of this association, and woody *Aesculus turbinata* and *Cercidiphyllum japonicum* grow sparsely in places.

Valley-bottom forest communities of the same structure and composition as this syntaxon occur in Hokuriku of northern Chubu, northern Kanto, and central and western Tohoku, where the climatic environment is that of the division and semi-division of the Sea of Japan according to Suzuki (1962) (Figs. 1, 2).

Structural analysis of the species composition was done by using the relevés from previous surveys in the regions. The results showed that this association is distinguished by such character and differential species as *Hydrangea macrophylla* (Thunb.) Ser. var. *angusta* (Franch. et Savat.) Hara, *Viburnum plicatum* Thunb. var. *glabrum* (Koidz.) Hara and *Diphylleia grayi* Fr. Schm. (Table 1). Some character species of the alliance *Pterocaryion*, a shrub *Cephalotaxus harringtonia* var. *nana* (Sea of Japan element, Hotta 1974) and ferns *Polystichum retroso-paleaceum, Matteuccia struthiopteris* (L.) Todaro and *Dryopteris monticola* (Makino) C. Chr., are dominant plants on the forest floor of this association. Therefore, this syntaxon and the *Polysticho-Aesculetum* are grouped together in a sub-alliance *Dryopteridenion monticolae*, based on the above-mentioned differential species (Table 1).

The upper vertical limit of the *Fagetea* region in northern Chubu and the Sea of Japan side of Tohoku is located at 1,600 m to 1,700 m above sea level and the lower limit at 400 m to 500 m (Miyawaki 1985, 1986). The *Polysticho-Pterocaryetum* is distributed in these regions in the *Fagetea* region (Fig. 6). In northern Tohoku, on the other hand, the upper limit of the *Fagetea* is located at about 1,000 m and the lower limit goes down near the coast (Miyawaki 1987). The distribution area of this association suddenly goes down in northern Tohoku, north of latitude 39°N (Table 1). This may be because it there enters the Köppen Df (subpolar rainy) climate, in the sub-boreal zone, with cold humid climatic conditions (Yoshino 1978). In addition, there are few mountains over 1,000 m in northern Tohoku, so the range must descend.

The above-mentioned two syntaxa belong to the sub-alliance *Dryopteridenion monticolae*, which is characterized by diagnostic taxa that are mainly Sea of Japan elements. It is also evaluated as an association group of the Sea of Japan region (Table 1; Fig. 2).

Fig. 6. Stand from the *Polysticho-Pterocaryetum* occurring on steep colluvial slopes in the montane belt of northern Chubu. In the herbaceous layer a strong growth of ferns is the first thing noticed

4.4.3.7 Valley-bottom forests in Hokkaido

In Hokkaido there is little influence of the seasonal rain front and of typhoons (Murata 1995). While Honshu is located in the temperate zone with a rainy climate (Köppen Cfa: temperate rainy climate), most of Hokkaido is included in the sub-boreal climate corresponding to the Df climate (Saito & Okitu 1987).

A so-called "pan mixed forest" (Tatewaki 1958), with conifers (*Abies sachalinensis* (Fr. Schm.) Masters var. *mayriana* Miyabe et Kudo and *Picea jezoensis* (Sieb. et Zucc.) Carr.) and deciduous broad-leaved trees (*Quercus crispula* Blume, *Acer mono* Maxim. var. *glabrum* (Lév. et Van.) Hara and *Tilia japonica* (Miq.) Simonkai), develops widely in Hokkaido under such a climatic environment, as a climatic climax forest. Thus, the vegetation landscape of Hokkaido and its climatic environment are greatly different from other regions of the Japanese Archipelago.

Because the northern limit of *Pterocarya* is in the Kuromatsunai lowland of the Oshima Peninsula of Hokkaido (Fig. 1), the valley-bottom forests there are mainly composed of *Cercidiphyllum japonicum* and *Ulmus laciniata*. In addition, *Ulmus davidiana* Planch. var. *japonica* (Rehder) Nakai and *Fraxinus mandshurica* Rupr. var. *japonica* Maxim, which are often component species of wetland forests, mix into the valley-bottom forests on gentle slopes along mountain streams. These valley-bottom *Ulmus-Cercidiphyllum* forests in Hokkaido are classified into the association *Ulmo laciniatae-Cercidiphylletum japonici*, which belongs to the *Pterocaryion rhoifoliae* (Miyawaki 1988).

The species composition of the valley-bottom forests of the *Pterocaryion* was compared with that of the wetland forests of the alliance *Ulmion davidianae* in various parts of Japan. The results showed that the character and differential species of this association are *Chrysosplenium ramosum* Maxim., *Aruncus dioicus* (Walt.) Fernald var. *tenuifolius* (Nakai) Hara, *Polystichum braunii* (Spenn.) Fee and *Scutellaria pekinensis* Maxim. var. *ussuriensis* (Regel) Hand.-Mazz.. In addition, this association was syntaxonomically moved from the *Pterocaryion* to the *Ulmion*, emphasizing the existence of *Acer mono* Maxim. var. *mayrii* (Schwerin) Sugimoto and *Syringa reticulate* (Bl.) Hara, which are character and differential species of the alliance. The association *Cacalio-Pterocaryetum* which represents valley-bottom forests in the Oshima Peninsula (Fig. 1) that had been attributed before to the *Pterocaryion* by Ohno (Miyawaki 1988), would be included in this *Ulmion* by comparing the species compositions.

The distribution area of the *Ulmo-Cercidiphylletum* in Hokkaido is mostly below 400 m above sea level. At lower altitude the slope of the river bed becomes gentle, and sedimentation exceeds erosion. Therefore, fine, wet soils accumulate on fluvial terraces. Such an accumulation of alluvial soil in the valley-levee areas facilitated the invasion of species of the wetland-forest element into the *Ulmo-Cercidiphylletum*, and that is probably a factor that makes this association belong to the *Ulmion*.

4.5 Relationship between the valley-bottom *Pterocarya* forests and other azonal vegetation

Ravine *Zelkova serrata* forests, wetland *Ulmus davidiana* var. *japonica* forests, pioneer scrub communities characterized by *Euptelea polyandra*, swamp forests dominated by *Alnus japonica* (Thunb.) Steud., and riparian groves with various *Salix* species represent azonal edaphic climax vegetation along streams and rivers, as well as the valley-bottom *Pterocarya* forests. The ecological characteristics of these vegetation types and phytosociological relationships with the valley-bottom *Pterocarya* forests are shown below.

4.5.1 Relationship to the ravine *Zelkova* forests

Ravine forest communities dominated by *Zelkova serrata* and valley-bottom *Pterocarya* forests differ especially in the micro-topography of their habitats. Ravine *Zelkova* forests mainly develop on shady steep slopes or in ravines and on talus in valleys of the submontane belt (Ohno 1983). There are two higher units in the syntaxonomical system of the ravine *Zelkova* forests. One of them is a sub-alliance *Zelkovenion* (Ohno 1983), belonging to the *Pterocaryion*, and the other is an alliance *Zelkovion* (Miyawaki 1977) that is independent of the *Pterocaryion* and belongs to the *Fraxino-Ulmetalia*. These ravine *Zelkova* forest communities are azonal vegetation that ranges between the *Fagetea* in the cool-temperate zone and the *Camellietea* in the warm-temperate zone. That is, they are so-called extra-orders independent of the two above-mentioned classes. Therefore, the author concludes that the syntaxa should probably be classified into the alliance *Zelkovion* and the order *Zelkovetalia*, independent of the *Fraxino-Ulmetalia*.

4.5.2 Relationship to the wetland *Ulmus* forests

The habitat of wetland *Ulmus* forests is on valley bottoms along upper streams as well as in the valley-bottom *Pterocarya* forests. However, there is a difference in each habitat, that is, the former occurs on fluvial terraces, while the latter develops on debris-flow terraces. In a cool-temperate zone at high latitude, such as Hokkaido, the wetland *Ulmus* forests occur on the natural levees where alluvial soils accumulated thickly in the lower course of river. Thus, both habitats are close to each other, and there are many common species between *Ulmus* and *Pterocarya* forests.

The wetland *Ulmus* forest communities are united in the alliance *Ulmion davidianae* (Suzuki 1954). This alliance belongs to the *Fraxino-Ulmetalia* as well as the *Pterocaryion* (Suzuki 1966).

4.5.3 Relationship to the pioneer scrub dominated by *Euptelea*

It is known that *Euptelea polyandra*, as a character species of the *Pterocaryion*, has a high reproductive and sprouting ability (Sakai et al. 1995). Therefore, this woody species promptly establishes itself on unstable shady valley slopes, where landslides often occur, and forms a pioneer scrub community. The association *Hydrangeo-Eupteleetum* that represents the *Euptelea* pioneer scrub has been described by Miyawaki et al. (1964) in the Fossa Magna region of central Honshu. In primary succession, this syntaxon is an early stage of the *Pterocarya* forests. Therefore, *Euptelea* growing in the *Pterocarya* forests may represent individuals alive for the remainder of the succession stage or living on unstable locations in the habitat.

Euptelea pioneer scrub can be seen in valley bottoms in various places to the south of Tohoku, which is the northern limit of its distribution (Horikawa 1972). Much of this pioneer scrub has been separated as syntaxa without character species. The *Euptelea* pioneer scrub communities in various regions of Japan are combined in the alliance *Eupteleion polyandrae* which belongs to the order *Fraxino-Ulmetalia* (Miyawaki 1983).

Another pioneer scrub community that is able to grow on the unstable slopes with intense movement of surface soils is the *Weigela-Alnus* pioneer scrub community in the order *Weigelo-Alnetalia firmae* (Ohba & Sugawara 1979). This syntaxon develops not only on shady slopes in valleys but also on sunny slopes with dry soil and frequent landslides.

4.5.4 Relationship to the *Alnus* swamp forests

There are distinct differences in habitat and species composition between the *Alnus japonica* swamp forests and the valley-bottom *Pterocarya* forests, althoughugh both are edaphic climax types of the same azonal vegetation. The *Alnus* swamp forests are generally found in spring-fed swamps in foot-hills and on alluvial fans, and in the back marshes along rivers with a higher groundwater level; their habitats are adjacent to sedge and reed marshes.

The species composition of the *Alnus* swamp forests is closer to that of the wetland *Ulmus* forests than that of the valley-bottom *Pterocarya* forests. An overlapping of species composition between the *Alnus* swamps and the *Ulmus* wetlands is observed through *Fraxinus mandshurica* var. *japonica* forests.

The *Alnus* swamp forest communities form an independent class *Alnetea japonicae* (Miyawaki et al. 1977a; Miyawaki et al. 1994), as an azonal vegetation, and their distribution area almost reaches from the lower side of the cool-temperate zone to the whole area of the warm-temperate zone.

4.5.5 Relationship to *Salix* riparian groves

The habitat and species composition of the *Salix* riparian groves and the valley-bottom *Pterocarya* forests have conspicuous differences, although they are both waterside vegetation. Habitat differences are mainly due to physical destruction by the rivers, and species differences depend on environmental factors such as soil and topography. Thus, the disturbance regime (Nakashizuka & Yamamoto 1987) in the *Salix* riparian groves and the valley-levee *Pterocarya* forests is quite different.

The *Salix* riparian groves are typical azonal vegetation that grows at the waterside from upper stream courses in the subalpine belt to lower river courses in the lowlands. The horizontal range of these groves reaches from the sub-boreal to the temperate zone. *Salix* riparian groves form a peculiar plant society, the class *Salicetea sachalinensis* (Ohba 1973; Miyawaki et al. 1994), beside the syntaxono-mic system of the valley-bottom *Pterocarya* forests.

4.5.6 Relationship to ravine and valley-bottom forests in central Europe

Although neither the ravine *Zelkova* forest nor the valley-bottom *Pterocarya* forest occurs in central Europe, "mixed Maple-Ash forest communities" found there have been evaluated as ecologically equivalent forests (Ellenberg 1978). The syntaxa corresponding to the valley-bottom forests include "brook-channel Ash forests: *Carici remotae-Fraxinetum*" and "slope-foot Maple-Ash forests: *Aceri-Fraxinetum*". Moreover, there is "ravine Ash-Maple forest: *Phyllitido-Aceretum*" as a syntaxon corresponding to the ravine *Zelkova* forests.

In the syntaxonomic system of central Europe, two (*Aceri-Fraxinetum* and *Phyllitido-Aceretum)* among the above-mentioned three associations are included in the alliance *Tilio-Acerion*, which belongs to the order *Fagetalia* within the class *Querco-Fagetea*. This alliance corresponds to the alliance *Zelkovion* (Miyawaki 1977) or the sub-alliance *Zelkovenion* (Ohno 1983) of Japan. Another association, *Carici remotae-Fraxinetum*, belongs to the alliance *Alno-Ulmion* that groups Elm and Ash woods occurring in floodplains of the alluvial lowlands in central Europe. This alliance, together with *Tilio-Acerion*, belongs to the *Fagetalia* (Ellenberg 1978). Moreover, this alliance corresponds to the alliance *Ulmion davidianae* (Suzuki 1954, 1966) of Japan.

Thus, forest vegetation that corresponds to the valley-bottom *Pterocarya* forests in Japan does not exist in central Europe. Geographic features in central Europe include generally gradual slopes, less precipitation than in Japan and fewer landslides. Therefore, debris-flow terraces that become the growth sites of the valley-bottom *Pterocarya* forests are not formed easily in central Europe.

4.6 Vegetation-geographic evaluations of the valley-bottom forests in Japan

This chapter reviewed the syntaxonomic system of the valley-bottom *Pterocarya* forests in Japan. That is, a new syntaxonomic system concerning the valley-bottom *Pterocarya* forests was established by using the method of combining the geographic unit and the concept of the extra-order, independent of the class group and the class group that Suzuki (1966) had advocated.

As a result, the valley-bottom *Pterocarya* forest communities were grouped together in the alliance *Pterocaryion rhoifoliae*, which consist of two associations on the Sea of Japan side of Honshu and four associations on the Pacific side of Honshu, Shikoku and Kyushu (Table 1; Fig. 2). This alliance does not occur in Hokkaido. The *Pterocaryion* is a peculiar syntaxon of higher rank occurring only in the Japanese Archipelago, where landslides and debris flows occur frequently in the heavy rain of typhoons and the rainy season and due to large earthquakes.

This alliance was included in the order *Fraxino-Ulmetalia* within the class *Fagetea crenatae* (Miyawaki et al. 1964), also in the past syntaxonomic system (Table 1). The *Fagetea crenatae* in the Japanese Archipelago and *Quercetea mongolicae* in the Korean Peninsula are climatic climax forests in the cool-temperate zone and form a class group in the regions surrounding the Sea of Japan (Table 1). The order *Fraxino-Ulmetalia* that grouped the valley-bottom *Pterocarya* forests and the wetland *Ulmus* forests as an edaphic climax forest in the cool-temperate zone was assigned to a so-called intra-order within the class group consisting of the two above-mentioned classes (Table 1). This class group is not a unit of the highest rank, called "division" by Westhoff and Maarel (1973) and Dierschke (1994) in the syntaxonomic system that Braun-Blanquet (1959) and other plant sociologists advocated, but rather a vegetation-geographic unit in the sense of Tüxen and Kawamura (1975) or Ohno (1994, 1999).

Various plant societies characterized by different geo-historical backgrounds and each with peculiar genealogy occur in the Japanese Archipelago, through the transition of various floras, according to the fluctuation of climate since the Tertiary Period (Suzuki 1966). The class group, as a vegetation-geographic unit, and the concept of the extra-order independent of the class group are effective methods not only for analyzing the characteristics of geographic distribution of vegetation but also the formative processes of the plant societies, including the factors of their differentiation and development. In addition, the above-mentioned analytical method enables a syntaxonomic systematization of the vegetation that has been called a forest of the intermediate temperate zone (Suzuki 1961) or a deciduous broad-leaved forest (Kira et al. 1976) of the warm-temperate zone, both of which extend over the cool-temperate zone (*Fagetea crenatae* region) in Japan and the warm-temperate zone (*Camellietea japonicae* region).

Acknowledgment

The author is grateful to Dr. Asako Ichisawa and Dr. Tadashi Ikeda for technical assistance, and also to Prof. Dr. Jong-Suk Song for excellent supporting in this research.

References

Braun-Blanquet J (1959) Grundfragen und Aufgaben der Pflanzensoziologie. Vistas in Botany 1:145-171 (in German)

Dierschke H (1994) Pflanzensoziologie. Ulmer, Stuttgart (in German)

Ellenberg H (1978) Vegetation Mitteleuropas mit den Alpen. Ulmer, Stuttgart (in German)

Horikawa Y (1972) Atlas of the Japanese Flora. Gakken, Tokyo

Horikawa Y, Sasaki Y (1959) Phytosociological studies on the vegetation of Geihoku-district (the Sandankyo gorge and its vicinity), Hiroshima Prefecture. In: Schientific Resarchies of the Sandankyo gorge and the Yawata highland. Hiroshima-ken, Hiroshima, pp 85-107 (in Japanese with English summary)

Hotta M (1974) Evolutionary boiology in plants III. History and geography of plants, Sanseido, Tokyo (in Japanese)

Ito K (ed) (1987) Vegetation of Hokkaido. Hokkaido Univ Press, Sapporo (in Japanese)

Kawanishi M, Ishikawa S, Miyake N, Ohno K (2005) Forest floor vegetation of *Pterocarya rhoifolia* forests in Shikoku, with special reference to micro-landform. Veg Sci 22:87-102 (in Japanese with English abstract)

Kawanishi M, Sakio H, Kubo M, Shimano K, Ohno K (2006) Effect of micro-landforms on forest vegetation differentiation and life-form diversity in the Chichibu Mountains, Kanto District, Japan. Veg Sci 23:13-24

Kawanishi M, Sakio H, Ohno K (2004) Forest floor vegetation of *Fraxinus platypda-Pterocarya rhoifolia* forest along Ooyamazawa valley in Chichibu, Kanto District, Japan, with a special reference to ground disturbance. Veg Sci 21:15-26 (in Japanese with English abstract)

Kikuchi T (2001) Vegetation and landforms. Univ Tokyo Press, Tokyo (in Japanese)

Kim J-W (1992) Vegetation of northeast Asia. On the syntaxonomy and syngeography of the oak and beech forests. Abteilung für Vegetationsökologie und Naturschutzforschung Institut für Pflanzenphysiologie. Universität Wien, Wien

Kira T, Shidei T, Numata M, Yoda K (1976) Vegetation of Japan. Kagaku 46:235-247 (in Japanese)

Kubo M, Shimano K, Ohno K, Sakio H (2001) Relationship between habitats of dominant trees and vegetation units in Chichibu Ohyamasawa riparian forest. Veg Sci 18:75-85 (In Japanese with English abstract)

Maekawa F (1949) Makinoesia and its bearing to Oriental Asiatic flora. J Jpn Bot 24:91-96 (In Japanese with English résumé)

Maekawa F (1961) Some problems on the plant geography. Chiri 6:1030-1035 (in Japanese)

Miyawaki A (ed) (1977) Vegetation of Toyama Prefecture. Toyama-ken, Toyama (in Japanese with German summary)

Miyawaki A (ed) (1979) Reale Vegetation der Prafektur Nagano. Nagano-ken, Nagano (in Japanese with German summary)

Miyawaki A (ed) (1981) Vegetation of Japan, Kyushu. Shibundo, Tokyo (in Japanese with German summary)

Miyawaki A (ed) (1982) Vegetation of Japan, Shikoku. Shibundo, Tokyo (in Japanese with German summary)

Miyawaki A (ed) (1983) Vegetation of Japan, Chugoku. Shibundo, Tokyo (in Japanese with German summary)

Miyawaki A (ed) (1984) Vegetation of Japan, Kinki. Shibundo, Tokyo (in Japanese with German summary)

Miyawaki A (ed) (1985) Vegetation of Japan, Chubu. Shibundo, Tokyo (in Japanese with German summary)

Miyawaki A (ed) (1986) Vegetation of Japan, Kanto. Shibundo, Tokyo (in Japanese with English summary)

Miyawaki A (ed) (1987) Vegetation of Japan, Tohoku. Shibundo, Tokyo (in Japanese with English summary)

Miyawaki A (ed) (1988) Vegetation of Japan, Hokkaido. Shibundo, Tokyo (in Japanese with English summary)

Miyawaki A, Fujiwara K, Mochizuki R (1977a) Vegetation of Ubayashiki in northern Honshu, Iwate-ken. Bull Yokohama Phytosociol Soc Jpn 7:1-82 (in Japanese)

Miyawaki A, Ohba T (1963) *Castanopsis sieboldii*-Walder auf den Amami-Inseln. Sci Rep Yokoahama Nat Univ Sec II, 9:31-48 (in German)

Miyawaki A, Ohba T, Murase N (1964) Phytosociological investigation on the vegetation of the Tanzawa Mountains. In: Kanagawa-ken (ed) Report of scientific research in Tanzawa-Ooyama. Knagawa-ken, Yokohama, pp 54-102 (in Japanese with English summary)

Miyawaki A, Okuda S, Fujiwara R (1994) Handbook of Japanese vegetation. Shibundo, Tokyo (in Japanese)

Miyawaki A, Suzuki K, Fujiwara K, Harada H, Sasaki Y (1977b) Vegetation of Yamanashi Prefecture. Yamanashi-ken, Kofu (in Japanese with German summary)

Murata G (1995) Flora and vegetation zones of Japan. Jpn J Historical Bot 3:55-60 (in Japanese with English abstract)

Nakamura F (1995) Forest and stream interactions in riparian zone. Jpn J Ecol 45:295-300 (in Japanese)

Nakashizuka T, Yamamoto S (1987) Natural disturbance and stability of forest community. Jpn J Ecol 37:19-30 (in Japanese with English synopsis)

Oberdorfer E (1968) Assoziation, Gebietsassoziation, georaphische Rasse. In: Pflanzensoziologische Systematik. Bericht über das Internationale Symposium (1964) der Internationalen Vereinigung für Vegetationskunde, Junk, Haag, pp 124-141 (in German with English summary)

Oberdorfer E (1977) Suddeutsche Pflanzengesellschaften. Teil I. Gustav Fischer, Stuttgart, New York (in German)

Ohba T (1973) On the vegetation of Kiyotsu valley, central Japan. In: Natural Conservation Society of Japan (ed) Scientific report of the conservation in Kiyotsu river dam planning. Nature Conservation Society of Japan. Tokyo, pp 57-102 (in Japanese with German synopsis)

Ohba T, Sugawara H (1979) Bemerkung uber die japanischen Vorwald-Gesellschaften. In: Miyawaki A, Okuda S (eds) Vegetation und Landschaft Japans. Bull Yokohama Phytosociol Soc Jpn, Yokohama, pp 267-288 (in German)

Ohno K (1981) Pflanzensoziologische Forschungen über die Schluchtwälder des Camellietea japonicae-Bereiches in Südwest-Japan. Hikobia Suppl 1:83-90 (in German with English synopsis)

Ohno K (1982) A phytosociological study of the valley forests in the Chugoku Mountains, southwestern Honshu, Japan. Jpn J Ecol 32:303-324

Ohno K (1983) Pflanzensoziologische Untersuchungen über Japanische Flußufer- und Schluchtwälder der Montanen Stufe. J Sci Hiroshima Univ Ser B Div. 2 (Bot), 18: 235-286 (in German)

Ohno K (1994) Studies on the analysis of forest communities. Through the phytosociological approach. For Sci 10:24-27 (in Japanese)

Ohno K (1999) Encouragement for multidimensional syntaxonomy. Actinia 12:95-102 (in Japanese with English abstract)

Ohno K, Ozeki S (1997) Vegetation of Tanzawa Mountains (especially on the vegetation occurring in the Fagetea crenatae region). In: Kanagawa-ken (ed) Report of scientific research on the natural environments in Tanzawa-Ooyama. Kanagawa-ken, Yokohama, pp 103-121 (in Japanese)

Saito S, Okitu S (1987) Inorganic environment in Hokkaido. In: Ito K (ed) Vegetation of Hokkaido. Hokkaido Univ Press, Sapporo (in Japanese)

Sakai A, Ohsawa T, Ohasawa M (1995) Adaptive significance of sprouting of *Euptelea polyandra,* a deciduous tree growing on steep slopes with shallow soil. J Plant Res 108:377-386

Sakio H (2002a) What is a riparian forest? In: Sakio H, Yamamoto F (eds) Ecology of riparian forests. Univ Tokyo Press, Tokyo, pp 1-19 (in Japanese)

Sakio H (2002b) Valley-bottom forests and montane riparian forest. In: Sakio H, Yamamoto F (eds) Ecology of riparian forests. Univ Tokyo Press, Tokyo, pp 21-60 (in Japanese)

Sasaki Y (1979) Der Verband Pterocaryion rhoifoliae in Japan. In: Miyawaki A, Okuda S, (eds) Vegetation und Landschaft Japans. Bull Yokohama Phytosociol Soc, Yokohama, pp 213-226 (in German)

Song JS (1988) Phytosociological study of the mixed coniferous and deciduous broad-leaf forests in South Korea. Hikobia 10:145-156 (in Japanese with English abstract)

Suzuki H (1962) The classification of Japanese climates. Geogr Rev 35:205-211 (in Japanese with English résumé)

Suzuki T (1949) Vegetation in the upper stream-area of River Tenryu. Gijutu Kenkyu 1:77-

91 (in Japanese with English résumé)

Suzuki T (1952) Vegetation of East Asia. Kokon Shoin, Tokyo (in Japanese)

Suzuki T (1954) Forest and bog vegetation within Ozeghara basin. In: Scientific Researches of the Ozegahara moor. Japan Society for the Promotion of Science, Tokyo, 8:1-12

Suzuki T (1961) A review of Japanese forest zones. Chiri 6:1036-1043 (in Japanese)

Suzuki T (1966) Preliminary system of the Japanese natural comm.unities. Forest Env 8:1-12 (in Japanese with English résumé)

Suzuki T, Yuki K, Ooki M, Kanayama T (1956) Vegetation of Mt Gassan. In: Scientific Researches of Mt. Gassan and Asahi Mountains. Yamagata-ken, Yamagata, pp 144-199 (in Japanese)

Takahashi H (1971) Fossa Magna element plants. Res Rep Kanagawa Pref Mus Nat His 2:1 63 (in Japanese with English summary)

Tatewaki M (1958) Forest ecology of the islands of the north Pacific Ocean. J Fac Agric Hokkaido Univ 50:371-471

Tüxen R (1970) Entwicklung, Stand und Ziele der pflanzensoziologischen Systematik (Syntaxonomy). Ber dt bot Ges 83:633-639 (in German)

Tüxen R, Kawamura Y (1975) Gesichtspunkte zur syntaxonomischen Fassung und Gliederung von Pflanzengesellschaften entwckelt am Beispiel des nordwestdeutschen Genisto-Callunetum. Phytocoenologia 2:87-99 (in German)

Uematu H (1951) Geobotanical studies on the northern part of Fossa Magna region. J Jpn Bot 26:33-40 (in Japanese with English résumé)

Westhoff V, Maarel Evd (1973) The Braun-Blanquet approach. In: Whittaker RH (ed) Handbook of vegetation science, part 5. Ordination and classification of communities. Junk, Hague, pp 619-726

Yamanaka T (1962) Deciduous forests in the cool temperate zone of Shikoku. Res Rep Kochi Univ (Nat Sci I) 11:9-14

Yamazaki T (1959) Distribution of Japanese Plants. Natural Science and Museum 26:1-19 (in Japanese)

Yoshino M (1978) Climatology. Taimeido, Tokyo (in Japanese)

Part 4

Riparian forests in headwater stream

5 Coexistence mechanisms of three riparian species in the upper basin with respect to their life histories, ecophysiology, and disturbance regimes

Hitoshi Sakio[1], Masako Kubo[2], Koji Shimano[3] and Keiichi Ohno[4]

[1]Saitama Prefecture Agriculture & Forestry Research Center, 784 Sugahiro, Kumagaya, Saitama 360-0102, Japan (*Present address*: Sado Station, Field Center for Sustainable Agriculture and Forestry, Faculty of Agriculture, Niigata University, 94-2 Koda, Sado, Niigata 952-2206, Japan)

[2]Yamanashi Forest Research Institute,2290-1 Saishoji, Masuho, Yamanashi 400-0502 Japan

[3]Faculty of Science, Shinshu University, 3-1-1 Asahi, Matsumoto, Nagano 390-8621 Japan

[4]Faculty of Environment and Information Science, Yokohama National University, 79-7 Tokiwadai, Hodogaya, Yokohama, Kanagawa 240-8501 Japan

5.1 Introduction

Forest vegetation in the upper basin was strongly dependent upon the topography. More specifically, landforms vary in the riparian zone, and the valley floors are mosaics that include active channels, abandoned channels, floodplains, terraces, and alluvial fans (Gregory et al. 1991; Sakio 1997). These complex mosaic structures of landforms result from various disturbance regimes in the riparian zone, and the high diversity of microsites and disturbance regimes may promote the coexistence of canopy trees in riparian zones.

Reproductive traits and responses to the environment differ widely among tree species. To understand tree coexistence, Nakashizuka (2001) pointed out the importance of investigating the entire life history and demography of coexisting tree species in a community. In particular, the characteristics of trees are most

Sakio, Tamura (eds) Ecology of Riparian Forests in Japan : Disturbance, Life History, and Regeneration
© Springer 2008

varied in the initial stages (e.g., flowers, seeds, and seedlings), and differences in reproductive traits are important factors contributing to the ability of tree species to coexist in a forest. Water and light are important factors for the growth of trees in the cool-temperate zone. Gap formation accelerates the growth and flowering of trees, and while trees require moderate humidity in soil, flooding may retard their growth.

In the Chichibu Mountains of central Japan, several canopy tree species dominate and coexist in a natural riparian forest (Maeda & Yoshioka 1952; Tanaka 1985). This forest is primarily dominated by *Fraxinus platypoda* Oliv. and *Pterocarya rhoifolia* Sieb. et Zucc., but *Cercidiphyllum japonicum* Sieb. et Zucc. sometimes dominates and coexists with these canopy species (Sakio et al. 2002). The Chichibu Mountains have a complex topography with steep slopes (>30°) and a network of mountain streams (Sakio 1997), and these riparian species must have adaptive strategies for responding to the environment and to natural disturbances throughout their life histories.

Many tree species coexist in natural riparian forest in the upper basin (Sakio et al. 2002; Suzuki et al. 2002). The role of niche partitioning versus chance events of establishment in determining the diversity of tree regeneration is unresolved (Brokaw & Busing 2000). In this chapter, we examine how the coexistence of three canopy tree species relates to tree life history, ecophysiology, and distur-bance regime of the riparian zone in the Chichibu Mountains of central Japan. We focus on the following three topics: the population structure of three riparian tree species, their reproductive strategies, and their responses to light and water environments.

5.2 Disturbance regime of the upper basin riparian zone

Natural disturbances are important to the regeneration of riparian forests, and the disturbance regime of the riparian zone varies in type, frequency, size, and magni-tude. In steeper montane landscapes, valley floor landforms are sculpted by fluvial processes and a variety of mass soil movement processes from tributaries and adjacent hillslopes (Gregory et al. 1991).

Natural disturbances in the riparian zone under study is divided into two types. One is the small disturbances that occur on the ground surface without the improvement of light conditions. In stream bars, sedimentation and erosion of sand and gravel by stream flow are repeated in the rain and typhoon season every year. These small disturbances prevent the establishment of seedlings while producing new safe sites for them (Sakio 1997). In the Chichibu Mountains, large typhoons accompanied by diurnal precipitation over 300 mm have occurred once every decade in the past century (Saitama Prefecture & Kumagaya Local Meteorological Observatory 1970). These heavy rains result in debris flow (Fig. 1), surface landslides, and channel movements without improvement in light

Fig. 1. Debris flow in the riparian zone. This occurred in 1982 due to heavy rain during a typhoon

Fig. 2. Large-scale landslide in which bedrock was broken

conditions. The other type of disturbance involves large areas with the destruction of canopy trees, resulting in improved light conditions. For example, large earthquakes or typhoons can cause large mass movements through landslides (Fig. 2).

High diversity of the disturbance regime is closely related to microtopographic variation, and each habitat is different in the various environments of, for example, soil, light, and water. In the active channel, the surface was covered with rock, gravel, and sand, while in the floodplain, some soil layers had formed due to repeated sedimentation. On a hillslope, a thick litter layer and humus layer was observed in the soil profile. Light conditions are altered by canopy gaps. In the summer season, the direct rays of the sun provide over 70,000 lux in single-tree gaps, and scattered light is less than 1000 lux beneath the canopy layer. As a result, a complex mosaic structure of habitats occurs in the riparian zone.

5.3 Population structure

5.3.1 Species composition

The study site (Ooyamazawa Riparian Forest Research Site: 35°57'N, 138°45'E) is a typical riparian forest in the cool-temperate deciduous broad-leaved forest of the Chichibu Mountains, central Japan. The area around the study plot is important in that human disturbances, such as logging or erosion control, have not yet affected it. This forest was classified as Cacalio yatabei–Pterocaryetum rhoifoliae

Table 1. Species composition of trees over 4 cm in diameter at breast height (DBH). Study plot is 4.71 ha in area and 1170 m in stream length.

Species	No. of tree	No.of canopy tree	(%)	DBH(cm) of canopy tree	
				Average	Maximum
Abies homolepis Sieb. et Zucc.	12	4	(0.8)	62.9	93.0
Acer amoenum Ohwi	3				
Acer argutum Maxim.	91				
Acer carpinifolium Sieb. et Zucc.	442				
Acer cissifolium K. Koch	1				
Acer distylum Sieb. et Zucc.	1				
Acer mono Maxim.	267	12	(2.4)	42.6	72.4
Acer mono f. *dissectum* Rehd.	3	1	(0.2)	21.5	21.5
Acer nikoense Maxim.	2				
Acer nipponicum Hara	12	1	(0.2)	28.7	28.7
Acer palmatum Thunb.	6				
Acer rufinerve Sieb. et Zucc.	11	1	(0.2)	33.6	33.6
Acer shirasawanum Koidz.	378	2	(0.4)	46.7	54.4
Acer tenuifolium Koidz.	7				
Acer tschonoskii Maxim.	1				
Actinidia arguta Planch. ex Miq.	13	1	(0.2)	14.5	14.5
Betula grossa Sieb. et Zucc.	3	3	(0.6)	36.2	50.8
Betula maximowicziana Regel	4	4	(0.8)	66.5	73.0
Carpinus cordata Blume	51				
Carpinus japonica Blume	1				
Celtis jessoensis Koidz.	1				
Cercidiphyllum japonicum Sieb. et Zucc.	59	47	(9.6)	76.8	153.4
Clethra barbinervis Sieb. et Zucc.	3				
Euonymus sieboldianus Blume	5				
Euptelea polyandra Sieb. et Zucc.	10				
Fagus crenata Blume	6	1	(0.2)	62.6	62.6
Fraxinus apertisquamifera Hara	7				
Fraxinus lanuginosa Koidz.	2	1	(0.2)	19.5	19.5
Fraxinus platypoda Oliv.	417	303	(61.6)	56.9	140.5
Hydrangea petiolaris Sieb. et Zucc.	1				
Kalopanax pictus Nakai	3	3	(0.6)	57.7	78.0
Phellodendron amurense Rupr.	1	1	(0.2)	35.5	35.5
Prunus buergeriana Miq.	1				
Prunus grayana Maxim.	1				
Pterocarya rhoifolia Sieb. et Zucc.	113	81	(16.5)	44.6	77.7
Pterostyrax hispida Sieb. et Zucc.	79				
Schizophragma hydrangeoides Sieb. et Zucc.	6				
Sorbus alnifolia C. Koch	4				
Stewartia pseudo-camellia Maxim.	2				
Swida controversa Soják	2	1	(0.2)	23.9	23.9
Tilia japonica Simonkai	9	5	(1.0)	60.5	93.9
Trochodendron aralioides Sieb. et Zucc.	2				
Tsuga sieboldii Carr.	2				
Ulmus laciniata Mayr	91	19	(3.9)	55.9	89.4
Viburnum furcatum Blume	4				
Vitis coignetiae Pulliat	4	1	(0.2)	5.7	5.7
Total	2144	492	(100.0)		

and it contains a total of 230 species of vascular plants (Kawanishi et al. 2006). The main herb species on the floor are *Mitella pauciflora* Rosend., *Asarum caulescens* Maxim., *Meehania urticifolia* Makino, and *Dryopteris polylepis* C. Chr. We found 2144 individual trees (DBH ≥ 4 cm) of 46 species in a 4.71-ha study plot (Table 1).   The plot contained 492 individual trees of 20 canopy species. The dominant canopy species were *F. platypoda*, *P. rhoifolia*, and *C. japonicum*, which accounted for 61.6, 16.5, and 9.6%, respectively, of the total number of canopy stems. Dominant sub-canopy species included *Acer shirasawanum* Koidz., *Acer mono* Maxim., *Carpinus cordata* Blume, and *Pterostyrax hispida* Sieb. et Zucc., while *Acer carpinifolium* Sieb. et Zucc. and *Acer argutum* Maxim. dominated the understory.

5.3.2 Size and spatial structure

Mean DBH of canopy tree species (Fig. 3) differed significantly among the dominant canopy tree species (Kruskal-Wallis test, $P < 0.0001$). The DBH (mean ± SD) of *F. platypoda* was 45.3 ± 25.5 cm with a maximum of 140.5 cm. *Pterocarya rhoifolia* measured 35.6 ± 17.8 cm with a maximum of 77.7 cm, and *C. japonicum* had a mean DBH of 64.9 ± 34.8 cm with a maximum of 153.4 cm. The maximum DBH of *P. rhoifolia* was about half that of the other species.

Fraxinus platypoda and *P. rhoifolia* showed two peaks (under 10 cm and 40 cm in DBH), but *C. japonicum* did not. This suggests that *F. platypoda* and *P. rhoifolia* have sapling banks, and synchronous regeneration as indicated by the peaks in the 40-cm class. The age class distribution of *F. platypoda* had a peak in the 200–220-year class corresponding with the 40–50-cm DBH range (Sakio 1997). This synchronous regeneration might have been caused by large distur-bances such as landsides, debris flows, and earthquakes.

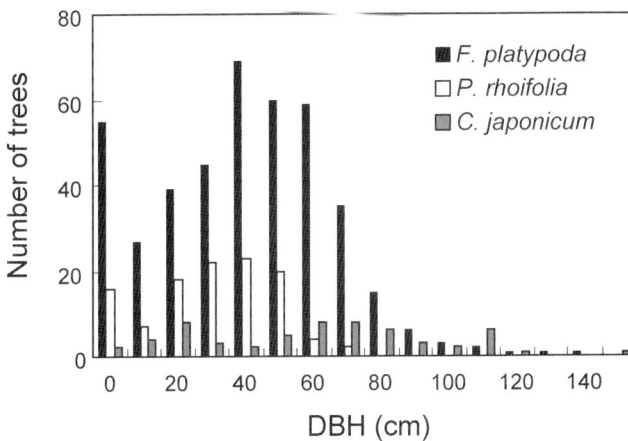

Fig. 3. DBH class distribution (10-cm interval) of three riparian tree species.

The spatial pattern of *F. platypoda* and *C. japonicum* showed a random distribution, but *P. rhoifolia* had an aggregated distribution in some clumps on the deposits of large-scale landslides (Fig. 4). Following the analysis of annual growth rings, we estimated the mean ages of *P. rhoifolia* trees in patch A, B, and C to be 88.0 ± 7.6 (mean $\pm$ SD) years, 88.7 ± 5.8 years, and 94.3 ± 7.9 years, respectively, which suggests that synchronous regeneration of *P. rhoifolia* in these three large-disturbance sites occurred about 90 years ago. The age of *C. japonicum*

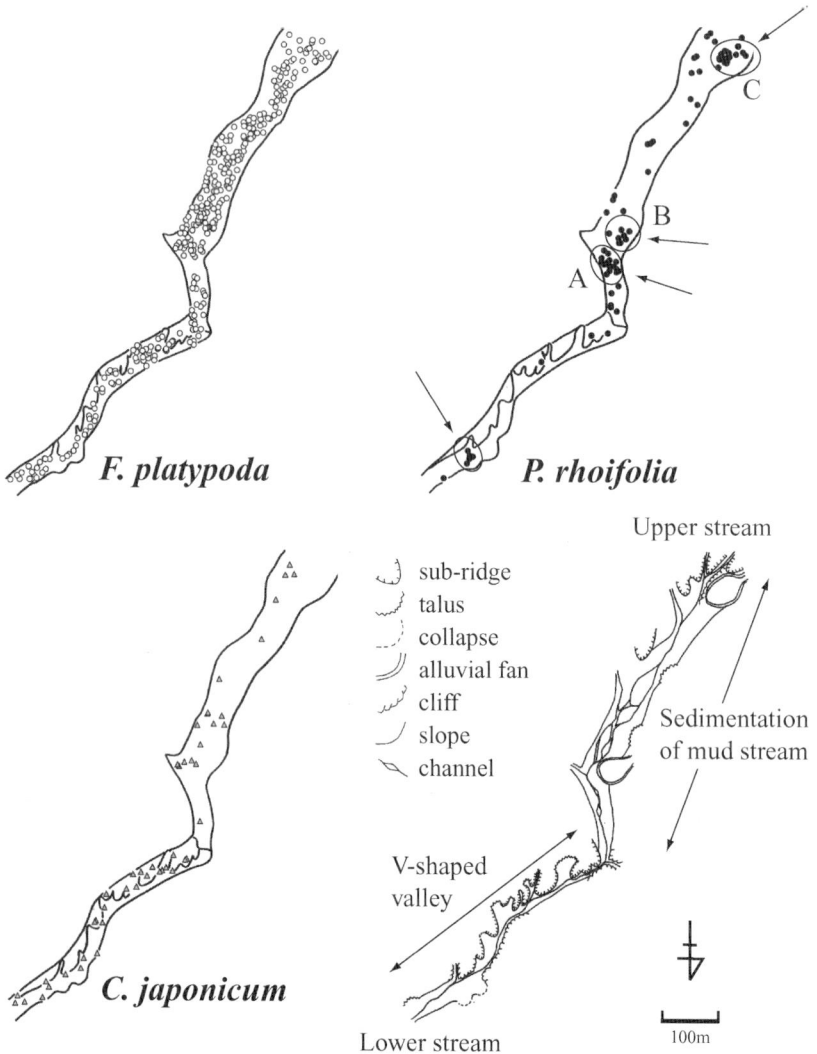

Fig. 4. Spatial distribution of three riparian canopy tree species and micro-topography (Kubo et al. 2001). Arrows indicate large landslides with canopy destruction

sub-canopy trees around these patches is the same as that of *P. rhoifolia*, indicating that these species invaded the large-disturbance site at the same time, and that *P. rhoifolia* grew faster than *C. japonicum*.

5.4 Reproductive strategies of riparian tree species

5.4.1 Seed characteristics and production

Fraxinus platypoda, *P. rhoifolia*, and *C. japonicum* produce wind-dispersed fruits with wings (Fig. 5). Oven-dried fruits of *F. platypoda*, *P. rhoifolia*, and *C. japonicum* (mean ± SD, $n = 20$) weighed 144 ± 24 mg, 90 ± 11 mg, and 0.82 ± 0.15 mg, respectively (Sakio et al. 2002). Fruits and seeds of *C. japonicum* were significantly lighter than those of the other species (Welch's *t*-test, $P < 0.001$) and were distributed over substantial distances. The maximum seed dispersal distance of *C. japonicum* was over 300 m as determined by microsatellite analysis (Sato et al. 2006). As the seed dispersal of *C. japonicum* begins in November after leaf fall of the canopy layer, seeds can flow over long distances without obstruction of the leaf layer. In contrast, the maximum seed dispersal distance of *F. platypoda* was about 20–30 m, and its seeds are dispersed at the same time as leaf fall. However, the seeds of *F. platypoda* suggest an adaptation to waterborne fruit dispersal and have a strategy of secondary seed dispersal by stream water (Fig. 6). The seed dispersal distance of *P. rhoifolia* is about the same as *F. platypoda*, and *P. rhoifolia* may exhibit hydrochorous dispersal.

In general, climax species have large seeds and mast years, and pioneer species have small seeds and regular fruiting. In general, these three species exhibited irregular fruiting behaviors (Fig. 7). *Fraxinus platypoda* had non-mast years in 1997 and 2001 and mast years in 1996, 1998, 2002, and 2004. In other words, this species showed irregular fruiting as in *Fraxinus excelsior* L. (Tapper 1992, 1996).

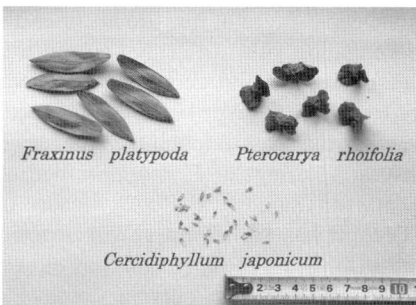

Fig. 5. Fruits of riparian tree species

Fig. 6. Dispersed fruits of *F. platypoda* by stream water

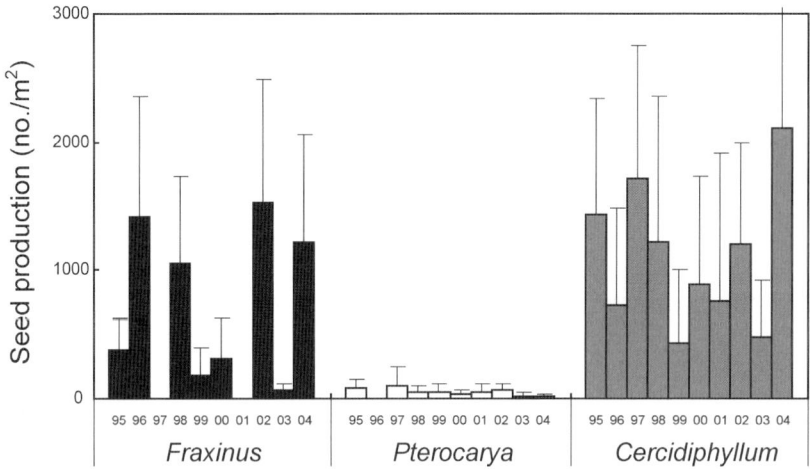

Fig. 7. Annual fluctuation of seed production in the three riparian tree species from 1995 to 2005

Cercidiphyllum japonicum produced a certain amount of seeds in every year. *Fraxinus platypoda* had the largest coefficient of variation (CV) in seed production (CV = 100.05); the corresponding CV of *P. rhoifolia* was 66.2, whereas *C. japonicum* (CV = 49.61) had the smallest variation of the three species.

Fraxinus platypoda exhibited large seeds, mast fruiting, and a short dispersal distance of seeds. In contrast, *C. japonicum* had small seeds dispersed over substantial distances and produced a lot of seeds every year. In *F. platypoda* and *C. japonicum*, a life-history trade-off existed between seed weight and seed number.

5.4.2 Seedling establishment

Germination and seedling establishment in the early stages of tree life histories are very important for forest regeneration. Germination sites differ among the tree species studied. *Fraxinus platypoda* and *P. rhoifolia* could germinate on gravel along an active channel and on forest floor with a litter layer because their seeds and seedlings are large enough to produce roots that pass through the litter layer and reach the ground. Most of the current-year seedlings of *P. rhoifolia* died on the forest floor litter layer during the current year (Sakio et al. 2002) because of low light intensity and drought, but the decline was slower on gravel deposits without litter. The decline in current-year seedlings of *F. platypoda* was slower than that of *P. rhoifolia*. The seedlings of *F. platypoda* can survive for several years, and seedling banks of this species occurred on gravel deposits along the channel (Sakio 1997). Their seedling banks can survive at least 10 years beneath canopy gaps, including those of *P. rhoifolia*.

In contrast, *C. japonicum* is adapted to germinate on fine soil without litter or on fallen logs (Kubo et al. 2000). Kubo et al. (2004) showed that the current-year seedlings of *C. japonicum* emerged from fine bare soil as opposed to soil with litter or gravel treatments in the nursery. Seedling emergence of small-seeded species was reduced by litter (Seiwa & Kikuzawa 1996). However, most of the current-year seedlings of *C. japonicum* that emerged on fine soil surfaces were washed away by the heavy rain during the rainy season or a typhoon. As a result, we did not see many seedlings of *C. japonicum* in this area, and found only two saplings >1 m tall in the plot (4.71 ha) at the Ooyamazawa Riparian Forest Research Site (Sakio et al. 2002).

5.4.3 Sprouts

Cercidiphyllum japonicum has more sprouts than *F. platypoda* and *P. rhoifolia* (Fig. 8). The number of *C. japonicum* sprouts was positively correlated with the DBH of the main stem ($r = 0.37$, $P < 0.05$) with a maximum of 60 sprouts, but no significant correlation was observed between the number of sprouts and DBH of the main stem in *F. platypoda* or *P. rhoifolia*. A typical *C. japonicum* stool is shown in Fig. 9. Two main stems stood at the center of the stool and several sprouts emerged from the main stems. Many species can sprout under the brighter light conditions following overstory removal by humans (Rydberg 2000; Louga et al. 2004).   Even in undisturbed stable forests, light conditions are important for sprouting (Koop 1987; Ohkubo et al. 1988, 1996; Peter & Ohkubo 1990). After the death of main stems, *C. japonicum* fills the ensuing gap with sprouts (Kubo et al. 2005), which apparently play a role in the maintenance of individuals.

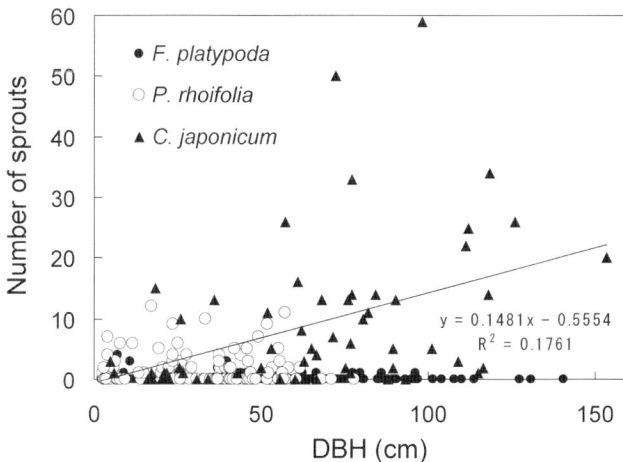

Fig. 8. Relationship between DBH of the main stem and the number of sprouts of each individual

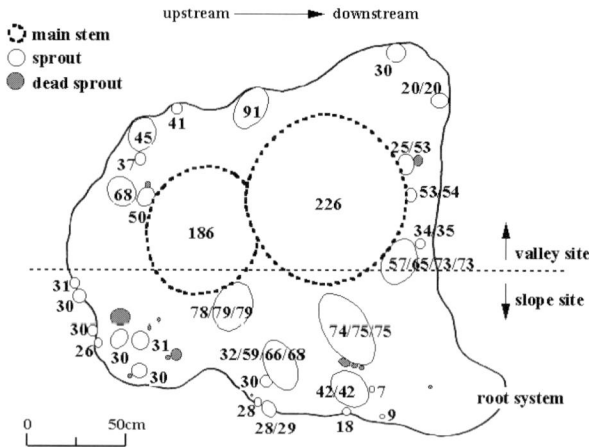

Fig. 9. Distribution of main stems and sprouts by age class in the *C. japonicum* stand (Kubo et al. 2005). Numbers indicate the age of stems and sprouts

5.5 Responses to light and water stress

5.5.1 Effects of the light environment on seedling growth

Canopy gaps play an important role in forest regeneration and the coexistence of plant species. Gap formation increases light intensity and accelerates the growth of tree saplings. In this riparian forest, many saplings of *F. platypoda* and *P. rhoifolia* occurred in abandoned channels and gravel deposits, but the sapling layer of *C. japonicum* consisted of only two individuals in a 4.71-ha plot (Sakio et al. 2002). The saplings of *F. platypoda* were found not only in canopy gaps but under the canopy, whereas *P. rhoifolia* was found only in canopy gaps (Sakio 1993). Figure 10 shows the height of current-year seedlings of the three riparian tree species by light treatments in the nursery (Sakio, unpublished data).

At 10% relative light intensity (RLI), we saw no difference in the height of current-year seedlings among the three species. The heights of current-year seed-lings of *P. rhoifolia* increased according to the increase in RLI and were about 30 cm over 40% RLI, which were higher than for the other species. In forest gaps, the growth of current-year shoots of *P. rhoifolia* was faster than that of *F. platypoda* (Sakio 1993). We found no difference in these values between *F. platypoda* and *C. japonicum*, but the growth of *C. japonicum* may be very fast considering that the fruit weight of *F. platypoda* is 175 times larger than that of *C. japonicum*. In the riparian zone, a wide range of gap sizes occurs from as small as a tree to more than 50 m in diameter. The larger forest gaps are captured by *P. rhoifolia* through competition among the tree species invading each gap (Ann & Oshima 1996).

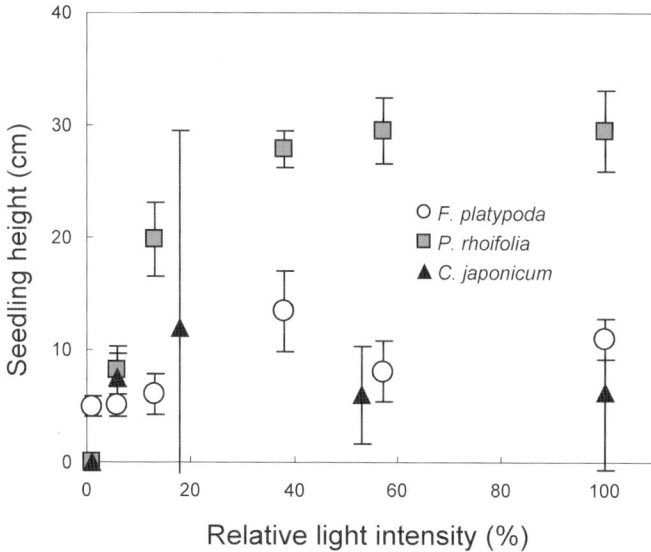

Fig. 10. Relative light intensity and height of current-year seedlings

5.5.2 Seedling growth responses to water stress

The growth of flooded trees is often retarded and they do not survive. Many reports have shown the physiological, morphological, and anatomical responses of tree seedlings to flooding in wetland tree species and coniferous trees (Yamamoto & Kozlowski 1987a, b; Yamamoto et al. 1995a, b). In the Chichibu Mountains, current-year seedlings and small-sized seedlings near the stream are covered by rising water during the rainy or typhoon season every year, with flooding lasting up to 10 days.

This flooding may have an effect on the seedlings of *F. platypoda* and *P. rhoifolia* distributed on sand or gravel deposits near the active channel. Flooding affected the growth of the 1-year-old seedlings of the three riparian species in this cool-temperate region, as well as wetland tree species (Sakio 2005). The flooding treatment during one growing season was initiated by raising the water level to the soil surface in the unglazed pots in which three species of seedlings were planted. The flooding treatment reduced the number of attached leaves, shoot elongation, dry weight increment, and the survival ratio. For all species, the mean values of the stem and branch dry weights were lower in flooded plants than in control plants. Dry weights did not differ significantly among control and flooded seedlings of *F. platypoda*, but they did in the gap-dependent riparian species, including *P. rhoifolia* and *C. japonicum* (Fig. 11). The survival ratio of flooded plants after the experiment differed by species. All *F. platypoda* seedlings survived, whereas 80 and 70% of the *P. rhoifolia* and *C. japonicum* seedlings, respectively, survived.

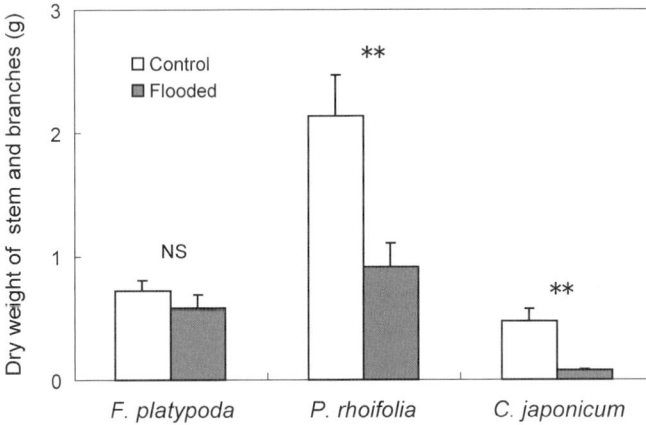

Fig. 11. Dry weights of flooded and control seedlings of three species. Vertical bars represent standard errors. Asterisks indicate the level of significance for differences from control seedlings. Double asterisks denote the 1% level; NS indicates not significant

Table 2. Flooding periods (days) and survival ratio (%) of current-year seedlings

Flooding periods	*F. platypoda*	*P. rhoifolia*	*C. japonicum*
0	90	90	100
2	100	100	80
4	100	70	80
10	100	70	10
20	80	20	0
30	0	0	0

In the case of current-year seedlings, the difference in the effect of flooding was more conspicuous among the three species. The survival rate of flooded plants after the experiment differed by species (Table 2). For *F. platypoda* current-year seedlings, over 80% survived a flooding treatment of 20 days, and all seedlings died after 30 days of flooding. For *P. rhoifolia* current-year seedlings, over 70% survived a flooding treatment of 10 days, 20% survived 20 days of flooding, and all seedlings died after 30 days of flooding, while among *C. japonicum* current-year seedlings, over 80% survived 4 days of flooding, 10% survived 10 days flooding, and all seedlings died after 20 and 30 days of flooding.

Fraxinus platypoda seedlings had relatively high tolerance to the flooding treatments. However, flooding seriously affected the growth of the gap-dependent riparian species (*P. rhoifolia* and *C. japonicum*). The effects of flooding differed among the three species, and *F. platypoda* had higher survival under flood conditions than the other species.

Fraxinus mandshurica Rupr. var. *japonica* Maxim., native to swampy areas, grew adventitious roots with many aerenchyma in the bark tissue and produced ethylene during flooding treatments (Yamamoto et al. 1995b; Nagasaka 2001). Sakio (2002) demonstrated that planted *F. platypoda* trees near the stream form adventitious roots on buried stems. Like *F. mandshurica*, this species may also be adapted to flooded environments. These results show one reason why *F. platypoda* is the dominant species in riparian zones in which flooding and movement of sand or gravel occur repeatedly. The responses of tree seedlings to flooding reflected species habitat and growth patterns (Sakio 2005).

5.6 Coexistence mechanisms of three riparian species

The three riparian species have different regeneration strategies. In the riparian zone, *F. platypoda*, a shade-tolerant species, regenerated not only in single-tree gaps in the canopy by way of advanced saplings, but also in large disturbance sites via dispersed seeds (Sakio 1997). Moreover, the saplings of *F. platypoda* were more tolerant of flooding than the other species. For these reasons, *F. platypoda* was a dominant canopy species in the riparian zone, which experienced more variable disturbances with regard to type, frequency, and magnitude.

Pterocarya rhoifolia had a high growth rate during the seedling and sapling stage in open sites. This species, having adapted its reproductive characteristics to unpredictable conditions, was dependent on large disturbances for regeneration.

Cercidiphyllum japonicum has exhibited sprouting from canopy stems for long-term maintenance of individuals, and has continued to wait for new establishment sites with the mass dispersal of seeds every year through the miniaturization of seeds.

Natural disturbance without destruction of canopy trees in the riparian zone occurred at high frequencies. Under these conditions, *F. platypoda*, a shade-tolerant species, is the dominant species because of adaptation to growth in small gaps. However, large disturbances with the destruction of some canopy trees allow the chance of regeneration for *P. rhoifolia* and *C. japonicum.* Improvement in light conditions and safe sites for germination allow the establishment and growth of these species. In these environments, these two species grow faster than *F. platypoda* and form the canopy layer.

Variety in type, frequency, and magnitude of natural disturbances in riparian zones result in various microhabitats for the establishment of the three species. Ecological characteristics differ among the three tree species at each life-history stage (Table 3), but the coexistence of canopy trees was maintained in riparian areas through some niche partitioning and through unpredictable large-scale disturbances. In the riparian zone, niche and chance are key factors in the coexistence of tree species.

Table 3. Ecological characteristics of the three riparian canopy species studied

Ecological characteristics		*Fraxinus*	*Pterocarya*	*Cercidiphyllum*
Canopy Tree	Number	61.6%	16.5%	9.6%
	Sprout	Few	Intermediate	Many
	Life span (Years)	<300	<150	<500
Sapling	Number	Many	Intermediate	Very few
	Shade tolerance	High	Low	Low
	Growth rate	Low	High	Intermediate
Seedling	Number	Many	Intermediate	Few
(current)	Shade tolerance	High	Low	Low
	Survival ratio	High	Low	Very low
	Establishment site	Gravel	Gravel	Fine soil
Seed	Seed size	Large	Large	Very small
	Dispersal pattern	Wind/Water	Wind/Water	Wind
	Dispersal distance	20–30 m (Wind)	20–30 m (Wind)	300 m
	Mast year	Yes	Yes	No ?
Flood tolerance	(1-year old)	High	Intermediate	Intermediate
	(current year)	High	Intermediate	Low

5.7 Conclusions

In this chapter, we examined forest structure, reproductive strategies, and responses to the light and water environments of three riparian tree species in the
Chichibu Mountains of central Japan. In riparian areas, the three dominant canopy
species are well adapted to disturbances throughout their life histories. These
species have different habitat requirements, but variation in the disturbance regime
results in habitat differentiation in the riparian zone. The coexistence of canopy
trees is maintained through niche partitioning and chance in the early life-history
stages. Chance events such as unpredictable large-scale disturbances affect the
regeneration of *P. rhoifolia* and *C. japonicum*. To determine the most important
factor, long-term studies that correlate the occurrence of large-scale disturbances
and the recovery of riparian forests are needed.

References

Ann SW, Oshima Y (1996) Structure and regeneration of *Fraxinus spaethiana - Pterocarya
    rhoifolia* forests in unstable valleys in the Chichibu Mountains, central Japan. Ecol Res
    11:363-370
Brokaw N, Busing RT (2000) Niche versus chance and tree diversity in forest gaps. Trend
    Ecol Evol 15(5):183-188

Gregory SV, Swanson FJ, Mckee WA, Cummins KW (1991) An ecosystem perspective of riparian zones: focus on links between land and water. BioScience 41(8):540-551

Kawanishi M, Sakio H, Kubo M, Shimano K, Ohno K (2006) Effect of micro-landforms on forest vegetation differentiation and life-form diversity in the Chichibu Mountains, Kanto District, Japan. Veg Sci 23:13-24

Koop H (1987) Vegetative reproduction of trees in some European natural forests. Vegetatio 72:103-110

Kubo M, Shimano K, Sakio H, Ohno K (2000) Germination sites and establishment conditions of *Cercidiphyllum japonicum* seedlings in the riparian forest. J Jpn For Soc 82(4):349-354 (in Japanese with English summary)

Kubo M, Shimano K, Ohno K, Sakio H (2001) Relationship between habitats of dominant trees and vegetation units in Chichibu Ohyamasawa riparian forest. Veg Sci 18: 75-85 (in Japanese with English summary)

Kubo M, Sakio H, Shimano K, Ohno K (2004) Factors influencing seedling emergence and survival in *Cercidiphyllum japonicum.* Folia Geobot 39:225-234

Kubo M, Sakio H, Shimano K, Ohno K (2005) Age structure and dynamics of *Cercidiphyllum japonicum* sprout based on growth ring analysis. For Ecol Manage 213:253-260

Luoga EJ, Witkowski ETF, Balkwill K (2004) Regeneration by coppicing (resprouting) of mimbo (African savanna) trees in relation to land use. For Ecol Manage 189: 23-35

Maeda T, Yoshioka J (1952) Studies on the vegetation of Chichibu Mountain forest.(2) The plant communities of the temperate mountain zone. Bull Tokyo Univ For 42:129-150+3pls (in Japanese with English summary)

Nagasaka A (2001) Effect of flooding on growth and leaf dynamics of two-year-old deciduous tree seedlings under different flooding treatments. Bull Hokkaido For Res Inst 38:47-55 (in Japanese with English summary)

Nakashizuka T (2001) Species coexistence in temperate, mixed deciduous forests. Trend Ecol Evol 16(4):205-210

Ohkubo T, Kaji M, Hamaya T (1988) Structure of primary Japanese beech (*Fagus japonica* Maxim.) forest in the Chichibu Mountains, central Japan, with special reference to regeneration processes. Ecol Res 3:101-116

Ohkubo T, Tanimoto T, Peters R (1996) Response of Japanese beech (*Fagus japonica* Maxim.) sprouts to canopy gaps. Vegetatio 124:1-8

Peters R, Ohkubo T (1990) Architecture and development in *Fagus japonica-Fagus crenata* forest near Mount Takahara, Japan. J Veg Sci 1. 499-506

Rydberg D (2000) Initial sprouting, growth and mortality of European aspen and birch after selective coppicing in central Sweden. For Ecol Manage 130:27-35

Saitama Prefecture & Kumagaya Local Meteorological Observatory (1970) Weather disaster of Saitama Prefecture. Saitama Prefecture & Kumagaya Local Meteorological Observatory. Saitama (In Japanese)

Sakio H (1993) Sapling growth patterns in *Fraxinus platypoda* and *Pterocarya rhoifolia.* Jpn J Ecol 43:163-167 (in Japanese with English synopsis)

Sakio H (1997) Effects of natural disturbance on the regeneration of riparian forests in a Chichibu Mountains, central Japan. Plant Ecol 132:181-195

Sakio H (2002) Survival and growth planted trees in relation to the debris movement on gravel deposit of a check dam. J Jpn For Soc 84(1):26-32 (in Japanese with English summary)

Sakio H, Kubo M, Shimano K, Ohno K (2002) Coexistence of three canopy tree species in a riparian forest in the Chichibu Mountains, central Japan. Folia Geobot 37:45-61

Sakio H (2005) Effects of flooding on growth of seedlings of woody riparian species. J For Res 10:341-346

Sato T, Isagi Y, Sakio H, Osumi K, Goto S (2006) Effect of gene flow on spatial genetic structure in riparian canopy tree *Cercidiphyllum japonicum* revealed by microsatellite analysis. Heredity 96:79-84

Seiwa K, Kikuzawa K (1996) Importance of seed size for the establishment of seedlings of five deciduous broad-leaved tree species. Vegetatio 123:51-64

Suzuki W, Osumi K, Masaki T, Takahashi K, Daimaru H, Hoshizaki K (2002) Disturbance regimes and community structure of a riparian and an adjacent terrace stand in the Kanumazawa Riparian Research Forest, northern Japan. For Ecol Manage 157:285-301

Tanaka N (1985) Patchy structure of a temperate mixed forest and topography in the Chichibu Mountains, Japan. Jpn J Ecol 35:153-167

Tapper PG (1992) Irregular fruiting in *Fraxinus excelsior*. J Veg Sci 3:41-46

Tapper PG (1996) Long-term patterns of mast fruiting in *Fraxinus excelsior*. Ecology 77(8):2567-2572

Yamamoto F, Kozlowski TT (1987a) Effect of flooding of soil on growth, stem anatomy, and ethylene production of *Cryptomeria japonica* seedlings. Scand J For Res 2:45-58

Yamamoto F, Kozlowski TT (1987b) Effects of flooding, tilting of stems, and ethrel application on growth, stem anatomy and ethylene production of *Pinus densiflora* seedlings. J Exp Bot 38:293-310

Yamamoto F, Sakata T, Terazawa K (1995a) Growth, morphology, stem anatomy, and ethylene production in flooded *Alnus japonica* seedlings. IAWA J 16:47-59

Yamamoto F, Sakata T, Terazawa K (1995b) Physiological, morphological and anatomical responses of *Fraxinus mandshurica* seedlings to flooding. Tree Physiol 15:713-719

6 Population dynamics and key stages in two Japanese riparian elements

Yuko KANEKO[1] and Takenori TAKADA[2]

[1]Lake Biwa Environmental Research Institute, 5-34 Yanagasaki, Otsu, Shiga 520-0022, Japan

[2]Graduate School of Environmental Earth Science, Hokkaido University, N10W5 Kita-ku, Sapporo 060-0810, Japan

6.1 Introduction

The vegetative landscapes in riparian forests differ from those in upper hill-slope forests (Hiroki 1987; Oshima et al. 1990; Yamanaka et al. 1993; Sugita et al. 1995; Suzuki et al. 2002). Riparian forests suffer more frequently from disturbance than do upper hill-slope forests because active water channels constitute the most powerful geomorphic processes in riparian areas (Nakamura 1990; Gregory et al. 1991; Ito & Nakamura 1994; Kaneko 1995). Riparian forests are suitable subjects with which to study the effects of natural disturbance regimes on plant populations. The effects of natural disturbance will vary among plant species in different habitats and may be reflected in the vital rates, i.e., plant growth rate, survival rate, and fecundity. We analyzed the demographic structure and population dynamics of two typical dominant riparian tree species based on census data collected for eight years, from 1989 through 1996, in a cool-temperate forest in Japan. Censuses were conducted of all individuals, from current-year seedlings to mature trees. The results of these long-term demographic studies and matrix analyses provide much new knowledge concerning the demographic parameters and population dynamics of two typical riparian species.

6.2 Riparian habitats and disturbance regimes

The study site was located in the Kyoto University Ashiu Experimental Forest, Kyoto Prefecture, Japan. The study plots were established in a typical cool-

Sakio, Tamura (eds) Ecology of Riparian Forests in Japan : Disturbance, Life History, and Regeneration
© Springer 2008

temperate deciduous broadleaved forest within the Mondoridani watershed in Ashiu (35°20' N, 135°44' E; 688.7–836.5 m above sea level; watershed area = 16.0 ha). Ashiu is characterized by a typical Japan Sea climate. The snow season lasts for five months, from December through April. Topographically, the Mondoridani valley consists of a non-riparian area (from ridges to upper hill slopes) and a riparian area (from lower hill slopes to terraces and floodplains) along a continuous transect of the valley system. A distinct feature of the riparian habitat is the frequent natural disturbance, which causes the formation of a mosaic of different micro-topographical and micro-environmental conditions. We divided the study area into three habitats based on topography: slope, terrace (high floodplain), and floodplain (low floodplain).

Natural disturbance regimes may create spatial variation because the effects vary depending on the topography. Natural disturbances are unpredictable and temporally variable. Differences among natural disturbance regimes also provide different patterns of gaps (i.e., areas in which the height of the regeneration canopy layer is < 10 m) that have different properties (i.e., spatial distribution, size, density). Gap dynamics create spatial and temporal variations caused by natural disturbance regimes. The disturbance and gap properties of the three habitats are shown in Table 1 (see Kaneko et al. 1999 for details of other environmental conditions). These differences in background environments may sharply determine the patterns of ecological distribution and structures of plant populations in riparian habitats that develop over an environmental gradient.

Flooding frequently disturbs floodplain habitats. There are a few large gaps in areas subjected to destructive flooding. Two or more gap makers often form such large gaps. The percentage of the total gap area is high on the floodplains. Canopy gaps that are formed by a physically destructive event such as a typhoon are usually accompanied by geomorphic disturbances, which often alter the forest floor entirely and create numerous new open sites. The slope habitat consists of lower and toe slopes, with gradients from 30° to 50°. On the slopes, one or few gap makers often

Table 1. Disturbance and canopy gap properties along a valley system

Topography (% of total area in the Mondoridani valley)	Riparian habitats			Non-riparian habitats		
	Floodplain 4.1	Terrace 6.3	Lower slope 22.5	Middle slope 24.5	Upper slope 17.4	Ridge 25.2
Agents of geomorphic disturbance	Flooding	Flooding/ Landslide	Landslide	Landslide	Landslide	Landslide
Frequency of geomorphic disturbance	10^0–10^1 year	10^2 year	10^0–10^2 year	10^0–10^2 year	10^0–10^2 year	10^2 year
Agents of other physical disturbances	Wind/Snow/ Treefall/ Branchfall	Wind/Snow/ Treefall/ Branchfall	Wind/Snow/ Treefall/ Branchfall	Wind/Snow/ Treefall/ Branchfall	Wind/Snow/ Treefall/ Branchfall	Wind/Snow/ Treefall/ Branchfall
Total gap area (% of habitat)	High 29.8	Low 14.6	Medium 20.0	Medium 18.3	Low 14.9	High 22.4
Mean gap size (m^2)	Large 145.7	Large 120.9	Medium 99.6	Medium 60.5	Medium 60.7	Small 47.8
Density of gaps (no./ha)	Low 14.8	Low 9.5	High 34.6	Medium 24.8	Medium 24.0	High 34.4

leads to the formation of many small gaps, although the percentage of the total gap area remains low. Canopy gaps that are created by small-scale treefalls or branchfalls are not usually accompanied by geomorphic disturbances of the forest floor. Such gap areas are safe sites for seeds and current-year seedlings to establish because a thick and mature forest soil develops on the slopes, even though large-scale geomorphic disturbances caused by landslides sometimes occur. The terrace habitat occurs between the floodplain and slope habitats and is characterized by a flat terrace surface with a thick and nutrient-rich forest soil. Terraces are usually quite stable habitats, although the effects of large-scale landslides on slopes or flooding by rivers can occasionally reach terraces and create gaps.

Most major disturbances in Japan are caused by typhoons. During the study period, geomorphic disturbances such as flooding, bank undercutting, and landslides occurred in the study area because of a major typhoon in September 1990, which released 287 mm of rain in two days. Such severe typhoons hit Ashiu at intervals of several years (Ando et al. 1989) and have occurred in 1990, 1998, and 2004. Some natural physical disturbances and effects on tree populations in the Mondoridani valley are shown in Table 2 (see Kaneko 1995 for details). There are two effects of disturbances: the temporary direct and destructive effect and the time-delayed, extended, and beneficial effect.

6.3 Study species and their ecological niches

Aesculus turbinata Blume (Japanese horse chestnut; hereafter referred to as *Aesculus*) and *Pterocarya rhoifolia* Sieb. et Zucc. (Japanese wingnut; hereafter referred to as *Pterocarya*) were selected for study. These are deciduous, dominant canopy species and are representative riparian elements in Japanese temperate forests. *Aesculus* is the only native species of Hippocastanaceae in Japan. It can live for 250 years and grows to approximately 25 m in height, with a diameter at breast height (DBH) of 150 cm. *Pterocarya* is the only volunteer species of its genus in Japan. It has a lifespan of about 150 years and grows to approximately 30 m in height, with a DBH of 100 cm. Demographic censuses of tagged individuals revealed size-dependent growth, mortality, and flowering patterns for both species (Kaneko et al. 1999; Kaneko & Kawano 2002). Populations of the two species can be divided into five stages based on their growth and mortality patterns (see Kaneko 2005): seeds, established current-year seedlings (S), monopodial juveniles (J_1), branching and pre-reproductive juveniles (J_2), and mature trees (M). Figure 1 shows a stage-classified life cycle diagram for each species (Kaneko 1998).

The two species have a persistent juvenile stage in which stunted juveniles, awaiting release from shading, survive close to the photosynthetic compensation point until a gap opens in the overtopping layer. Neither species has a dormant seed pool in the soil. The waiting juveniles in the study area were divided into two stage groups: juveniles with a monopodial stem and those with branches. The waiting span of *Aesculus* is longer than that of *Pterocarya*. The two species have life-history characteristics intermediate to those of typical climax and pioneer species.

Table 2. Effects of natural physical disturbances on tree populations in a mountainous riparian area

Category	Agent	Cause	Events affecting trees	Temporary direct effects on individual trees	Main factors causing time-delayed, extended effects
Geomorphic	Flood	Heavy rain	Submergence of valley bottom or floodplain	Damage to a tree	Transport of forest floor layer
				Transport of a tree	Disturbance to surface soil
		Excess precipitation	Scour riverbank	Damage to a tree	Appearance of treefalls and canopy gaps
				Uprooting	
			Deposition of sedimentation	Part or whole of a tree buried	Formation of bare sites on forest floor
	Slope landslide	Heavy rain	Landcreep of topsoil	Damage to a tree	Formation of bare sites on forest floor
				Transport of a tree	
		Excess precipitation	Deposition of colluvial soil	Uprooting	Formation of bare sites in a canopy gap
		Continuous precipitation	Fully saturated soil	Part or whole of a tree buried	Formation of bare sites on forest floor
		Snowmelt		Uprooting	Formation of bare sites in a canopy gap
Physical or other than geomorphic	Wind	Violent storm	Violent storm	Damage to a tree	Change in micrometeorology
				Uprooting	Formation of bare sites in a canopy gap
	Snow	Snow cover	Snow loads	Damage to a tree	Change in micrometeorology
			Pressure by moving snow on slopes	Roots rise to surface	Transport of forest floor layer
		Snowmelt	Sheet erosion	Uprooting	Disturbance to surface soil
	Treefall	Violent storm	Falling trees	Damage to a tree	Change in micrometeorology
		Continuous precipitation		Crushed by a falling tree	Appearance of treefalls
				Uprooting	Formation of bare sites in a canopy gap
	Branch-fall	Violent storm	Falling branches	Damage to a tree	Change in micrometeorology
		Snow load		Crushed by a falling branch	Appearance of branchfalls

Pterocarya is considered more adapted to stochastic environments and has less shade-tolerant characteristics than *Aesculus*. We emphasize the importance of the waiting juvenile stages of long-lived tree species in understanding factors limiting the ecological distribution patterns and species composition of the plant community.

Differences in the disturbance regime often cause patterns of intra- and interspecific competition (Alvarez-Buylla & Garcia-Barrios 1993; Oshima & Takeda 1993; Takada & Nakashizuka 1996). It is also very important to understand the demographic structure of each riparian species in relation to the mechanisms of coexistence within a multispecies system over an environmental gradient, and furthermore, to estimate the potential persistence of a particular riparian species in changing habitats.

Aesculus (left):
ovules (pollinated by bumble bees) → immature fruits
mature seeds → attacked seeds
viable seeds (no dormant period) ← (dispersed by wood mice) the other sites
emergent seedings
S
J₁
J₂
M
Death

Pterocarya (right):
ovules (pollinated by wind) → immature fruits
mature seeds → attacked seeds
viable seeds (no dormant period) ← (dispersed by wind) the other sites
emergent seedings
S
J₁
J₂
M
Death

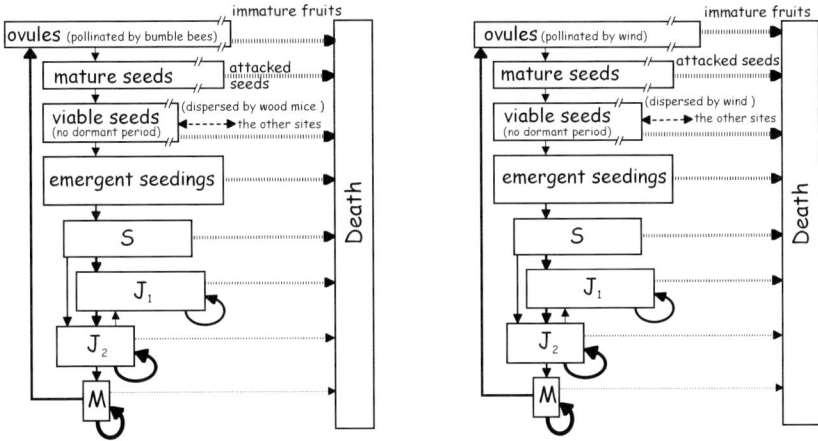

Fig. 1. Stage-classified life cycle diagrams for populations of *Aesculus* (left) and *Pterocarya* (right).

Table 3. Ecological distribution of major woody elements and vegetation properties (> 10 cm in diameter at breast height)

Topography		Riparian habitats			Non-riparian habitats		
		Floodplain	Terrace	Lower slope	Middle slope	Upper slope	Ridge
	Pterocarya	83.1	0.3	0.0	0.0	0.0	0.0
Dominance	*Aesculus*	0.2	44.0	7.3	0.7	0.0	0.0
(% of total	*Quercus*	0.0	25.7	10.9	7.1	6.0	5.5
basal area)	*Fagus*	2.5	17.2	34.4	25.6	15.0	7.5
	Cryptomeria	0.0	1.8	27.3	46.6	66.9	73.1
Stem density (no./ha)		Low	Low	Medium	Medium	High	High
		155.9	122.6	277.0	498.3	640.2	784.4
Total basal area of stems		Small	Small	Medium	Medium	Large	Large
(m²/ha)		12.8	15.4	26.2	38.4	50.4	67.7

Earlier vegetation studies of the Mondoridani valley demonstrated that distinct ecological distributions of woody species occur over an environmental gradient, forming a continuum consisting of several predominant species (Oshima et al. 1990; Yamanaka et al. 1993). There is characteristic niche partitioning among these major woody elements, reflecting a background environment (e.g., availability of resources such as soil moisture, nutrients, available gap sites, light, etc.) provided under a localized disturbance regime for each topographic habitat and the intensity of biotic interactions (Table 3). Within the non-riparian area, the dominant canopy layer trees are *Fagus crenata* Blume and *Cryptomeria japonica* (L. fil.) D. Don var. *radicans* Nakai, both of which are representative temperate forest elements in Japan. From the lower hill slope to the terrace within the riparian area, the canopy layer trees are gradually replaced by *Aesculus* and *Quercus crispula* Blume. *Aesculus* is abundant in slope and terrace habitats and is often accompanied by a thick understory of dwarf bamboo, *Sasa senanensis* (Franch. et Savat.) Rehder, on the forest floor. *Pterocarya* is dominant on the floodplain. *Aesculus* and *Pterocarya* co-occur in the riparian area because of different microhabitat selection.

Fitness is measured by the population growth rate. The relative importance of the two species' local subpopulations in fitness was given in Kaneko et al. (1999) and Kaneko and Kawano (2002). Elasticity analysis using a topographically combined projection matrix showed that the sum of the elasticity values was zero in the floodplain subpopulation of *Aesculus* and in the slope subpopulation of *Pterocarya*. Considering the metapopulation concept (Hanski & Gilpin 1997), this clearly indicates that the floodplain subpopulation constitutes an island sink population, whereas the slope and terrace subpopulations constitute the mainland source population for *Aesculus*, providing the main source of offspring recruitment. In *Aesculus*, the effect of the slope subpopulation on the population growth rate is 10.1 times as great as that of the terrace subpopulation. In *Pterocarya*, the slope subpopulation constitutes an island sink population, whereas the floodplain and terrace subpopulations constitute the mainland source populations. The effect of the floodplain subpopulation on the population growth rate is 5.8 times as great as that of the terrace subpopulation.

6.4 Spatio-temporal variations in population growth rate under typhoon disturbances along a riparian environmental gradient

Recent spatio-temporal analyses of census data under different disturbance regimes have shown that vital rates can fluctuate differently in neighboring plant subpopulations, depending on localized disturbance regimes in the different habitats (Damman & Cain 1998; Martinez-Ramos & Samper 1998; Kaneko et al. 1999; Kaneko & Kawano 2002; Garcia 2003; Kwit et al. 2004). Disturbance regimes produce the environment that affects the vitality of individuals. How do disturbance regimes affect population growth? It is likely that local subpopulations have different population dynamics. Spatio-temporal analyses, not only of vitality of individuals, but also of population statistics, are necessary to study the effect of spatial differences in disturbance regimes and unpredictable temporal patterns of disturbances on the maintenance mechanisms of local populations. The effects of disturbances on vitality are usually diverse and stage specific, and vital rates can be collected into population projection matrices. Matrix population models provide a tool with which to synthesize these effects into population statistics that can be calculated from projection matrices and that quantify the effects of disturbances at the population level. The population growth rate λ is the most important and most frequently used statistic because λ measures fitness of the species. Life table response experiments (LTRE) were carried out to compare the demographic responses of the two species to different disturbance regimes at the population level. LTRE analysis is a form of retrospective population perturbation analysis and can

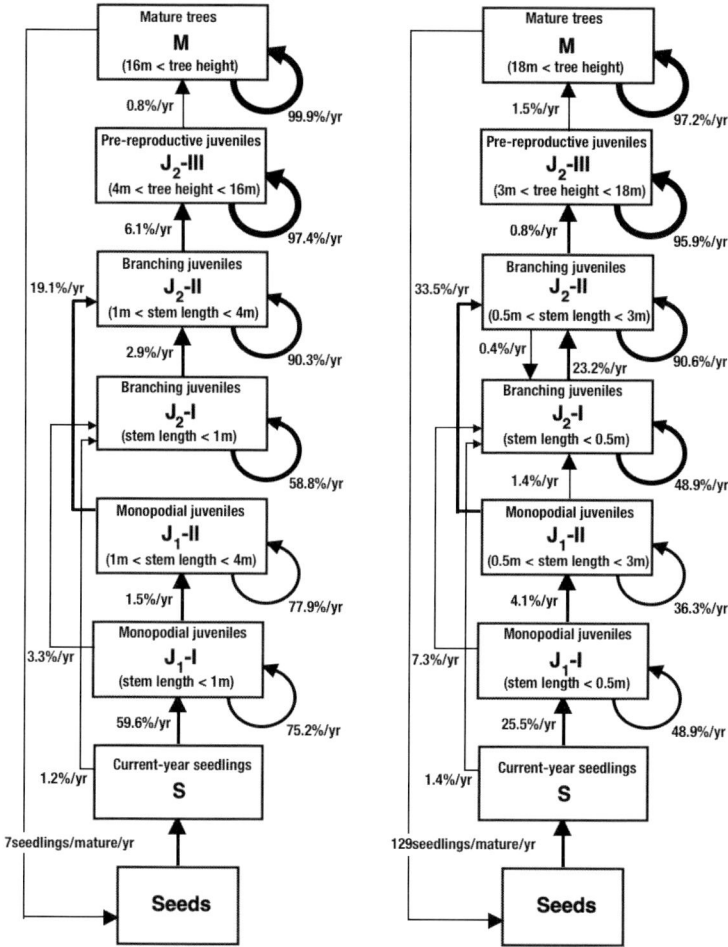

Fig. 2. Flow charts indicating the life cycles of *Aesculus* (left) and *Pterocarya* (right).

be classified by their design in analogy to analysis of variance (Caswell 2001).

Fluctuations in population growth rates were analyzed using the fixed factorial design of LTRE analysis based on data from 1989 to 1992 for natural *Aesculus* populations (Kaneko & Takada 2003, unpublished data). A stage-classified projection matrix was made according to Kaneko et al. (1999). The first matrix was developed by pooling all data from the source subpopulations over a four-year period to give a reference matrix. Values for each matrix entry in the reference matrix are shown in Figure 2. Six projection matrices were then made separately for the two subpopulations (slope and terrace in *Aesculus*, and terrace and floodplain in *Pterocarya*) for each year period. When all data were pooled, the whole population showed an increase in size, with population growth rates of the six matrices show-

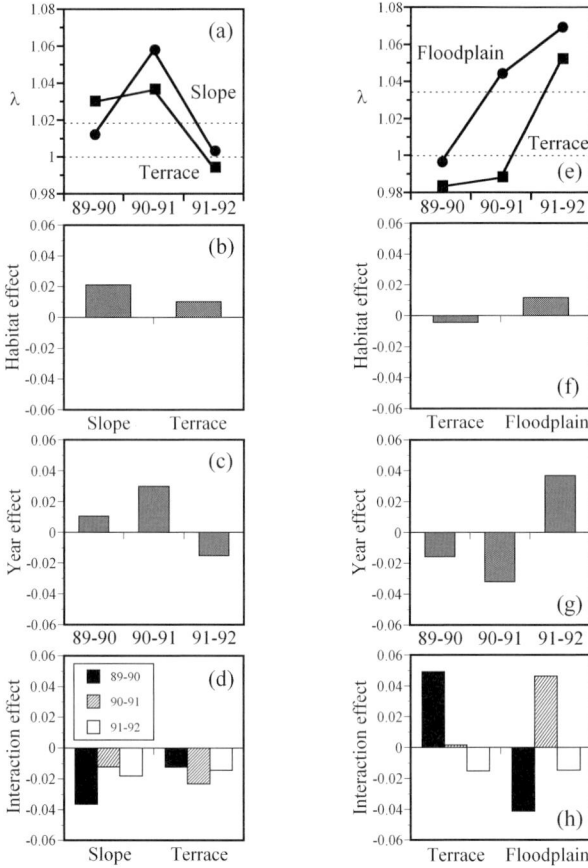

Fig. 3. (a) and (e): Population growth rate in each year; (b) and (f): main effects of habitat; (c) and (g): main effects of year; (d) and (h): interaction effects for *Aesculus* (left) and *Pterocarya* (right).

ing spatio-temporal variations from 0.994 to 1.058. The population growth rate of the reference matrix was 1.0195. The population growth of *Aesculus* was largest in the slope habitat in 1990–1991 (Fig. 3a). The population growth of *Aesculus* was also larger in the slope habitat than in the terrace habitat, except in the typhoon year of 1989–1990. The slope habitat had a larger positive effect than that of the terrace habitat (Fig. 3b). The year 1989–1990 had a larger negative effect on the slope habitat than on the terrace habitat (Fig. 3d). As mentioned in Section 3, the slope subpopulation is the mainland source population and has the largest prospective effect on population growth rates (Kaneko et al. 1999).

Fluctuations in population dynamics with respect to typhoon disturbance were analyzed based on data from 1989 to 1992 for natural *Pterocarya* populations (Kaneko & Takada 2003, unpublished data). A stage-classified projection matrix was made according to Kaneko and Kawano (2002). A reference matrix and six projection matrices were made in the same way as for *Aesculus*. The values of each matrix entry in the reference matrix are shown in Figure 2. The whole population showed an increase in size, despite the effects of the major typhoon in 1990. The population growth rate of *Pterocarya* was smallest in 1989–1990 and has since increased (Fig. 3e). The population growth rate of the reference matrix was 1.0344. The year 1989–1990 had the largest negative effect, and the year 1990–1991 had the largest positive effect on the floodplain (Fig. 3h). The total *Pterocarya* population decreased in size in 1990, but increased rapidly after 1991. The population growth of *Pterocarya* was greater in the floodplain habitat than in the terrace habitat in all year periods. The floodplain habitat showed a positive effect, whereas the terrace showed a negative effect on the population growth (Fig. 3f). As mentioned in Section 3, the floodplain subpopulation is the mainland source population and has the largest prospective effect on the population growth rate (Kaneko & Kawano 2002).

The population growth rate may have been reduced because mortality increased, growth was impaired, or reproduction was limited. Are these causes all equally responsible for the effects on the population growth rate or can some of the effects be attributed to each cause? To answer these questions requires a decomposition of the effect of the disturbance regime on population growth into elements arising from the effect on each stage-specific value.

6.5 What are the key stages for population dynamics?

To what extent does each matrix entry that measures stage-specific vitality contribute to the population growth rate? The contributions obtained from the fixed factorial LTRE method are shown in Tables 4 and 5 for *Aesculus*. Figure 4 represents the results in Tables 4 and 5 (Koop & Horvitz 2005). The means of the absolute values of the contributions of each matrix entry ($a_{i,j}$) to the habitat effects are given in Table 4. The most influential entries in determining habitat effects were: $a_{3,2}$ (growth of monopodial J_1-I juveniles to branching J_1-II juveniles); $a_{7,6}$ (growth of pre-reproductive J_2-III juveniles to mature M trees); and $a_{2,1}$ (growth of current-year seedlings, S, to J_1-I juveniles). Spatial variations in the *Aesculus* population growth rate are caused by the survival rates of juveniles. Mortality from biotic causes (e.g., starvation, consumption by herbivores, diseases) was tree size-dependent, whereas mortality from disturbances (e.g., geomorphic processes, wind) occurred in all size classes (Kaneko et al. 1999). Mortality from disturbances was highest in floodplain habitats and lowest in terrace habitats because the subpopulations on terrace sites were not affected by geomorphic disturbances (Kaneko et al. 1999). The differences between the three habitat types with varying disturbance regimes overrode year-to-year variation in environmental stresses (type, scale, and frequency of disturbance). Mortality from disturbances is

Table 4. Contribution matrix for stage-specific contributions to habitat effects on *Aesculus* population growth rate

Stage-class at Year $t+1$	Stage-class at Year t						
	S	J_1-I	J_1-II	J_2-I	J_2-II	J_2-III	M
S	0	0	0	0	0	0	*0.00247*
J_1-I	0.00618 [3]	**0.00265**	0	0.00066	0	0	0
J_1-II	0	0.00954 [1]	**0.00147**	0	0	0	0
J_2-I	0.00073	0.00044	0	**0.00210**	0	0	0
J_2-II	0	0	0.00213	0.00324 [4]	**0.00058**	0	0
J_2-III	0	0	0	0	0.00151	**0.00146**	0
M	0	0	0	0	0	0.00778 [2]	**0.00268** [5]
Total	0.00690	0.01263	0.00360	0.00599	0.00209	0.00925	0.00516

Table 5. Contribution matrix for stage-specific contributions to year effects on *Aesculus* population growth rate

Stage-class at Year $t+1$	Stage-class at Year t						
	S	J_1-I	J_1-II	J_2-I	J_2-II	J_2-III	M
S	0	0	0	0	0	0	*0.01440* [1]
J_1-I	0.00138	**0.00190**	0	0.00041	0	0	0
J_1-II	0	0.00886 [2]	**0.00296**	0	0	0	0
J_2-I	0.00062	0.00071	0	**0.00051**	0	0	0
J_2-II	0	0	0.00639 [4]	0.00284	**0.00300** [5]	0	0
J_2-III	0	0	0	0	0.00234	**0.00206**	0
M	0	0	0	0	0	0.00671 [3]	**0.00275**
Total	0.00200	0.01147	0.00935	0.00376	0.00534	0.00877	0.01715

Bold letters show the diagonal elements. Italic letter shows contibution of fecundity element. Little letters show the rank

site-dependent. For example, the terrace environment was almost four times better for the growth of J_2-III to M stages of *Aesculus* than the slope environment (Kaneko et al. 1999). On the slope, all survivors of the S stage advanced to the J_1-I stage, but on the terrace and floodplain, some survivors advanced to the J_2-I stage. The fates of J_2-I juveniles also varied between the sites. On the slope, 83.3% of the survivors remained in the J_2-I stage, whereas on the terrace, 50.0% of the survivors regressed to the J_1-I stage. Individuals that regressed to the J_1-I stage were forced to survive at that stage, with its higher mortality risk.

The population growth rate of *Aesculus* was greatest in 1990–1991 because 1990 was a mast year in slope habitats. The typhoon year of 1989–1990 had a larger negative effect on population growth rate in the slope habitat than in the terrace habitat because the terrace is a more stable habitat for *Aesculus*. The large contributions to the year effects were from $a_{1,7}$ (fecundity), $a_{3,2}$ (growth of J_1-I to J_1-II juveniles), and $a_{7,6}$ (growth of J_2-III to mature trees) (Table 5, Fig. 4). Annual changes in fecundity are mainly caused by annual changes in seed production. LTRE fixed analysis identified the growth of monopodial waiting juveniles and pre-reproductive juveniles and fecundity as important contributions to both habitat and

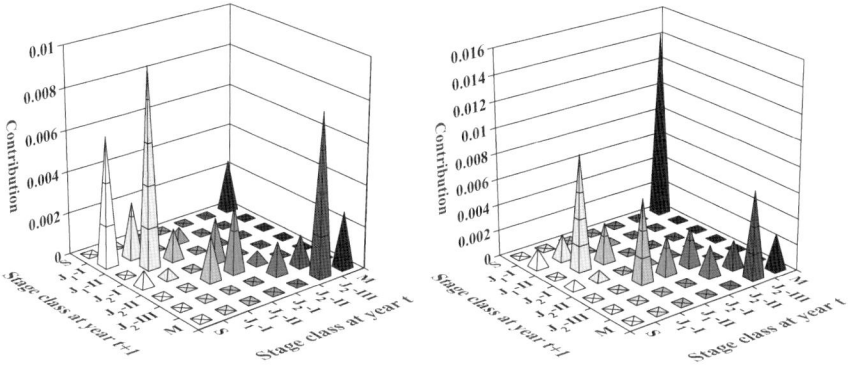

Fig. 4. Mean of the absolute value of the contributions of each matrix element to the habitat effect (left) and to the year effect (right) for *Aesculus*

Table 6. The indices (i,j) of the four most important matrix entries for *Aesculus*

Rank	Prospective analysis	Retrospective analysis (LTRE fixed effects)		
	Elasticity	Habitat	Year	Habitat-by-Year
1	7, 7	3, 2	1, 7	1, 7
2	6, 6	7, 6	3, 2	7, 7
3	5, 5	2, 1	7, 6	3, 3
4	3, 3	5, 4	5, 3	5, 3

year effects. The fates of a monopodial persistent juvenile stage and a pre-reproductive stage are identified as key events for *Aesculus*, both in predicting the outcome of hypothetical perturbations and in explaining the observed variation in population growth rates (Table 6).

The contributions obtained from the fixed factorial LTRE method are shown in Figure 5 for *Pterocarya*. The most influential entries determining habitat effects were: $a_{1,7}$ (fecundity), $a_{7,7}$ (survival of mature trees), and $a_{6,5}$ (growth of branching J_2-II juveniles to J_2-III). Spatial variation in the *Pterocarya* population growth rate was more affected by the mortality of mature trees (not caused by typhoon disturbances) than of juveniles (caused by typhoon disturbances). The large contributions to the year effects were from $a_{1,7}$, $a_{7,7}$, and $a_{5,5}$ (stasis of J_2-II juveniles). Since the major typhoon in 1990, several geomorphic disturbances have occurred that have caused not only physical destructive effects on vital rates, but also extended beneficial effects in surviving individuals or the establishment of newly recruited seedlings. The first mast year after the typhoon of 1990 occurred in 1991; this, together with the extended beneficial effect of the typhoon, led to the largest positive effect occurring in 1991–1992. Retrospective analyses showed that the most influential entry in determining year effects in both species was fecundity. This

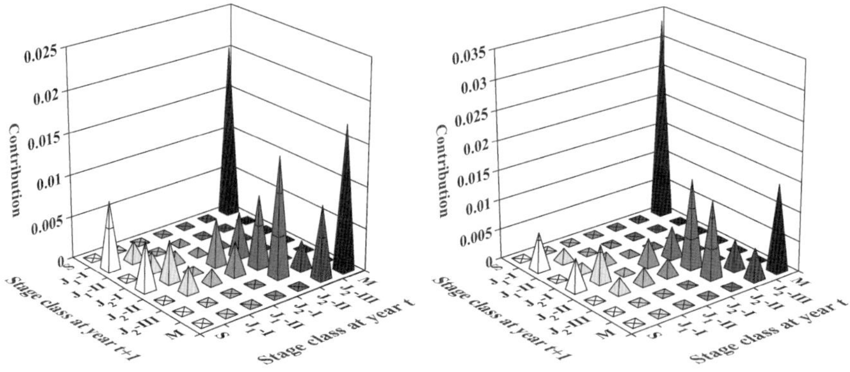

Fig. 5. Mean of the absolute value of the contributions of each matrix element to the habitat effect (left) and to the year effect (right) for *Pterocarya*

Table 7. The indices (i,j) of the four most important matrix entries for *Pterocarya*

Rank	Prospective analysis	Retrospective analysis (LTRE fixed effects)		
	Elasticity	Habitat	Year	Habitat-by-Year
1	7, 7	1, 7	1, 7	6, 5
2	6, 6	7, 7	5, 5	2, 1
3	5, 5	6, 5	7, 7	4, 1
4	6, 5	7, 6	6, 5	4, 4

clearly reflects the direct and extended beneficial effect of the typhoon and the effect of temporal variation in seed production. However, unlike *Aesculus*, yearly flux in the fecundity of *Pterocarya* was mainly related to annual changes in survival from seed dispersal to seedling establishment in the following year.

The LTRE fixed analysis identified branching waiting juveniles, mature trees, and fecundity as important to both habitat and year effects, implying that, instead of growing rapidly or producing more seeds, the best strategy for *Pterocarya* is to regulate growth and decrease the risk of mortality in the J_2-II and M stages. In *Pterocarya*, the transition to, and fates of, branching persistent juveniles and mature individuals are key to predicting the outcome of hypothetical perturbations and in explaining observed variations in the population growth rate (Table 7).

6.6 Conclusion

Natural disturbances play important roles in triggering regeneration in many types of forest. The site-dependent patterns of advantages and disadvantages for individuals may determine the population dynamics of riparian tree species. The topography that is most beneficial to some stages may not be good for other stages. In the studied species, not only was the survival at each stage affected by the habitat, but similar variations in stage classes were observed in the fate of individual trees, i.e., in how much they regressed, stayed the same, or grew. As mentioned above, geomorphic disturbances such as flooding may have physically destructive effects on the vital rates of individuals in a local population, but at the same time, have beneficial effects on the surviving individuals or the establishment of newly recruited seedlings through the development of gaps on the forest floor. Therefore, for a riparian species, long-term spatial and temporal analyses of demographic processes are necessary to understand the effects of disturbance regimes on the maintenance mechanisms of their populations.

The elasticity matrices for the two woody plant populations indicated that the survival rates of individuals at the same stage, especially the larger size stages of M, J_2-III, and J_2-II, contributed more to the population growth rate than did the other investigated parameters. The maintenance of individuals at the pre-reproductive and reproductive stages would have the largest effect on λ and is the most important factor for population maintenance. These two stages are also the most efficient for maintaining the fitness of the species. Fecundity, on the other hand, had little effect on λ according to the prospective analysis. Prospective analyses indicate the most vulnerable factors for a population, i.e., where it may be attacked or protected, and provide the best guidance for species conservation (Caswell 2001).

However, prospective analyses cannot reveal how vital rates varied in the past, are varying now, or might vary in the future. Retrospective analysis, in contrast, is concerned with precisely this kind of variation and how it contributes to actual variation in λ (Caswell 2000). The results of the LTRE fixed analysis highlighted the most important factors in the population dynamics. For *Aesculus*, these were the fates of monopodial persistent juveniles and pre-reproductive juveniles and fecundity; for *Pterocarya*, these were the fates of mature trees and branching persistent juveniles and fecundity. The magnitude of variance in fecundity was greater than 10^4 and 10^7 times as large as that of the other elements, a_{ij}, in *Aesculus* and *Pterocarya* populations, respectively. Nevertheless, the LTRE fixed contribution of fecundity was not large because the sensitivity value of fecundity was small. The LTRE contribution of fecundity was only 2.3 times as great as that of the transition probability from J_1-II to J_2-II in the *Aesculus* population. In the *Pterocarya* population, the LTRE fixed contribution of the survival of mature trees to the variance of λ was 8.1 times as great as that of fecundity. This clearly indicates that evaluating the long-term growth processes of juveniles and plants in the reproductive phase is important and necessary, not only for understanding population maintenance mechanisms and life-history strategies, but also for the conservation of canopy tree species.

References

Alvarez-Buylla ER, Garcia-Barrios R (1993) Models of patch dynamics in tropical forests. Trend Ecol Evol 8:201-204

Ando M, Noborio H, Kubota J, Kawanabe S (1989) Analysis of weather station data at Ashiu Experimental Forest (I). Bull Kyoto Univ For 61:25-45 (in Japanese)

Caswell H (2000) Prospective and retrospective perturbation analyses: their roles in conservation biology. Ecology 81(3):619-627

Caswell H (2001) Matrix Population Models – Construction, Analysis, and Interpretation. Second edition. Sinauer Associates, Inc. Sunderland, MA, USA

Damman H, Cain ML (1998) Population growth and viability analyses of the clonal woodland herb, *Asarum canadense*. J Ecol 86:13-26

Garcia MB (2003) Demographic viability of a relict population of the critically endangered plant, *Borderea chouardii*. Conserv Biol 17(6):1672-1680

Gregory SV, Swanson FJ, McKee WA, Cummins KW (1991) An ecosystem perspective of riparian zones – Focus on links between land and water. BioScience 41:540-551

Hanski IA, Gilpin ME (1997) Metapopulation biology. Ecology, genetics, and evolution. Academic Press, New York, NY, USA

Hiroki S (1987) The difference in regeneration between *Aesculus turbinata* and *Pterocarya rhoifolia* in the Migimata valley of Mt. Hodaka. In: Papers on plant ecology and taxonomy in memory of Dr. Satoshi Nakanishi, The Kobe Geobotanical Society, pp 319-323 (in Japanese with English summary)

Ito S, Nakamura F (1994) Forest disturbance and regeneration in relation to earth surface movement. J For Environ 36(2):31-40 (in Japanese with English summary)

Kaneko Y (1995) Disturbance regimes of a mountainous riparian forest and effects of disturbance on tree population dynamics. Jpn J Ecol 45:311-316 (in Japanese)

Kaneko Y (1998) Demography and Matrix Analyses of Two Japanese Riparian Elements. D Sc Thes Kyoto Univ

Kaneko Y (2005) Life-history strategies of *Aesculus turbinata* and *Pterocarya rhoifolia* in a riparian forest. 111-136. In: The society for the study of species biology (ed) Science of plant biology. Bun-ichi, Tokyo (in Japanese)

Kaneko Y, Takada T, Kawano S (1999) Population biology of *Aesculus turbinata* Blume.: a demographic analysis using transition matrices on a natural population along a riparian environmental gradient. Plant Species Biol 14:47-68

Kaneko Y, Kawano S (2002) A demographic and matrix analysis on a natural *Pterocarya rhoifolia* population developed along a mountain stream. J Plant Res 115:341-354

Kaneko Y, Takada T (2003) Spatiotemporal analyses of population growth rates on tree populations using mathematical models. Abstracts of the 50th annual meeting of the ecological society of Japan, pp 252 (in Japanese)

Koop AL, Horvitz CC (2005) Projection matrix analysis of the demography of an invasive, nonnative shrub (*Ardisia elliptica*). Ecology 86(10):2661-2672

Kwit C, Horvitz CC, Platt W (2004) Conserving slow-growing, long-lived tree species: input from the demography of a rare understory conifer, *Taxus floridana*. Conserv Biol 18(2):432-443

Martinez-Ramos M, Samper KC (1998) Tree life-history patterns and forest dynamics: a conceptual model for the study of plant demography in patchy environments. J Sustain For 6:85-125

Nakamura F (1990) Analyses of the temporal and spatial distributions of floodplain deposits. J Jpn For Soc 72:99-108 (in Japanese with English summary)

Oshima Y, Yamanaka N, Tamai S, Iwatsubo G (1990) A comparison of the distribution properties of two dominant species, *Aesculus turbinata*, *Pterocarya rhoifolia*, in the natural riparian forest of Kyoto University Forest in Ashiu. Bull Kyoto Univ For 62:15-27 (in Japanese with English summary)

Oshima Y, Takeda H (1993) Effects of topographic properties on mortality of seedlings of some dominant tree species in a riparian forest. Trans Kansai Branch Jpn For Soc 2:131-132 (in Japanese)

Sugita H, Shimomoto H, Narimatsu M (1995) Spatial patterns and size structures of tree species in the Ohtakizawa research site, Omyojin Experimental Forest of Iwate University. Bull Iwate Univ For 26:115-130 (in Japanese)

Suzuki W, Osumi K, Takahashi K, Daimaru H, Hoshizaki K (2002) Disturbance regimes and community structures of a riparian and an adjacent terrace stand in the Kanumazawa Riparian Research Forest, northern Japan. For Ecol Manage 157:285-301

Takada T, Nakashizuka T (1996) Density-dependent demography in a Japanese temperate broad-leaved forest. Vegetatio 124:211-221

Yamanaka N, Matsumoto A, Oshima Y, Kawanabe S (1993) Stand structure of Mondori-Dani watershed, Kyoto University Forest in Ashiu. Bull Kyoto Univ For 65:63-76 (in Japanese with English summary)

7 Rodent seed hoarding and regeneration of *Aesculus turbinata*: patterns, processes and implications

Kazuhiko HOSHIZAKI

Department of Biological Environment, Akita Prefectural University, 241 Kaidoh-bata-Nishi, Shimoshinjo-Nakano, Akita 010-0195, Japan

7.1 Introduction

Tree regeneration is often limited by seed abundance, herbivore activities, and microhabitat suitability (Crawley 2000). In montaneous riparian habitats, tree seedling establishment often depends on particular microhabitats associated with the disturbance regime (e.g. gravel debris and mineral soil exposure; Sakio 1997; Kubo et al. 2000; Masaki et al. 2007). However, the Japanese horsechestnut, *Aesculus turbinata* Blume (Hippocastanaceae), is a tree species that is common in montane riparian forests but does not appear to have particular requirements for establishment. The reproductive traits of this species are characterized by conspic-uously large inflorescences and extremely large seeds (21g in fresh weight and 6.2 g in dry weight). In general, large seeds provide a greater chance of seedling establishment through persistence under shade and resistance to disease (Westoby et al. 1992).

In riparian heterogeneous environments, seed dispersal may play important roles in enhancing the success of regeneration. Considering the large amounts of seed reserves, seedlings of large-seeded trees would be to some extent tolerant of physical conditions in particular environments such as germination substrates, and water and nutrient availability (Westoby et al. 1992). Therefore, regeneration success may depend on how wide the seeds are dispersed, to e.g. various topo-graphic units and patches separated by fluvial disturbances. Actually, large seeds of *Juglans ailanthifolia* Carr., also dispersed by rodents and squirrels, are often delivered to floodplains isolated by channels irrespective of microtopography, and contribute to riparian dominance for the species (Goto & Hayashida 2002).

The regeneration of large-seeded species is often affected by small mammals (Crawley 2000), especially rodents which are often responsible for the regener-

Sakio, Tamura (eds) Ecology of Riparian Forests in Japan : Disturbance, Life History, and Regeneration
© Springer 2008

ation, in opposite ways. First, since large seeds and seedlings are attractive foods for herbivores, rodents may act as antagonistic seed/seedling predators (Janzen 1971; Hulme 1993; Crawley & Long 1995). On the other hand, rodents rigorously collect, transport and hoard more seeds than they immediately consume (caching behavior); they can also be mutualistic seed dispersers (Vander Wall 1990, 2001). Although a number of studies have examined the ecological role of seed dispersal (e.g. Nakashizuka et al. 1995), little attention has been paid to when and how those seeds die through rodent seed-hoarding processes and its consequences for the plants (Vander Wall & Joyner 1998; Hoshizaki & Hulme 2002).

In this chapter, I review studies on the regeneration of *A. turbinata* (Hoshizaki et al. 1997, 1999; Hoshizaki & Hulme 2002). Emphasis is placed primarily on its seed and seedling demography. First, I describe patterns and processes of seed hoarding by rodents. In the second part, the seedling ecology of *A. turbinata* is reviewed. Finally, the ecological significance of seed hoarding for *A. turbinata* is explained. I also discuss whether the relationship among *A. turbinata* and rodents is mutualistic or antagonistic. Implications of rodent-mediated regeneration in riparian habitat structures and for life history are presented.

7.2 Methods

7.2.1 Study forest

The studies were undertaken in the Kanumazawa Riparian Research Forest (4.71 ha). The forest contains ca. 2.8 ha of undisturbed riparian area consisting of a diverse array of tree species (Suzuki et al. 2002; Masaki et al. this volume). *A. turbinata* is one of the dominant tree species in the riparian area, comprising 18.5% of total basal area (Suzuki et al. 2002). A 0.4-ha plot (50 × 80 m) and a 1-ha plot (100 × 100 m) were used in the forest. The former was established to track the fate of *A. turbinata* seeds removed by rodents, and 5 of 8 adult (i.e. reproductive) trees were investigated (Hoshizaki et al. 1999). In the 1-ha plot, seedfall and seedling censuses of major tree species has been investigated extensively for years (Hoshizaki et al. 1997; Masaki et al. 2007). Twenty adults of *A. turbinata* are included.

7.2.2 Seed tracking and final destinations

Among several methods used to track rodent-dispersed seeds, thread- or wire-tagging is the simplest way and can be applied most effectively to relatively large seeds. It is advantageous in that tagged seeds can be tracked longer: from seedfall until seedling emergence. By installing seeds tagged with a 40-cm-long wire during 1995-1997 in the 0.4-ha plot, the complete fate of 337 seeds of *A. turbinata* that had fallen from the 5 adults selected as source trees was followed (Hoshizaki & Hulme 2002).

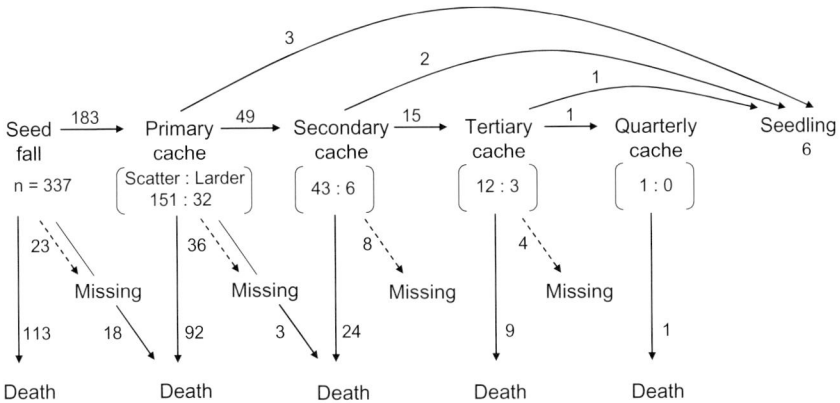

Fig. 1. Seed fate pathways for *Aesculus turbinata*. Numbers along each arrow indicate the seed numbers pooled for 1995-1997. Data reanalyzed from Hoshizaki & Hulme (2002)

The final destination of a seed is the site of seedling emergence. Since *A. turbinata* shows hypogeal germination, the seed coat remains attached to the emerging seedling (Hoshizaki & Miguchi 2005). Using this characteristic feature, the destination of fallen seeds of *A. turbinata* was investigated (Hoshizaki et al. 1999). A total of 1731 seeds were marked with ink on the seed coat and placed under 5 source trees in the 0.4-ha plot. During the season of seedling emergence, the hypogeal cotyledons of all the emerging seedlings were checked.

7.2.3 Spatial pattern of seedfall and seedling census

Seedfall was monitored using 121 seed traps with 10-m regular spacing in the 1-ha plot. Emerged seedlings were sampled throughout the plot, and their locations recorded (Hoshizaki et al. 1997). Survival was checked at least once a month, with any herbivore damages recorded.

7.3 Seed dynamics: cache generation, retrieval and consumption

7.3.1 How often are the seeds hoarded?

In research on seed dispersal by seed-hoarding animals, a primitive question is how often the seeds are removed, hoarded, and consumed (Forget 1990; Isaji & Sugita 1997). Rodents removed 95.8% of the seeds marked (n = 337 in total). They removed only intact seeds. Those infested by insects were not taken. Among the seeds removed, 183 (56.7%) were cached (primary cache; Fig. 1), which is

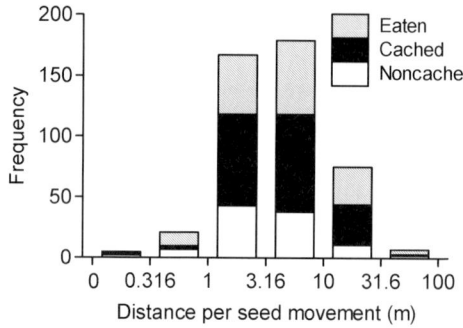

Fig. 2. Frequency distribution of seed transportation distances by rodents. Note that distance classes were categorized on a log scale, by every seed status: eaten, newly cached, or only transported (noncache)

Fig. 3. *Apodemus speciosus* having an *Aesculus turbinata* seed in its mouth. The mouse is just entering a 'camera trap' from above. Scales indicate 1 cm for dots and 5 cm for black lines. Photo taken with a far-infrared sensor, by K. Hoshizaki

1.85-times the number consumed immediately after removal (99 seeds). Primary caches were placed mostly within 10-20 days (depending on year) after the seed-fall.

Caches experienced several rounds of retrieval and recaching. Each seed was handled by rodents on average 2.2±1.2 times after seedfall until its ultimate fate (death or seedling emergence). Forty nine seeds experienced secondary caches, 15 experienced tertiary caches, and one experienced quarterly cache (Fig. 1). Seeds were sometimes left on the forest floor ('noncache' status) after being transported.

Rodents handled seeds at relatively short intervals; a handling interval of 2 days was most frequent, and 75.7% of all seed handlings occurred within 8 days.

7.3.2 Seed transportation and the dispersal agent

A rodent carries a single *A. turbinata* seed in its mouth when transporting it (Fig. 3). Camera-trapping revealed that the large Japanese field mouse *Apodemus speciosus* is the legitimate seed disperser (Hoshizaki 1999). At each carry, these rodents moved seeds a wide range of distances, from only 10 cm to 50.4 m. In most cases, the distance moved ranged from 1 to 10 m (Fig. 2). There seemed no difference in distances moved among cache status (eaten, newly cached, or noncache) in these dominant classes. In contrast, rodents tended to cache seeds after transporting them as far as 30 m or more, but this comparison seems premature because sample sizes are small.

7.3.3 Scatter- vs. larder-hoarding

Rodents placed mostly scatterhoard over larderhoard caches. Larderhoard caches occupied only 17.4% (n=32) of the primary caches, 12.2% of the secondary caches, and 20.0% of the tertiary caches. In 1996, when rodent density was highest among the three years studied, primary caches included more larder hoards (52.5%) than in the other two years (7.7%, average of 1995 and 1997). The fate of larderhoarded seeds was death in all cases. These results suggest that when the density of seed hoarders is high, rodents become more eager to quickly acquire their own food, free from cache pilferage, than to store foods for later use (Vander Wall & Joyner 1998).

Scatterhoard caches usually contained a single seed, with from 1 to 3 seeds per cache (Hoshizaki & Hulme 2002). All the seedlings from the marked seeds emerged from scatterhoard caches. Nevertheless, seedlings were sometimes found in large clumps of >10 individuals (see below), indicating that larger caches can eventually emerge. Thus, rodents typically transport the seeds several times from cache site to cache site. These relocations would benefit single-seeded caches to increase the distance between caches, probably contributing to a spatially even distribution of seedlings (see below).

7.4 End-points of seed dispersal

7.4.1 Dispersal distance

How far can rodents move seeds away from the source tree? As mentioned above, seed dispersal distance can be directly measured by tracking seeds at the time of seedling emergence whose seed coat was marked after seedfall. The result revealed that the seeds had been moved considerable distances during autumn and winter. The seeds were sometimes moved beyond the crown projection of neighboring fruiting trees. The frequency distribution of seed dispersal distances showed that the tail of the seed shadow reaches up to 115 m from the source tree (12.2-44.7 m for means, n = 2 years; Fig. 4). This is comparable to or greater than other large seeds, *Quercus* spp. and *Fagus crenata* (Jensen & Nielsen 1986; Miguchi 1994; Iida 1996).

Seed dispersal events where the distance from source trees is >100 m indicate that long-distance dispersal (LDD) can occur via seed hoarding by rodents. The seed dispersal curves of *A. turbinata* suggest that LDD is a relatively minor event in this species, and any unusual behavior in rodents can cause LDD (Higgins et al. 2003). However, LDD might play an important role in establishing a new habitat after perturbations and in the historical spread of populations after climate change (e.g. post-glacial spread; Clark et al. 1998).

7.4.2 Spatial patterns of seedfall and seedling emergence

The distribution of seedlings that emerged (i.e. seedling shadows) in the 1-ha plot was quite different from the seed shadows. In the case of a seedling cohort in 1993, prior seedfalls concentrated (87 %) under the crowns of *A. turbinata*, whereas few seeds fell under other species and in gaps. Despite that, more seedlings emerged in gaps and under the canopy of other species than expected from seedfall densities (Figs. 5, 6a). This tendency was consistent for most cohorts (Hoshizaki et al. 1997).

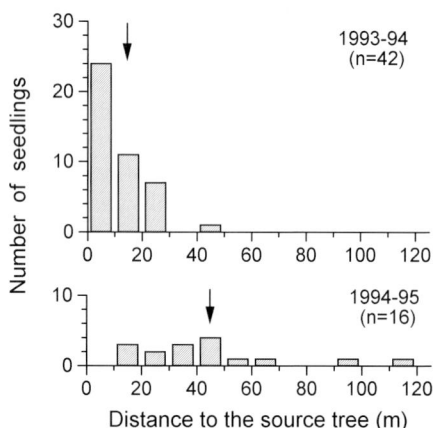

Fig. 4. Seed dispersal curves in *A. turbinata*. Data for seedlings whose source trees were identified by seed-marking in 1993 (upper) and 1994 (lower). The arrow indicates mean dispersal distance for each year (12.2 m for 1993-94 and 44.7 m for 1994-95). After Hoshizaki et al. (1999).

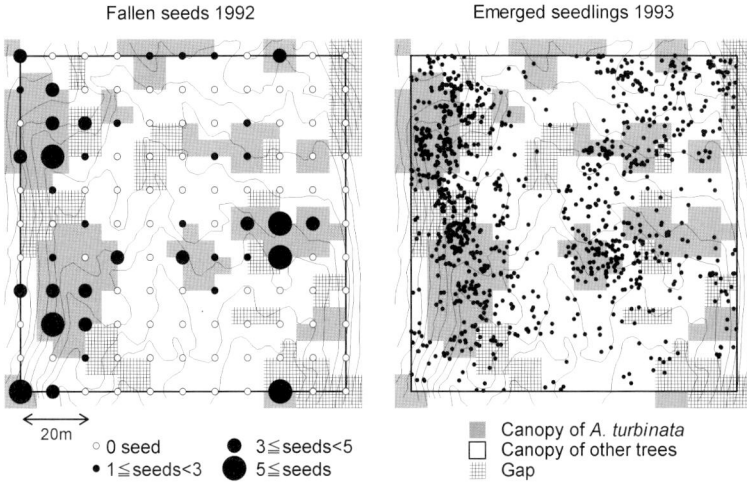

Fig. 5. Seedling spatial distribution in a "successful dispersal" year, after enlargement of seed shadows via rodent hoarding activities. Note that seedfall concentrates beneath *A. turbinata* canopies (left) but that seedlings in the following spring emerged over the 1-ha plot (right). Modified from Hoshizaki et al. (1997), with permission

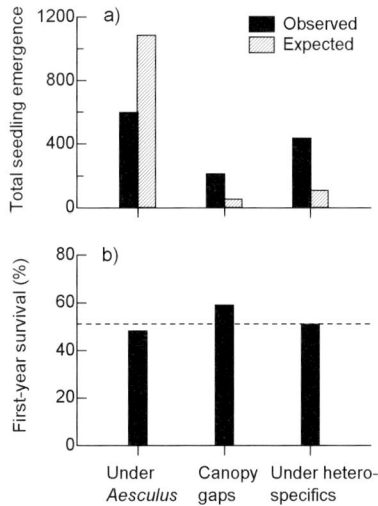

Fig. 6. A test of the colonization hypothesis for the adaptive significance of seed dispersal in *A. turbinata*. Expected values, calculated under the assumption that (a) emergence and (b) survival of seedlings are, respectively, proportional to seedfalls and emergence, are shown with (a) hatched bars and (b) the broken line. Data from Hoshizaki et al. (1997)

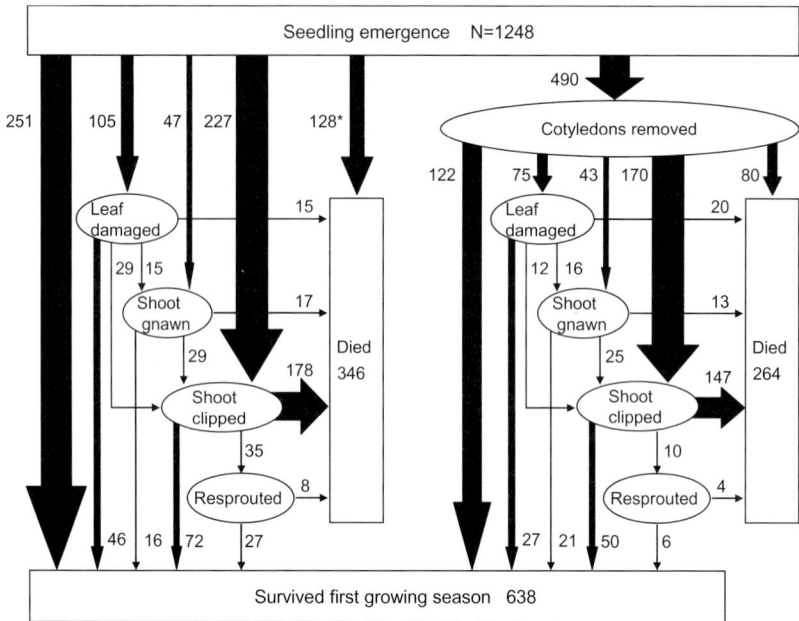

Fig. 7. Fate pathways of first-year seedlings. Numerals represent the number of seedlings for each transition. Width of arrows indicates the importance of each transition. The transition with an asterisked number represents deaths due to only pathogenic fungi and/or light shortage (see also text). Redrawn from Hoshizaki et al. (1997), with permission.

7.5 Seedling regeneration

7.5.1 Impacts of herbivory

What kinds of hazards are there during the first year of seedling emergence, and how does a large seed mass influence seedling survival? A detailed observation of the fate of current-year seedlings was continued for 4 years (1993-1997). Seedlings experienced various types of herbivory including shoot clipping and cotyledon removal by rodents. Many seedlings suffered from different types of herbivory. Therefore fates of the first-year seedlings had complex pathways (Fig. 7). In the case of a large cohort that emerged in 1993 (1248 seedlings/ha), 39% of seedlings suffered cotyledon removal (n = 490), almost the same fraction experienced shoot clipping (n = 492), and 207 experienced both types of herbivory.

Shoot clipping had the largest effect on seedling mortality, followed by cotyledon removal (Hoshizaki et al. 1997). In contrast, leaf herbivory rarely resulted in death. Although the detection of mortality factors is difficult, 58% of the seedlings suffering rodent herbivory ultimately died (n = 467). This represents

77% of all deaths (Fig. 7). On the other hand, the mortality of intact seedlings, mostly caused by pathogenic fungi, was lower (34%) and accounted only for 21% of total deaths. These results indicate the importance of rodent herbivory and high tolerance to shade and/or pathogens during the first growing season (Hoshizaki et al. 1997).

7.5.2 Seedling resistance and the roles of a large seed mass

Seedlings show high resistance to herbivory. In the case of the above cohort, there was no herbivory type that caused 100% mortality, and 32% (n=155) of clipped seedlings survived. The large seed reserves probably contributed to the enhanced tolerance. Furthermore, in *A. turbinata*, because remaining seed reserves are still rich in hypogeal cotyledons even one month after emergence, some seedlings were able to re-sprout after the shoot was clipped. Stored reserves in the hypogeal cotyledons play a pivotal role in resprouting (Hoshizaki et al. 1997; Hoshizaki & Miguchi 2005). The large seed mass of *A. turbinata* thus serves a risk spreading against herbivory (Forget 1992; Harms & Dalling 1997).

Another role of the large seed mass of *A. turbinata* may lie in becoming tall in the riparian understory. It has been suggested that strong pressure of competition with neighboring plants for light on the forest floor can favor a large seed mass (Westoby et al. 1992). Growth of the first year seedling of *A. turbinata* is characteristic in that it quickly reaches ca. 40 cm tall and expands all leaves within 2-3 wk after emergence, but its shoot does not elongate further until the second year (Seiwa & Kikuzawa 1991; Hoshizaki & Miguchi 2005). The seedlings show higher biomass allocation to aboveground parts (3.9-fold larger than to roots), which is much higher than in other large-seeded trees, *Q. crispula* and *F. crenata* (by a factor of 1.8 and 0.53, respectively). Because the understory of riparian forests often has a dense cover of herbaceous/fern species (Hoshizaki & Akiyoshi, unpublished), this distinct feature of seedling morphology may suggest that *A. turbinata* seedlings are competitive in riparian forest floor environments.

Thus the large seed size of *A. turbinata* plays important roles in resistance to damage from herbivores and competition with other plants in the dense understory during the seedling stage.

7.5.3 Influence of canopy gaps and density

Canopy gaps provide increased light levels, making them the most crucial microhabitat for regeneration in many species (e.g. Nakashizuka 1983; Schupp et al. 1989). In *A. turbinata*, seedlings survived significantly well in gaps (Fig. 6b, Hoshizaki et al. 1997). Saplings also showed higher growth rates in sites with higher light levels (Hoshizaki et al. 1999). The frequency of herbivory did not differ among canopy conditions, irrespective of the type of damage (Hoshizaki et al. 1997). Other microhabitats (microtopography, rooting substrate, etc.) did not affect seedling survival for *A. turbinata* (Hoshizaki et al. unpublished).

Seedling survival is often negatively affected by density, and this has been emphasized for demography and population dynamics (e.g. Augspurger & Kitajima 1992) and for species richness of forest communities (Janzen 1970). In *A. turbinata*, seedling survival was affected by the local density at emergence; when the density was low enough (1 seedling per 25 m^2), the survival rate was significantly increased. The causal factor was not obvious, however. Mortality from fungal attacks increased at higher seedling densities, but those from herbivory, the most important factor (Sect. 5.1), did not vary as a function of density (Hoshizaki et al. 1997).

Thus regeneration of *A. turbinata* is strongly limited by herbivores (sensu Crawley 2000) and enhanced in canopy gaps. The little effect of microsites and seed abundance (see below) indicates an advantage of the large seed size of this species.

7.6 Ecological roles of seed dispersal by rodents

7.6.1 Testing hypotheses regarding adaptive significance

One may expect that seed dispersal involves a selective advantage when dispersed offspring show enhanced performance compared to non-dispersed offspring (Willson 1992; Nakashizuka et al. 1995). Three hypotheses have been proposed regarding adaptive significance: colonization, escape, and directed dispersal (Howe & Smallwood 1982; Willson 1992). To date, comprehensive tests for small mammal-dispersed plants have been scarce, compared to other dispersal systems (Hanzawa et al. 1988; Augspurger & Kitajima 1992; Jordano & Schupp 2000). Moreover, there has been a bias toward the colonization and escape hypotheses in earlier studies (e.g. Janzen 1970, 1971; Schupp et al. 1989; but see Nakashizuka et al. 1995). In *A. turbinata*, all three hypotheses have been tested (Hoshizaki et al. 1997, 1999).

The simplest way of testing was shown for the colonization hypothesis. As mentioned above, the seedlings showed higher performance in canopy gaps, suggesting that gaps are suitable sites for successful regeneration in *A. turbinata*. Second, densities of seedling emergence were similar between in canopy gaps and beneath the canopy of other species (Fig. 6a). Therefore the colonization hypothesis was supported (Hoshizaki et al. 1997). The dispersal distance curve (Fig. 4) suggests that seeds of *A. turbinata* have the capability of reaching canopy gaps.

In the escape hypothesis, detection of positive density-dependent mortality should be a prerequisite for the test, but the causes of the density dependence should also be specified to interpret the plant-enemy interactions (Janzen 1970, 1971). A substantial amount of evidence has been found in support of this hypothesis (e.g. Augspurger & Kitajima 1992; Peres et al. 1997). In *A. turbinata* seedlings, the density dependence of mortality from fungal attacks suggests support for this hypothesis. However, because this mortality was a minor hazard, as mentioned above, the importance of this hypothesis seems weak (Hoshizaki et al. 1999).

The directed dispersal hypothesis explains the advantage of non-random seed transportation by animals to a particular patch type suitable for offspring establishment (Howe & Smallwood 1982; Wenny 2001). Several examples suggest its applicability to the dispersal by seed-caching rodents (e.g. Vander Wall 1993), but critical testing should be based on demographic data (Hanzawa et al. 1988). Testing needs to demonstrate both that seeds are dispersed disproportionately to a particular patch type and that the patch is more suitable for establishment than sites at random (Nakashizuka et al. 1995). In *A. turbinata*, changes in location of the seeds through secondary dispersal were investigated, as well as the survival and growth of seedlings at their destinations (Hoshizaki et al. 1999). Although light was most important in determining the survival and growth of seedlings, light levels at seed destinations did not differ from those at the locations of seeds before dispersal nor those on the surrounding forest floor. Therefore, this hypothesis is rejected in *A. turbinata* (Hoshizaki et al. 1999).

These three hypotheses are not mutually exclusive, so evaluating the relative importance of the hypotheses is also important for evolutionary implications for seed dispersal (Wenny 2001). Hoshizaki et al. (1999) examined the relative importance of the colonization and escape hypotheses using a statistical model, and showed that light level was a more important determinant of seedling survival than seedling density. Therefore, the colonization hypothesis is more important in this case. Herbivory did not show any density dependence, suggesting also that the significance of the escape hypothesis may be weak. Thus, the role of rodents in dispersing large seeds of *A. turbinata* is most important for finding suitable sites with more light, merely by enlarging the seed shadow.

7.6.2 Are rodents mutualistic dispersers?

While rodents are effective seed dispersers, they also impose heavy damage on the *A. turbinata* population through seed predation and seedling herbivory. Are they really mutualistic seed dispersers, more than antagonistic predators? First of all, the burying of seeds by rodents may be beneficial for *A. turbinata*, because the seeds are prone to desiccation (Katsuta et al. 1998); seeds may lose vigor without being hoarded underground. This effect might be subtle, considering that most of the fallen seeds were hoarded. Nonetheless it may be suggestive evidence of mutualism (Steele & Smallwood 2002).

In another view, a comparison of seedling spatial patterns may help to answer the above question. In autumn 2002, the legitimate seed dispersers (*Apodemus speciosus*) were much sparser (ca. 2 mice/ha) than usual (20-50 mice/ha). In the following spring, >80% of emerged seedlings were beneath the canopy of conspecific adults (Fig. 8); the extent of seed-shadow enlargement was much less than in 'normal' years, despite that seedlings were especially more abundant than usual (4070 vs. 100-2000 seedlings/ha). Therefore, *A. turbinata* seems not to be able to

a) 2000 (rodent densities: normal)   b) 2003 (*A. speciosus* almost absent)

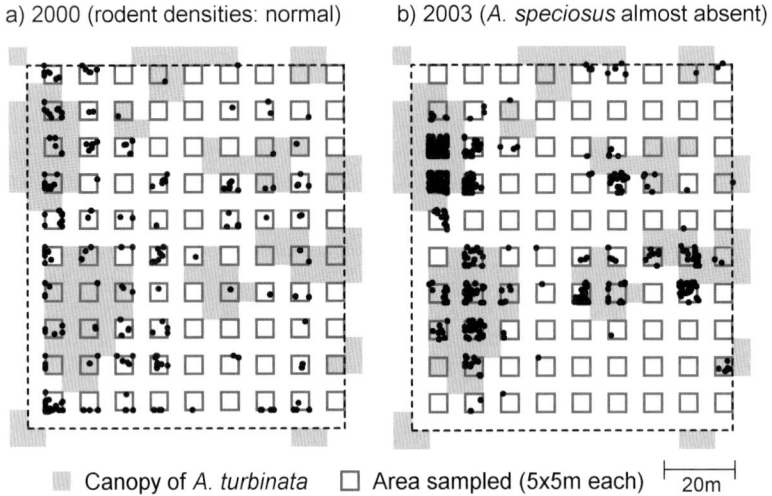

Canopy of *A. turbinata*   ☐ Area sampled (5x5m each)   ⊢——⊣ 20m

Fig. 8. Contrasting spatial distributions of seedling emergence under conditions with and without effective dispersal agents. Patterns for years (a) when rodent composition and densities in the preceding autumn were at 'normal' levels and (b) when the autumn density of the legitimate seed disperser, *Apodemus speciosus*, was extremely low (see also text). Seedling densities were 1120 seedlings/ha in 2000 and 4070 in 2003. The same 0.9-ha areas in the 1-ha plot are shown with a broken frame.

enlarge the seed shadow in the absence of legitimate seed dispersers. Forest voles such as *Eothenomys andersoni* have intermediate food habits between granivores and herbivores (Shimada & Saitoh 2006), so that they also cache *A. turbinata* seeds in this forest. Although they were abundant in autumn 2002 (27 voles/ha), the seedling spatial pattern suggests their inferiority in moving seeds beyond source trees (Vander Wall 1990). Considering the strong impact of rodent herbivory (Sect. 5.1), *E. andersoni* are regarded as antagonistic predators for both seeds and seedlings. Thus regeneration of *A. turbinata* may rely on scatter-hoarding by *A. speciosus*, suggesting a mutualistic relationship between them.

7.7 Annual variation

Regeneration of *A. turbinata* was highly variable among years. This may reflect complicated indirect interactions with rodents and a co-occurring large-seeded tree, *Fagus crenata*, that shares seed predators/dispersers with *A. turbinata*. The nature of the indirect interactions and its consequences for *A. turbinata*'s regeneration have been overviewed in previous papers (Hoshizaki & Hulme 2002; Hoshizaki & Miguchi 2005), and so only the demographic features of regeneration is briefly mentioned here.

The population produced seeds in 12 of 13 years studied (1992-2004), and its magnitude varied little among years (CV = 74%) (Hoshizaki et al. 1997; Hoshizaki & Hulme 2002 and unpublished data). In contrast, seedling numbers showed a greater among-year variation, ranging from 17 to 4070 seedlings/ha (CV = 116%). These patterns suggest that the regeneration of *A. turbinata* is rarely seed-limited. The regeneration may be determined not only by seed dispersal but also by the balance among abundances of seedfall, the legitimate seed-disperser (*A. speciosus*), and the antagonistic herbivores (*E. andersoni*). Masting of *F. crenata* causes a marked increase in rodent populations and, in turn, influences *A. turbinata* regeneration; too-high and too-low levels of rodent population under a given seed abundance may reduce regeneration success (Hoshizaki & Hulme 2002; Hoshizaki & Miguchi 2005).

7.8 Implications for life history in the riparian habitats

Although the regeneration was strongly limited by rodents, dispersal via rodent seed hoarding mitigates the negative impact, and seedlings are, to some extent, resistant to herbivory in *A. turbinata*. The large seed appears essential to establish a seedling bank in the understory of riparian habitats, especially under rich herbaceous vegetation. These features are particularly characteristic to *A. turbinata*, showing a remarkable contrast to other co-occurring riparian trees in which regeneration is mostly microsite-limited (Masaki et al. 2007).

Rodents may play further roles in riparian habitats. First, seeds of *A. turbinata* are often delivered to various habitats including upper slopes as well as riparian microhabitats (Hoshizaki et al. 1999). This suggests that rodents have the ability to deliver seeds across various habitats associated with the complex disturbance regime including long-distance dispersal (LDD). Second, seed burial via caching may also be effective not only in protecting from drought as mentioned above but also in preventing the seeds from being swept away with floods (Goto & Hayashida 2002). These features may support the stable population structure for *A. turbinata* in riparian forests (Kaneko et al. 1999; Suzuki et al. 2002).

The regenerative interactions reviewed here are not characteristic only of riparian forests. Nonetheless, *A. turbinata* populations are disproportionately distributed in riparian areas (Kaneko et al. 1999; Suzuki et al. 2002). Clearly, later life stages (i.e. saplings and other immature stages; Kaneko et al. 1999) seem important in *A. turbinata*, as a representative riparian species, to link the high capabilities of habitat colonization and of seedling-bank formation and the topographic bias of the population.

Acknowledgements

I am grateful to Dr. H. Sakio for invitation to this volume. This work was supported partly by the Ministry of Agriculture, Forestry and Fisheries, Japan (Bio-Cosmos Project) and partly by the Ministry of Education, Culture, Sports, Science and Technology of Japan (Grant-in-Aid for Scientific Research; No. 18770020).

References

Augspurger CK, Kitajima K (1992) Experimental studies of seedling recruitment from contrasting seed distributions. Ecology 73:1270-1284

Clark JS, Fastie C, Hurtt G, Jackson ST, Johnson C, King GA, Lewis M, Lynch J, Pacala S, Prentice C, Schupp EW, Webb T, Wyckoff P (1998) Reid's paradox of rapid plant migration: dispersal theory and interpretation of paleoecological records. BioScience 48:13-24

Crawley MJ (2000) Seed predators and plant population dynamics. In: Fenner M (ed) Seeds: The Ecology of Regeneration in Plant Communities, 2nd Edn. CAB International, Wallingford, UK, pp 167-182

Crawley MJ, Long CR (1995) Alternate bearing, predator satiation and seedling recruitment in *Quercus robur* L. J Ecol 83:683-696

Forget PM (1990) Seed-dispersal of *Vouacapoua americana* (Caesalpiniaceae) by caviomorph rodents in French Guiana. J Trop Ecol 6:459-468

Forget PM (1992) Regeneration ecology of *Eperua grandiflora* (Caesalpiniaceae), a large seeded tree in French Guiana. Biotropica 24:146-156

Goto S, Hayashida M (2002) Seed dispersal by rodents and seedling establishment of walnut trees (*Juglans ailanthifolia*) in a riparian forest. J Jpn For Soc 84:1-8 (In Japanese with English summary)

Hanzawa FM, Beattie AJ, Culver DC (1988) Directed dispersal: demographic analysis of an ant-seed mutualism. Am Nat 131:1-13

Harms KE, Dalling JW (1997) Damage and herbivory tolerance through resprouting as an advantage of large seed size in tropical trees and lianas. J Trop Ecol 13:617-621

Higgins SI, Nathan R, Cain ML (2003) Are long-distance dispersal events in plants usually caused by nonstandard means of dispersal? Ecology 84:1945-1956

Hoshizaki K (1999) Regeneration dynamics of a sub-dominant tree *Aesculus turbinata* in a beech-dominated forest: Interactions between large-seeded tree guild and seed/seedling consumer guild. D Thes Kyoto Univ, Kyoto

Hoshizaki K, Hulme PE (2002) Mast seeding and predator-mediated indirect interactions in a forest community: evidence from post-dispersal fate of rodent-generated caches. In: Levey DJ, Silva WR, Galetti M (eds) Seed Dispersal and Frugivory: Ecology, Evolution and Conservation. CAB International, Wallingford, Oxfordshire, UK., pp 227-239

Hoshizaki K, Miguchi H (2005) Influence of forest composition on tree seed predation and rodent responses: a comparison of monodominant and mixed temperate forests in Japan. In: Forget PM, Lambert JE, Hulme PE, Vander Wall SB (eds) Seed Fate: Predation, Dispersal and Seedling Establishment. CAB International, Wallingford, UK, pp 253-267

Hoshizaki K, Suzuki W, Nakashizuka T (1999) Evaluation of secondary dispersal in a large-seeded tree *Aesculus turbinata*: a test of directed dispersal. Plant Ecol 144:167-176

Hoshizaki K, Suzuki W, Sasaki S (1997) Impacts of secondary seed dispersal and herbivory on seedling survival in *Aesculus turbinata*. J Veg Sci 8:735-742

Howe HF, Smallwood J (1982) Ecology of seed dispersal. Annu Rev Ecol Syst 13:201-228

Hulme PE (1993) Post-dispersal seed predation by small mammals. Symp Zool Soc Lond 65:269-287

Iida S (1996) Quantitative analysis of acorn transportation by rodents using magnetic locator. Vegetatio 124:39-43

Isaji H, Sugita H (1997) Removal of fallen *Aesculus turbinata* seeds by small mammals. Jpn J Ecol 47:121-129 (In Japanese with English summary)

Janzen DH (1970) Herbivores and the number of tree species in tropical forests. Am Nat 104:501-528

Janzen DH (1971) Seed predation by animals. Annu Rev Ecol Syst 2:465-492

Jensen TS, Nielsen OF (1986) Rodents as seed dispersers in a heath - oak wood succession. Oecologia 70:214-221

Jordano P, Schupp EW (2000) Seed disperser effectiveness: the quantity component and patterns of seed rain for *Prunus mahaleb*. Ecol Monogr 70:591-615

Kaneko Y, Takada T, Kawano S (1999) Population biology of *Aesculus turbinata* Blume: A demographic analysis using transition matrices on a natural population along a riparian environment gradient. Plant Species Biol 14:47-68

Katsuta M, Mori T, Yokoyama T (1998) Seeds of Woody Plants in Japan - Angiospermae. Japan Forest Tree Breeding Association, Tokyo (In Japanese)

Kubo M, Shimano K, Sakio H, Ohno K (2000) Germination sites and establishment conditions of *Cercidiphyllum japonicum* seedlings in the riparian forest. J Jpn For Soc 82:349-354 (In Japanese with English summary)

Masaki T, Osumi K, Takahashi K, Hoshizaki K, Matsune K, Suzuki W (2007) Effects of microenvironmental heterogeneity on the seed-to-seedling process and tree coexistence in a riparian forest. Ecol Res 22:724-734

Miguchi H (1994) Role of wood mice on the regeneration of cool temperate forest. In: Kobayashi S, Nishikawa K, Danilin IM, Matsuzaki T, Abe N, Kamitani T, Nakashizuka T (eds) Proceedings of NAFRO seminar on sustainable forestry and its biological environment. Japan Society of Forest Planning Press, Tokyo, Japan, pp 115-121

Nakashizuka T (1983) Regeneration process of climax beech (*Fagus crenata* Blume) forests III. Structure and development processes of sapling populations in different aged gaps. Jpn J Ecol 33:409-418

Nakashizuka T, Iida S, Masaki T, Shibata M, Tanaka H (1995) Evaluating increased fitness through dispersal: A comparative study on tree populations in a temperate forest, Japan. Ecoscience 2:245-251

Peres CA, Schiesari LC, Dias-Leme CL (1997) Vertebrate predation of Brazil-nuts (*Bertholletia excelsa*, Lecythidaceae), an agouti-dispersed Amazonian seed crop: a test of the escape hypothesis. J Trop Ecol 13:69-79

Sakio H (1997) Effects of natural disturbance on the regeneration of riparian forests in a Chichibu Mountains, central Japan. Plant Ecol 132:181-195

Schupp EW, Howe HF, Augspurger CK, Levey DJ (1989) Arrival and survival in tropical treefall gaps. Ecology 70:562-564

Seiwa K, Kikuzawa K (1991) Phenology of tree seedlings in relation to seed size. Can J Bot 69:532-538

Shimada T, Saitoh T (2006) Re-evaluation of the relationship between rodent populations and acorn masting: a review from the aspect of nutrients and defensive chemicals in acorns. Popul Ecol 48:341-352

Steele MA, Smallwood PD (2002) Acorn dispersal by birds and mammals. In: McShea WJ, Healy WM (eds) Oak Forest Ecosystems: Ecology and Management for Wildlife. Johns Hopkins Univ Press, Baltimore, MD, USA, pp 182-195

Suzuki W, Osumi K, Masaki T, Takahashi K, Daimaru H, Hoshizaki K (2002) Disturbance regimes and community structures of a riparian and an adjacent terrace stand in the Kanumazawa Riparian Research Forest, northern Japan. For Ecol Manage 157:285-301

Vander Wall SB (1990) Food hoarding in animals. Univ Chicago Press, Chicago, Illinois, USA

Vander Wall SB (1993) Cache site selection by chipmunks (*Tamias* spp.) and its influence on the effectiveness of seed dispersal in Jeffrey pine (*Pinus jeffreyi*). Oecologia 96:246-252

Vander Wall SB (2001) The evolutionary ecology of nut dispersal. Bot Rev 67:74-117

Vander Wall SB, Joyner JW (1998) Recaching of Jeffrey pine (*Pinus jeffreyi*) seeds by yellow pine chipmunks (*Tamias amoenus*): potential effects on plant reproductive success. Can J Zool 76:154-162

Wenny DG (2001) Advantages of seed dispersal: a re-evaluation of directed dispersal. Evol Ecol Res 3:51-74

Westoby M, Jurado E, Leishman M (1992) Comparative evolutionary ecology of seed size. Trend Ecol Evol 7:368-372

Willson MF (1992) The ecology of seed dispersal. In: Fenner M (ed) Seeds: The ecology of regeneration in plant communities. CABI, Wallingford, UK, pp 61-85

8 Longitudinal variation in disturbance regime and community structure of a riparian forest established on a small alluvial fan in warm-temperate southern Kyushu, Japan

Hiroka ITO and Satoshi ITO

Division of Forest Science, Faculty of Agriculture, University of Miyazaki, 1-1 Gakuen Kibanadai Nishi, Miyazaki 889-2192, Japan

8.1 Introduction

Riparian forests maintain diverse vegetation structure through the influence of various types, intensities, and frequencies of disturbance promoted by fluvial and geomorphic processes (Nilsson et al. 1989; Baker 1990; Ito & Nakamura 1994; Sakio 1997; Suzuki ct al. 2002). These fluvial and geomorphic disturbances vary longitudinally from headwater streams to low-gradient alluvial rivers and result in different types of plant communities according to longitudinal location (Nakamura & Inahara 2007). For each location, many studies have revealed the coexistence or habitat segregation of tree species, showing how tree life history is linked to variations in disturbance regime and heterogeneous micro-landforms (Johnson et al. 1976; White 1979; Kovalchik & Chitwood 1990; Swanson & Sparks 1990; Gregory et al. 1991; Malanson & Kupfer 1993; Stewart et al. 1993; Loehle 2000). In Japan, many studies (e.g., other chapters of this issue) have reported distinct structures and dynamics of mountainous riparian forests compared to those of forests situated on adjacent hill slopes. These findings highlight how fluvial and geomorphic processes in sedimentation-dominated positions such as wide head-water streams or alluvial fans consisting of small terraces are important for the maintenance of regional plant diversity. However, most of these studies have been conducted in cool-temperate central or northern Japan (Sato 1992, 1995; Kaneko

Sakio, Tamura (eds) Ecology of Riparian Forests in Japan : Disturbance, Life History, and Regeneration
© Springer 2008

1995; Ann & Oshima 1996; Hara et al. 1996; Sakio 1997; Kaneko et al. 1999; Sakai et al. 1999; Sakio et al 2002; Sakio & Yamamoto 2002; Suzuki et al. 2002). In warm-temperate parts of the country, the relationship between species distribution and micro-landform or disturbance regime has been reported only for forest communities established on mountainous slopes (Hara et al. 1996; Sakai et al. 1999; Enoki 2003; Hattori et al. 2003; Enoki & Abe 2004) or for riparian forests along low-gradient alluvial rivers (Ishikawa 1988; Sakio & Yamamoto 2002). However, Ito and Nogami (2005) indicated that comparatively little information is available on the coexistence of trees and species diversity patterns in riparian forests in this region, including in the sedimentation zone in mountainous areas such as alluvial fans or the small floodplains of headwater streams.

In this chapter, we describe habitat segregation and diversity patterns of tree species (DBH>=3 cm) in relation to small-scale longitudinal variations in site conditions and fluvial and geomorphic disturbances along a sedimentation-dominated mountainous stream. The data referred to are from our recent study conducted in a warm-temperate region of Japan (Ito et al. 2006).

8.2 General description of the case study site

The study site was located on an alluvial fan in a volcanic caldera in the Koike lake in the Kirishima Mountains, southern Japan (Fig. 1). The forest is situated at 330–400 m elevation. The annual mean temperature is 15°C and average annual precipitation is 2260 mm. The climatic-vegetation zone of the forest is warm-temperate lucidophyllous (warm-temperate evergreen natural) forest dominated by *Quercus gilva* Blume and *Machilus thunbergii* Sieb. et Zucc. Several deciduous species associated with the cool-temperate zone, such as *Ulmus davidiana* Planch.

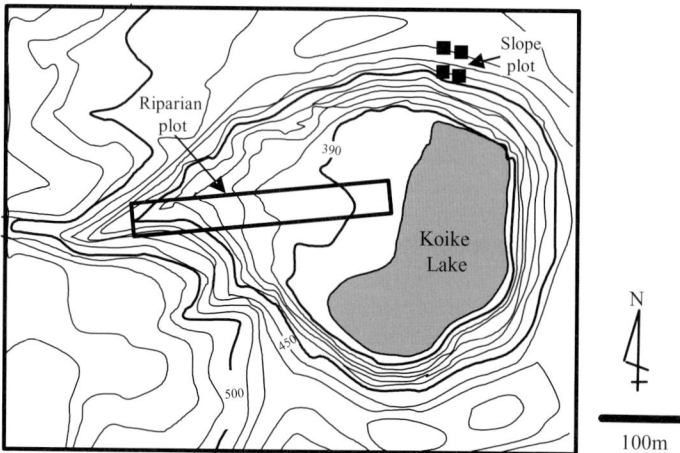

Fig. 1. Location of the study plots of riparian and slope forests (After Ito et al. 2006)

var. *japonica* (Rehder) Nakai, *Sapindus mukurossi* Gaertn. and *Morus bombycis* Koidz., were found at the site.

The survey plots of the riparian area were arranged to represent four geomorphic zones, with one plot per zone. The plots were 40 m wide and positioned end-to-end in the direction of the slope gradient (Fig. 1, see also Fig. 2 and Fig. 5 for detail). The four zones sampled by the plots were: V-shaped valley zone (VV, 40 m × 60 m), upper fan zone (UF, 40 m × 100 m), middle fan zone (MF, 40 m × 80 m) and lower fan zone (LF, 40 m × 80 m). We also investigated the vegetation on the hill slope (SL, 10 m × 40 m × 4) to compare the riparian forests with typical lucidophyllous forests established on relatively stable ground (Fig. 1). Each geomorphic zone was distinguished from the adjacent zone by slope gradient and observed geomorphology and soil surface characteristics.

8.3 Disturbance regime and site conditions along the stream gradient

The study site shows a clear gradient in disturbance regime and site conditions in relation to fluvial and geomorphic processes as summarized in Table 1. Along the longitudinal gradient of streams, from V-shaped valley to lower fan, the disturbance type represents erosion- to sedimentation-dominated disturbance, and the micro-site conditions vary from heterogeneous to homogeneous. Within the sedimentation-dominated riparian zones (UF-, MF- and LF-zones), the intensity of disturbance decreases along the longitudinal stream gradient from the UF-zone to the LF-zone, but the frequency of disturbance increases. The same trends in disturbance intensity and frequency were reported for an alluvial fan in a temper-

Table 1. Geomorphic and soil characteristic, disturbance regime and site condition in each riparian geomorphic zone.

	Geomorphic zone			
	V-shaped valley (VV)	Upper fan (UF)	Middle fan (MF)	Lower fan (LF)
Longitudinal slope gradient (%)	15.8	28	21.1	9.7
Ground surface undulation (m) [1]	1.06	0.68	0.27	0.11
Average thickness of each sediment layer (cm)	-	50.3	13.2	11.6
Average surface rock diameter (cm)	12.88	13.06	7.67	5.80
Disturbance regime				
Disturbance type	erosion	erosion and sedimentation		sedimentation
Site condition				
Stability of soil surface	low	high		low
Heterogeneity of micro-site	high	very high		low

[1] Ground surface undulation was calculated as the standard deviation of elevations measured on transverse grid lines at 20 m intervals in each plot.

ate riparian forest in central Japan (Ito & Marutani 1993). Sakio (1997) also reported that disturbance by the deposition of sand and gravel was more frequent than disturbance by debris flows in a cool-temperate *Fraxinus platypoda* Oliv. - *Pterocarya rhoifolia* Sieb. et Zucc. forest in central Japan.

8.3.1 V-shaped valley

In the V-shaped valley (VV-zone), erosion disturbances predominate. While the terraces formed by debris flows are exposed to frequent erosion, small-scale landslides also occur on the steep bank slope as a feature of the disturbance regime of a "lower-sideslope", as described by Tamura (1987). The ground surface along sections orthogonal to the flow direction displays strong undulations because the valley bottom is narrow and the bank is steep (Fig. 2), indicating a high hetero-geneity of micro-site conditions.

8.3.2 Upper fan and middle fan zone

The upper fan zone (UF-zone) and middle fan zone (MF-zone) appear to have

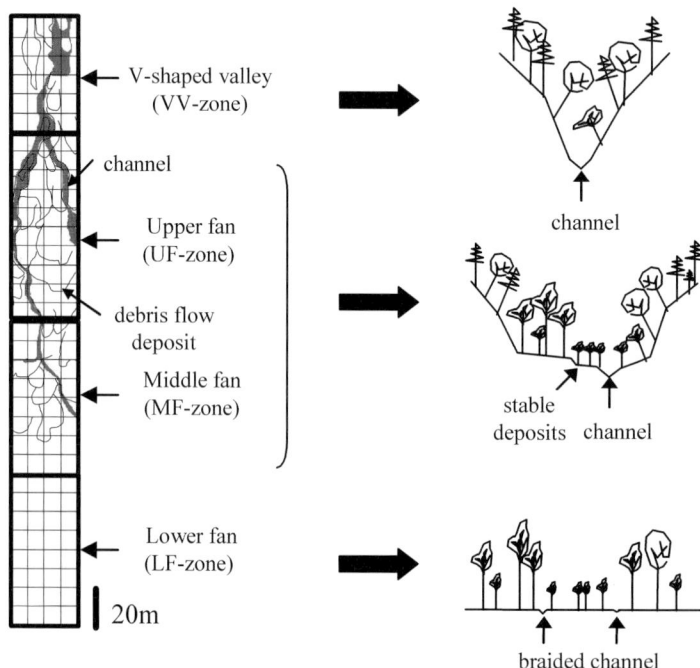

Fig. 2. Schematic diagrams showing the longitudinal variation in transverse sections in sequence from the V-shaped valley to the lower fan

sedimentation-dominated disturbances (evidenced by abundant debris flow deposits), but erosion disturbances also occur in these zones as evidenced by the
existence of clear channels. Large rocks (mainly originating from debris flows)
and thick sediment in these two zones indicate a high intensity and presumably
low frequency of disturbance.

The ground surface of the terraces in the UF- and MF-zones could be considered more stable than the other zones due to their relatively high position above
the channel (Fig. 2). Generally, terraces formed by the deposition of large debris
flows are stable for long periods of time (Ito & Nakamura 1994). Considering the
presence of both stable deposits and frequently disturbed channels (Fig. 2), the
UF- and MF-zones provide a more heterogeneous habitat than the other zones.

8.3.3 Lower fan zone

In the lower fan zone (LF-zone), the type of disturbance is believed to be predo-
minantly sedimentation due to the minimal soil surface undulations (i.e., lack of
clear channel formation). The unclear channels were presumably a result of
subsurface flow in this zone due to the low gradient of the riverbed and permeable
sandy sediments, forming a homogeneous micro-site condition (Fig. 2). The thin
sediment layer in the LF-zone is characterized by tractive deposits comprised of
fine rocks and sands. Thus, the sedimentation disturbance of this zone is the least
intensive but the most frequent, resulting in an unstable soil surface. With the
distribution of fine rocks and sands, flat ground surface, and high sedimentation
frequency, the LF-zone had the most differentiated habitat condition from the hill
slopes within the sedimentation-dominated riparian zones (UF-, MF- and LF-
zones).

8.4 Habitat segregation and species diversity pattern

According to our analysis of tree distribution to detect the geomorphic guilds
(Table 2), which were distinguished by having the same "preferred site" in a
specific zone by bootstrap methods, the studied geomorphic zones were segregated from each other as representing different habitats for tree species. The
number of individuals belonging to each geomorphic guild dominated in the
defining geomorphic zone (Fig. 3) and the proportion of individuals in each
geomorphic guild gradually declined or increased along the longitudinal gradient,
suggesting the continuous variation of species composition among the geomorphic
zones. Another analysis of species composition using DCA ordination of 36 stands
collected from the five zones based on a matrix of the absolute number of trees
(DBH>=3 cm) as the dominance values for each species in each stand (Fig. 4) also
indicated the species composition varying along the longitudinal stream gradient

(along the first axis), which might reflect the significant habitat segregation detected by the guild analysis. These variations of the species composition were associated with the pattern of species richness (Fig. 6). The community structure of each zone can be summarized as follows.

Table 2. Stem density of main component species in each geomorphic zone. Species are ordered in each guild type excluding infrequent species

Guild[1] type	Species	Stem density (No./ha)				
		SL	VV	UF	MF	LF
SL	Camellia japonica L.	344	60	50	25	-
	Distylium racemosum Sieb. et Zucc.	188	70	29	34	-
	Cinnamomum japonicum Siebold ex Nakai	169	8	17	13	3
	Castanopsis sieboldii (Thunb.) Schottky var. sieboldii (Makino) Nakai	100	13	25	16	-
	Castanopsis cuspidata (Thunb.) Schottky	56	-	4	22	-
VV	Machilus thunbergii Sieb. et Zucc.	194	143	396	306	41
	Aucuba japonica Thunb.	-	55	79	28	-
UF	Machilus japonica Sieb. et Zucc.	256	88	442	159	50
	Camellia sasanqua Thunb.	81	8	467	241	41
	Cleyera japonica Thunb.	194	63	338	169	9
	Actinodaphne longifolia (Blume) Nakai	138	40	263	94	31
	Illicium religiosum Sieb. et Zucc.	25	-	58	16	-
MF	Quercus gilva Blume	56	20	383	316	109
	Swida controversa (Hemsl.) Soják	-	13	54	50	-
LF	Neolitsea aciculata (Blume) Koidz.	81	30	108	66	41
	Cephalotaxus harringtonia (Knight) K. Koch	13	8	46	66	66
	Ficus erecta Thunb.	13	-	17	47	25
	Ilex rotunda Thunb.	13	3	38	9	16
GEN	Eurya japonica Thunb.	50	8	42	31	13
	Quercus salicina Blume	31	8	29	25	-
	Neolitsea sericea (Blume) Koidz.	38	-	25	13	13
	Celtis sinensis Pers. var. japonica (Planch.) Nakai	-	8	29	25	3
	Styrax japonica Sieb. et Zucc.	-	5	13	19	6
	Ternstroemia gymnanthera (Wight et Arn.) Sprague	6	-	21	9	6
	All species	2062	1129	1885	1931	541

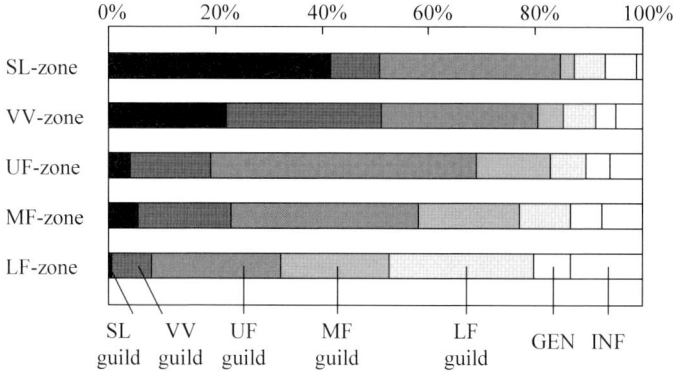

Fig. 3. The proportion of individuals by guild type on the five different geomorphic zones (Modified from Ito et al. 2006). Generalist (GEN) species were defined as those that had no bias toward any topographic zone. Infrequent species (INF) were defined as species which had a minimum expected value of less than 10 occurrences

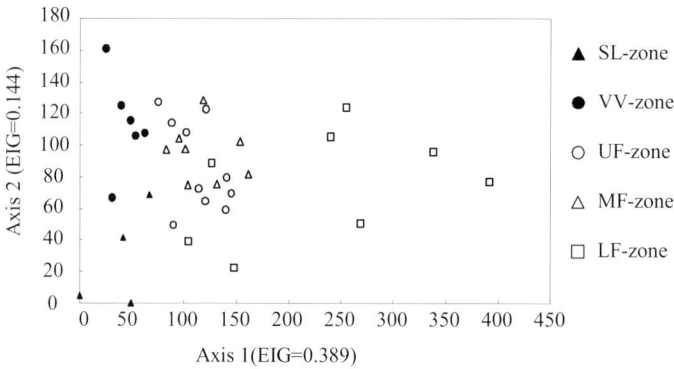

Fig. 4. Scatter plot of 36 stands for the first two axes of the detrended correspondence analysis (DCA) ordination (After Ito et al. 2006)

8.4.1 V-shaped valley

The VV- zone had the most individuals of lucidophyllous species (e.g. *Machilus thunbergii, Machilus japonica* Sieb. et Zucc. and *Distylium racemosum* Sieb. et Zucc.) (Table 2 and Fig. 5), which were also common on the hill slope (SL-zone), indicating that the species composition of the VV- zone was similar to typical lucidophyllous vegetation. DCA ordination of species composition (Fig. 4) also demonstrated that the VV-zone vegetation was the most alike of the four riparian zones to the hill slope (SL-zone). These results suggested that the VV-zone, which would experience erosion-dominated disturbance, was not clearly segregated from

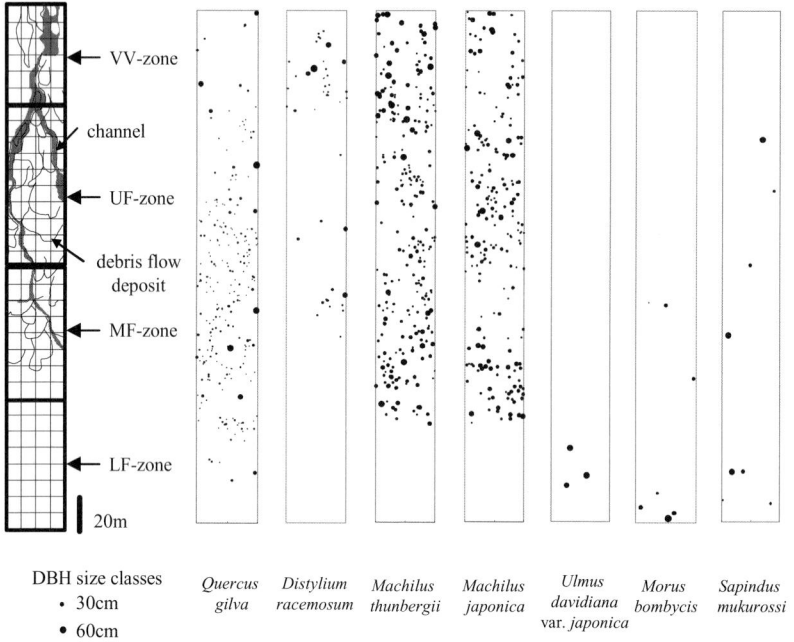

Fig. 5. Spatial distribution of the main component species and infrequent deciduous species in the geomorphic zones in the riparian area

the SL-zone as tree habitat. The cumulative increase with area in species number in the VV- zone was gradual (Fig. 6), resulting in low species richness among the geomorphic zones. This implies that despite the high heterogeneity of site conditions presented by the steep slopes of this zone, species diversity is not particularly high compared with the hill slopes.

8.4.2 Upper fan zone and middle fan zone

The UF-zone and MF-zone were dominated by *Quercus gilva*, which is not common in the VV-zone (Table 2 and Fig. 5). These zones consisted of a stable ground surface of alluvial terraces, which presumably have better soil moisture conditions than the upper slopes because of the location close to the channel (c.f., Nagamatsu & Miura 1997; Kikuchi 2001). The relatively stable sediments with moist soil conditions in these zones would provide suitable habitat for *Q. gilva*, which prefers flat and moist sites with fertile soil (Miyawaki 1981). DCA ordination did not differentiate between the species composition of the UF- and MF- zones, reflecting the similar site conditions and disturbance regimes of these two zones.

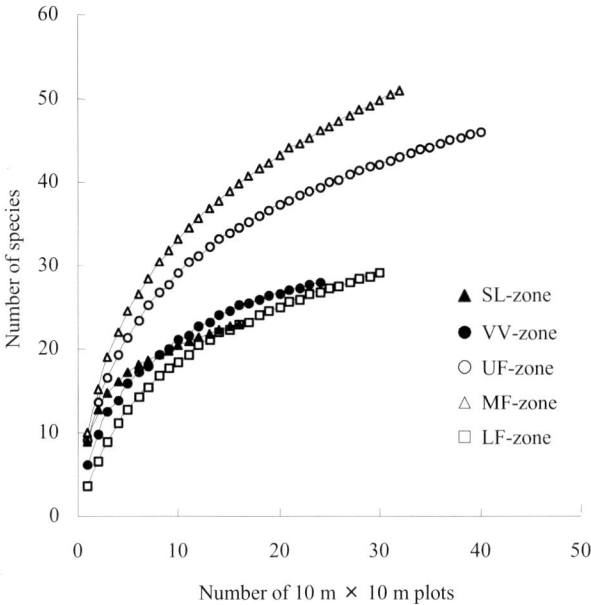

Fig. 6. Species-area curves for the five different geomorphic zones (After Ito et al. 2006)

The UF- and MF-zones were also characterized by high tree density and species richness (particularly of infrequent species with less than 10 individuals across all zones). These characteristics would be related to the stable ground surface of these high-elevation deposits. On the other hand, the formation of clear (relatively deep) channels contributes to the establishment of infrequent species that are dependent on frequent disturbance. The relative stability and high heterogeneity were thought to be major factors in determining the high species diversity in the UF- and MF- zones.

8.4.3 Lower fan zone

Of the four riparian zones, the LF-zone is likely to have the habitat that is most differentiated from the SL-zone as shown by guild composition (Fig. 3). Ordination of vegetation also indicated the strong specialization of species in this zone compared to the others (Fig. 4). This vegetation differentiation would be a function of the disturbance regime in this zone which produces a flat but unstable soil surface resulting from frequent sedimentation of fine rocks and sands. We suggest that this disturbance regime (Table 1) also has a significant influence on the low tree density of this zone (Table 2).

The species distinguished as LF-guild or infrequent species that were observed more commonly in the LF-zone (Table 2) could be expected to have life histories that are well adapted to less intensive but frequent disturbances rather than to intensive disturbances such as the destruction of canopy trees by debris flows (c.f., Sakio 1997; Nakamura & Inahara 2007). These species typical of the LF-zone were uncommon in the other zones. In addition, the spatial distribution of several infrequent species such as *Ulmus davidiana* var. *japonica, Morus bombycis* and *Sapindus mukurossi*, which are presumably dependent on frequent sedimentation, showed large individuals occurring in the LF- zone (Fig. 5).

The species richness of this zone was low compared with the UF- and MF-zones (Fig. 6), presumably because of the homogeneity of micro-site conditions. However, the disturbance regime and site conditions of the LF-zone would contribute towards maintaining the diversity of the regional flora by providing habitat for infrequent species.

8.5 Occurrence of infrequent species

The occurrence of infrequent species and their characteristics also provided funda-mental information for understanding the ecological importance of riparian forests of the region. In our case study, 44 of the 68 species were infrequent (less than 10 individuals within the whole study plot). Each of the riparian zones (VV-, UF-, MF- and LF-zones) had more infrequent species than the hill slope (Fig. 3), as reported for cool-temperate mountainous forests (Sakio 1997; Sakio et al. 2002;

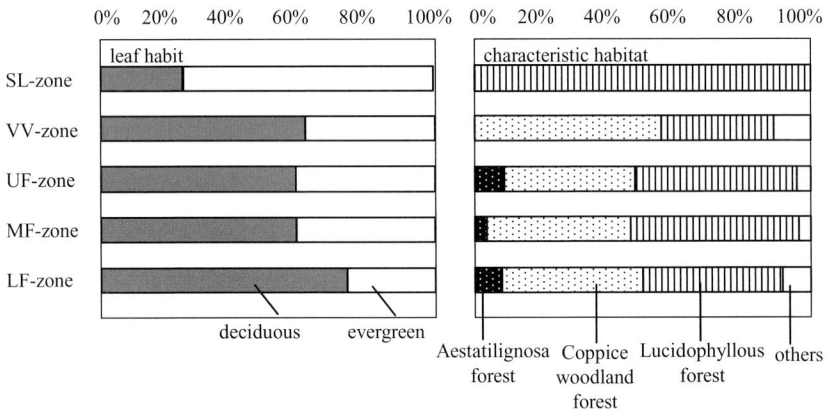

Fig. 7. The proportion of the number of individuals of infrequent species by leaf habit and characteristic habitat on the five different geomorphic zones (Redrawn from Ito et al. 2006). Aestatilignosa natural forest species are cool-temperate, summer-green forest species. Coppice woodlands species are related to severe disturbance. Lucidophyllous forest species are warm-temperate evergreen forest species

Suzuki et al. 2002). We found large individuals of infrequent species in the sedimentation-dominated riparian zones (UF-, MF- and LF-zones). This was most apparent in the LF-zone. The distribution of infrequent species by characteristic habitat (representing the species' typical habitat) showed that the riparian zones were characterized by more occurrences of aestatilignosa species (typically associated with the cool-temperate zone, e.g,. *Phellodendron amurense* Rupr. and *Ulmus davidiana* var. *japonica*) and coppice woodland species (related to severe disturbances, e.g., *Morus bombycis* and *Sapindus mukurossi*) compared to the predominance of lucidophyllous species (originally common in warm-temperate zone, e.g., *Ligustrum japonicum* Thunb. and *Daphniphyllum macropodum* Miq.) in the SL-zone (Fig. 7). This suggested that the habitats of the riparian zones were basically segregated from that of the hill slopes not only by their defining guilds, but also by the occurrence of infrequent species that are not common in the region. The occurrence of infrequent coppice woodland species in the riparian zone implies the influence of severe disturbances. We observed the stem base of most trees, including large individuals of *Ulmus davidiana* var. *japonica, Morus bombycis* and *Sapindus mukurossi*, buried under the sediments of the fine rocks and sand in the LF-zone. Thus, these deciduous species observed in the LF-zone could be expected to be tolerant to burial as suggested by their life histories of adaptation to less intensive but frequent sedimentation. This might be one of the major causes of the differentiation of the species composition of the LF-zone from those of the other zones.

8.6 Conclusions

This chapter described the patterns of site conditions, fluvial and geomorphic disturbance regimes, and habitat segregation of tree species along the longitudinal stream gradient in a warm-temperate mountainous riparian forest. The type (erosion or sedimentation) and regime (intensity and frequency) of disturbance varied continuously along the longitudinal stream gradient. In our case study site, this continuous variation in disturbance regime forms diverse site conditions. This feature is strongly related to the tree distribution and diversity pattern combined with the effects of micro-site heterogeneity promoted by erosion and sedimentation.

Sakio (1995) reported that the disturbance regime of the riparian zones is more complex than that of the hill slope, influencing the regeneration of tree species in the riparian zones. In our study site, the habitats for trees of the riparian zones (VV-, UF-, MF- and LF-zones) were obviously different from that of the hill slope (SL-zone) because of their diverse habitat conditions formed by complex disturbance regimes. In the analysis of guild structure, the riparian zones commonly had a low proportion of SL-guild individuals and a high proportion of infrequent species (Fig. 3), indicating the different habitat conditions. Suzuki et al. (2002) reported, for a site in a cool-temperate region, clear differences in species composition between the slope and riparian zone reflecting disturbance by flooding in the floodplain. Our case study of a warm-temperate riparian forest suggested the importance of sedimentation-dominated disturbance regime in the alluvial fan, resulting in a distinctly different forest structure than that resulting from the erosion-dominated disturbance regime of the V-shaped valley.

The site conditions promoted by the sedimentation-dominated disturbance regime would contribute to the maintenance of regional flora by supporting many infrequent species that are practically non-existent on the hill slopes. This was most apparent in the L F-zone. Also, in the UF- and MF-zones, many other deciduous infrequent species were observed along the channel or terrace walls, indicating their dependence on severe disturbance. Thus, we conclude that the riparian forests of this small alluvial fan in the warm-temperate region could support many deciduous species due to the continuously varying site conditions and disturbance regimes along the stream gradient.

References

Ann SW, Oshima Y (1996) Structure and regeneration of *Fraxinus spaetbiana-Pterocarya rhoifolia* forests in unstable valleys in the Chichibu Mountains, central Japan. Ecol Res 11:363-370

Baker WL (1990) Species richness of Colorado riparian vegetation. J Veg Sci 1:119-124

Enoki T (2003) Microtopography and distribution of canopy trees in a subtropical evergreen broad-leaved forest in the northern part of Okinawa Island, Japan. Ecol Res 18:103-113

Enoki T, Abe A (2004) Saplings distribution in relation to topography and canopy openness in an evergreen broad-leaved forest. Plant Ecol 173:283-291

Gregory SV, Swanson FJ, Mckee WA, Cumminks KW (1991) An ecosystem perspective of riparian zones: focus on links between land and water. BioScience 41:540-551

Hara M, Hirata K, Fujihara M, Oono K (1996) Vegetation structure in relation to micro-scale landform in an evergreen broad-leaved forest on Amami Ohshima Island, south-west Japan. Ecol Res 11:325-337

Hattori T, Asami K, Kodate S, Ishida H, Minamiyama N (2003) Distribution of the lucidophyllous elements and species richness of lucidophyllous forest along the micro-scale geomorphic condition in Kawanaka, Aya, Miyazaki Prefecture. Veg Sci 20:31-42 (in Japanese with English summary)

Ishikawa S (1988) Floodplain vegetation of the Ibi river in central Japan: I. distribution behavior and habitat conditions of the main species of the riverbed vegetation developing on the alluvial fan. Jpn J Ecol 38:73-84 (in Japanese with English summary)

Ito H, Ito S, Matsui T, Marutani T (2006) Effect of fluvial and geomorphic disturbances on habitat segregation of trees species in a sedimentation-dominated riparian forest in warm-temperate mountainous region in southern Japan. J For Res 11:405-417

Ito S, Marutani T (1993) Disturbance regime of debris flows and structural diversity in flood plain forests. Trans Jpn Forestry Soc 104:743-744 (in Japanese)

Ito S, Nakamura F (1994) Forest disturbance and regeneration in relation to each surface movement. Jpn J For Environ 36(2):31-40 (in Japanese)

Ito S, Nogami K (2005) Species composition and environments of riparian forests consisting of *Lagerstroemia subcostata* var. *fauriei*, an endangered species on Yakushima Island. Veg Sci 22:15-23 (in Japanese with English summary)

Johnson WC, Burgess RL, Keammerer WR (1976) Forest overstory vegetation and environment on the Missouri River floodplain in North Dakota. Ecol Monogr 46:59-84

Kaneko Y (1995) Disturbance regimes of mountainous riparian forest and effect of disturbance on tree population dynamics. Jpn J Ecol 45:311-316 (in Japanese)

Kaneko Y, Takada T, Kawano S (1999) Population biology of *Aesculus turbinata* Blume: A demographic analysis using transition matrices on a natural population along a riparian environmental gradient. Plant Species Biol 14:47-68

Kikuchi T (2001) Vegetation and Landform. Univ Tokyo Press, Tokyo, 220p (in Japanese)

Kovalchik BL, Chitwood LA (1990) Use of geomorphology in the classification of riparian plant associations in the mountainous landscapes of central Oregon, USA. Forest Ecol Manage 33/34:405-418

Loehle C (2000) Strategy space and the disturbance spectrum: a life-history model for tree species coexistence. Am Nat 156:14-33

Malanson GP, Kupfer JA (1993) Simulated fate of leaf litter and large woody debris at a riparian cut bank. Can J For Res 23:582-590

Miyawaki A (1981) Vegetation of Japan-Kyushu. Shibundo, Tokyo (in Japanese with English abstract)

Nagamatsu D, Miura O (1997) Soil disturbance regime in relation to micro-scale landforms and its effects on vegetation structure in a hilly area in Japan. Plant Ecol 133: 191-200

Nakamura F, Inahara S (2007) Fluvial geomorphic disturbances and life history traits of riparian tree species. In: Johnson EA, Miyanishi K (eds) Plant Disturbance Ecology: The Process and the Response. Elsevier, pp 283-310

Nilsson C, Grelsson G, Johosson M, Sperens U (1989) Patterns of plant species richness along riverbanks. Ecology 70:77-84

Sakai T, Tanaka H, Shibata M, Suzuki W, Nomiya H, Kanazashi T, Iida S, Nakashizuka T (1999) Riparian disturbance and community structure of a *Quercus-Ulmus* forest in central Japan. Plant Ecol 140:99-109

Sakio H (1995) Dynamics of riparian forest in mountain region with respect to stream disturbance and life-history strategy of trees. Jpn J Ecol 45 (3):307-310 (in Japanese)

Sakio H (1997) Effects of natural disturbance on the regeneration of riparian forests in the Chichibu Mountains, central Japan. Plant Ecol 132:181-195

Sakio H, Yamamoto F eds. (2002) Ecology of Riparian Forests (in Japanese). Univ Tokyo Press, Tokyo, Japan, 206p

Sakio H, Kubo M, Shimano K, Ohno K (2002) Coexistence of three canopy tree species in a riparian forest in the Chichibu Mountains, central Japan. Folia Geobot 37:45-61

Sato H (1992) Regeneration traits of saplings of some species composing *Pterocarya rhoifolia* forest. Jpn J Ecol 42:203-214 (in Japanese with English summary)

Sato H (1995) Studies on the dynamics of *Pterocarya rhoifolia* forest in southern Hokkaido. Bull Hokkaido For Res Inst 32:55-96 (in Japanese with English summary)

Stewart GH, Basher LR, Burrows LE, Runkle JR, Hole GMJ, Jackson RJ (1993) Beech-hardwood forest composition, landforms, and soil relationships, north Westland, New Zealand. Vegetatio 106:111-125

Suzuki W, Osumi K, Masaki T, Takahashi K, Daimaru H, Hoshizaki K (2002) Disturbance regimes and community structures of a riparian and an adjacent stand in the Kanumazawa Riparian Research Forest, northern Japan. For Ecol Manage 157:285-301

Swanson FJ, Sparks RE (1990) Long-Term Ecological Research and invisible space. BioScience 40:502-508

Tamura T (1987) Landform-soil features of the humid temperate hills. Pedologist 31: 29-40 (in Japanese)

White PS (1979) Pattern, process and natural disturbance in vegetation. Bot Rev 45:229-299

Part 5

Riparian forests on wide alluvial fan

9 Structure and composition of riparian forests with reference to geomorphic conditions

Shun-ichi KIKUCHI

Laboratory of Earth Surface Processes and Land Management, Graduate School of Agriculture, Hokkaido University, Kita 9 Nishi 9 Kita-ku, Sapporo 060-8589, Japan

9.1 Introduction

A riverbed is a route through which materials from hillslopes are swept downstream by water flow. During a flood event, large amounts of materials are transferred downstream without deposition. In contrast, when the power of the water flow has weakened after a flood, bed load materials are deposited in the riverbed. Therefore, riverbeds are dynamic sites in which the surface and vegetation are disturbed frequently and simultaneously (Kikuchi 2001). Hydrogeomorphic processes (e.g., erosion and deposition of sediment) and associated channel changes are the predominant disturbances that allow vegetation to become established along riverbeds (Nakamura et al. 1997). Consequently, the structure and species composition of riparian forest stands reflect the intensity and frequency of disturbance.

Disturbance regimes in riparian zones are classified into two types: debris flow, which is a flowing mixture of water-saturated debris that moves downslope under the force of gravity, and flooding. Debris flow generally occurs in low-order streams in mountain regions, while flooding generally occurs in alluvial fans. Unfortunately, most streams and rivers in Japan have been altered by development, so primary fluvial processes and natural riparian forests are disappearing. However, some undisturbed riparian zones remain in Hokkaido, northern Japan. Using data obtained in these natural areas, we can speculate about riparian forest dynamics. This chapter deals with the relationship between geomorphic site conditions and the dynamics of riparian forests on alluvial fans and floodplains in Hokkaido.

Sakio, Tamura (eds) Ecology of Riparian Forests in Japan : Disturbance, Life History, and Regeneration
© Springer 2008

9.2 Overview of the Tokachi River system

The Tokachi River is one of the largest catchments in Hokkaido, covering an area of 9,010 km^2. We investigated the shores of this river and the alluvial fan at its confluence with one tributary (Aruga et al. 1996; Nakamura et al. 1997). These rivers are located on the eastern side of Mount Tokachi (43°25' N, 142°41' E). Metamorphic rock and welded tuff dominate this region, and shallow landslides are frequently observed on the toeslopes. Floodplains measure about 500 m in maximum width. Parts of the floodplains and terraces are used for agriculture. River channels display a braided pattern, and the channel gradient is about 0.6%. The riverbed is composed mainly of sand, cobbles, and boulders, which can exceed 30 cm in diameter. The average annual precipitation between 2003 and 2006 at the nearest meteorological observatory (Tokachi Dam, 43°14'30" N, 142°56'30" E) was 1,237 mm; however, this area is a mountainous region that experiences heavy snowfalls. According to data gathered at the Nukabira meteorological observatory (540 m above sea level) from 2002 to 2006, the average annual snowfall was over 530 cm (The Japan Meteorological Agency 2007). The seasonal change in river discharge of the Tokachi River is typical for northern Japanese rivers; predictable snow floods occur from late April to late June, followed by low and stable discharge in subsequent months, and rare large floods due to typhoons or low pressure weather systems in August or September.

The riparian forests and adjacent toeslopes have different structures and species compositions. The riparian forests are composed of *Picea jezoensis* (Siebold et Zucc.) Carrière, *Abies sachalinensis* (F. Schmidt) Mast., *Populus maximowiczii* A. Henry, *Salix cardiophylla* Trautv. & Mey. var. *urbaniana* (Seemen) Kudo, formerly *Toisusu urbaniana* (Seemen) Kimura, *Alnus hirsuta* Turcz., and *Salix caprea* L., formerly *Salix bakko* Kimura. In addition, *Ulmus davidiana* Planch. var. *japonica* (Rehder) Nakai and *Fraxinus mandshurica* Rupr. var. *japonica* Maxim. occur sporadically in developed riparian stands. The dominant species of broad-leaved trees, *A. hirsuta*, *S. cardiophylla* var. *urbaniana*, and *P. maximowiczii*, are all wind-dispersed pioneer species, and they occasionally form almost pure stands. The dominant conifers, *P. jezoensis* and *A. sachalinensis*, are mixed with broad-leaved trees in young stands and form pure coniferous stands at maturity. The toeslope forests are composed of a mixture of broad-leaved and coniferous trees. The dominant conifers are *P. jezoensis*, *A. sachalinensis*, and *P. glehnii* (F. Schmidt) Mast., and the main broad-leaved species are *U. davidiana* var. *japonica*, *U. laciniata* (Trautv.) Mayr, *Acer mono* Maxim., *Phellodendron amurense* Rupr., *Tilia japonica* (Miq.) Simonk., and *Quercus crispula* Blume. The toeslope forests exhibit relatively uniform and high species richness.

9.3 Variation in site conditions on the floodplain in comparison to the hillslope

Various environmental factors can explain the variety in species composition and structure of forest stands. Previous studies have suggested a strong connection between the species composition of riparian vegetation and elevation above the channel bed (Bell 1980; Hupp & Osterkamp 1985) and its correlates, i.e., flooding frequency (Bell & del Moral 1977; Bell 1980; Hupp & Osterkamp 1996) and soil moisture (Adams & Anderson 1980). With regard to soil texture, previous studies have focused primarily on the physical characteristics of sediment with particles of less than 2 mm in diameter (Johnson et al. 1976; Hupp & Osterkamp 1985; Niiyama 1987, 1989; Ishikawa 1988), even though alluvium in mountain streams contains large amounts of coarse fragments, such as cobbles and boulders. It is obvious that these coarse fragments play an important role in stabilizing geomorphic surfaces in streams, and they represent shear stress that is a substitute for flooding intensity. Furthermore, some floodplains are repeatedly covered by thick deposits of sediment brought by successive floods (Scott et al. 1996; Friedman et al. 1996). Riparian trees are able to adapt to burial by sprouting adventitious roots from their buried trunks (Sigafoos 1964; Araya 1971). Therefore, coarse fragments in the surface layer where trees become established should be taken into account when examining the physical characteristics of sediment.

The grain size distributions of the substratum in a floodplain and a toeslope are shown in Figure 1. These data were obtained from six sites each in the floodplain

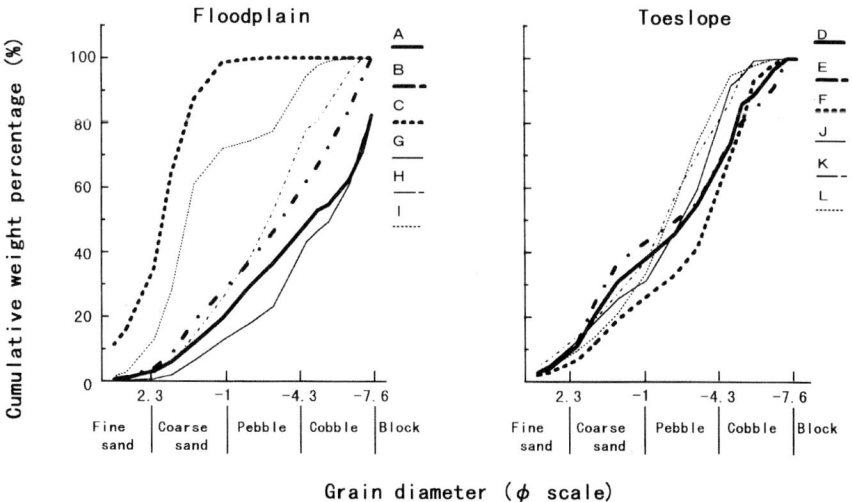

Fig. 1. Grain size distributions of the substratum in the floodplain and the toeslope (Aruga et al. 1996). The taxonomy of grain size follows the international system

and the toeslope. For the six sites in the floodplain, Sites A and G were closest to the riverbed and had the highest percentage of coarse particles, such as cobbles and blocks. Sites B and H had a similar distribution, dominated by pebbles. Sites C and I, which were far from the channel, had a higher percentage of fine and coarse sand than the other sites. In contrast, all toeslope sites had similar distributions of particles, ranging from coarse sand to cobbles.

Principal component analysis was used to determine trends in the substrate variables, including the variables for particle size distribution. The first principal component (PC-1) accounted for 39.6% of the variation and was highly positively correlated with mean particle size, water content, and organic matter content. These substrate variables were also compared between the floodplain and the toeslope (Table 1). Water content and organic matter content were significantly lower in the floodplain sites than at the toeslope (Mann-Whitney U test; $p<0.05$). The variance in mean particle size of the substratum and the PC-1 scores were significantly larger

Table 1. Differences in substrate characteristics between floodplain and toeslope (Aruga et al. 1996). Variables are expressed as mean ± SD. Values in parentheses indicate the minimum and maximum. [†] 'φ' indicates φ scale (log2-transformation of particle diameter [mm]). * indicates $p<0.05$

	Floodplain	Toeslope	Significance	
			F-test	U-test
Mean(φ)[†]	-2.37±2.55	-2.03±0.48	0.002*	0.485
	(-4.99, 1.69)	(-2.90, -1.53)		
Standard deviation(φ)[†]	2.58±0.59	2.86±0.33	0.222	0.394
	(1.52, 3.20)	(2.41, 3.34)		
Skewness(φ)[†]	0.33±0.51	0.55±0.28	0.198	0.589
	(-0.64, 0.87)	(0.13, 0.89)		
Kurtosis(φ)[†]	2.71±0.66	2.42±0.49	0.539	0.699
	(2.06, 3.82)	(1.74, 3.02)		
Water content (%)	1.48±1.01	4.17±0.78	0.581	0.004*
	(0.67, 3.37)	(3.25, 5.23)		
Organic matter content (%)	3.88±4.67	14.15±2.71	0.257	0.015*
	(1.22, 4.77)	(11.48, 18.58)		
PC-1 score	-0.53±1.19	0.53±0.31	0.009*	0.065
	(-1.45, 1.69)	(0.01, 0.90)		

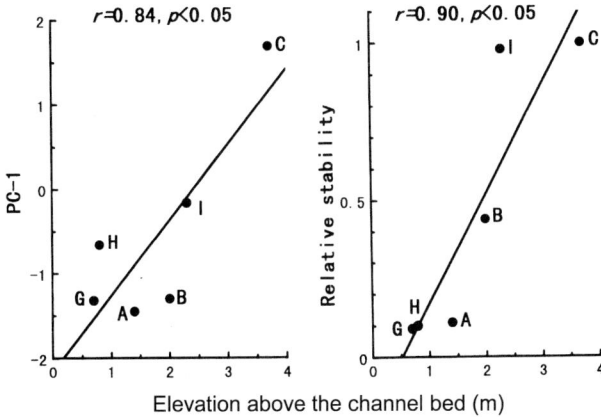

Fig. 2. Variation in PC-1 score and relative stability of floodplain sites with respect to elevation above the channel bed (Aruga et al. 1996). Stand age was determined as the average of ages of dominated trees in each site and relative site stability was calculated on the basis of the maximum stand age

in the floodplain sites than at the toeslope (F test; $p<0.05$). In addition, in the floodplain sites, the PC-1 scores and relative stability values were strongly and positively correlated with elevation above the channel bed (PC-1: $r=0.84$, $p<0.05$; relative stability: $r=0.90$, $p<0.05$; Fig. 2).

The attributes of riverbed sediment that directly affect vegetation growth are water retention, nutrient content, hardness, and air permeability (Ishikawa 1988). Grain size distribution corresponds to water retention and nutrient content (Johnson et al. 1976). As nutrient content is closely related to litter accumulation and humification, it is thought that organic content should be nearly equivalent to nutrient content. Therefore, PC-1, representing grain size distribution, water content, and organic content, was defined as an appropriate variable representing substrate characteristics.

The PC-1 scores in the floodplain sites were more variable than those at the toeslope and were highly correlated with elevation above the channel bed. The elevation above the channel bed is closely correlated with the frequency of fluvial disturbance and is associated with the particle size of scoured or deposited sediment (Hupp & Osterkamp 1985). Areas close to the channel generally have lower elevations. Thus, they endure repeated floods due to rainfall in summer and snow melt in spring, resulting in frequent deposition and scouring associated with lateral migrations of channel courses. Hence, organic matter and fine sediment are constantly being flushed out, and only coarse fragments remain on the surface of the riverbed. As a result, fine sediment, soil moisture, organic content, and site stability are quite low in these areas. In contrast, sites far from the channel are high above the bottom of the riverbed and are rarely inundated. Even when these areas are covered by a major flood, the water is usually shallow and tree trunks act as large roughness elements, reducing the stream power of the water (Nakamura 1995). Organic and inorganic materials tend to be deposited rather than scoured downstream. Site conditions are therefore characterized by high levels of fine sediment, soil moisture, and organic content, which are due mainly to high site stability.

Toeslopes are rarely disturbed by fluvial processes but are disturbed by colluvial processes. Shallow landslide scars were often observed around the study sites, although the PC-1 scores and relative site stabilities were similar among the toeslope sites. Homogeneous conditions may have been created by shallow landslides larger in scale than the actual size of each study site. In addition, litter accumulation and humification tends to progress on hillslopes without fluvial disturbances, and consequently, soil-forming processes, which result in high moisture retention and high nutrient content, have progressed for a relatively longer period. Thus, site conditions in floodplains and toeslopes are regulated by their respective predominant disturbance regimes.

9.4 Site conditions for dominant tree species

Riparian forests in the Tokachi River catchment consist of a number of elongated patches that extend parallel to the stream on its wide valley floor. Which tree species are predominant in each patch? What are the site conditions required by the dominant species? Which site conditions govern species composition? To answer these questions, 19 quadrats representing a wide range of elevations of geomorphic surfaces and distances from the adjacent stream channel were established along a 7-km reach of the Tokachi River (Nakamura et al. 1997).

An overview of species composition and structure of riparian forests was provided above. We used gradient analysis to examine the distribution of tree species on the valley floor. The dominant species used for the gradient analysis were three pioneer species (*A. hirsuta*, *S. cardiophylla* var. *urbaniana*, and *P. maximowiczii*) and two conifer species (*P. jezoensis* and *A. sachalinensis*). The two conifers were analyzed together because of their restricted occurrence in the quadrats. The relative dominance of each species across elevation and distance gradients was fit to a Gaussian curve, and correlation coefficients and their sig-nificances were compared. All species were significantly distributed along both gradients, with the exception of *P. maximowiczii* on the distance gradient. The correlation coefficients for elevation were higher than those for distance. Thus, elevation appeared to be the primary gradient that controlled the distribution of dominant species.

The changes in relative dominance along the elevation gradient differed among species (Fig. 3). *Alnus hirsuta* dominated low geomorphic surfaces, with a dramatic decrease on floodplains at least 100 cm above the riverbed. *Salix cardiophylla* var. *urbaniana* dominated low floodplains, and its dominance peaked at 100 cm. *Populus maximowiczii* is also a pioneer species, but it dominated floodplains at heights from 150 to 250 cm. The dominance of conifers gradually increased with elevation, and they achieved maximum dominance on sites at least 300 cm above

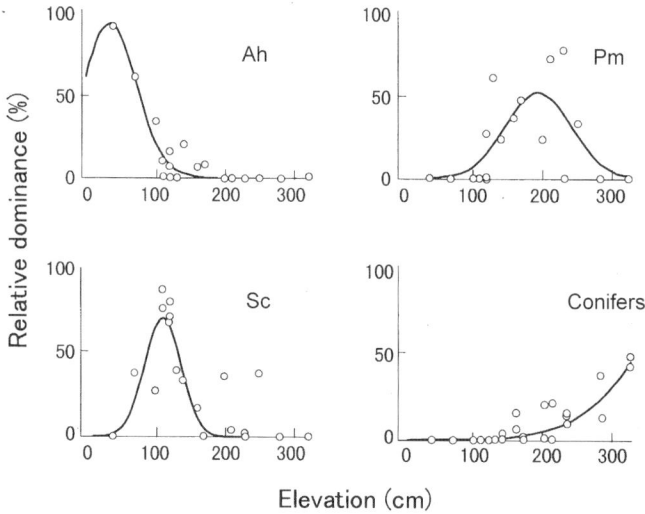

Fig. 3. Changes in relative dominance along the elevation gradient (Nakamura et al. 1997). Species abbreviations: Ah, *Alnus hirsuta*; Sc, *Salix cardiophylla* var. *urbaniana*; Pm, *Populus maximowiczii*; Conifers, *Picea jezoensis* and *Abies sachalinensis*

the riverbed. Not only did relative dominance shift from low to high elevation in the order *A. hirsuta, S. cardiophylla* var. *urbaniana, P. maximowiczii*, and conifers, but the amplitude of the curves also decreased in that order. Furthermore, the dispersion of the distributions seemed to increase with increasing elevation.

The three pioneer species and the two conifers in the study sites are all wind-dispersed and possibly water-dispersed species (Cappers 1993). The frequency and intensity of floods and sediment texture influence the establishment of seedlings, and resistance to fluvial disturbances is necessary for species to grow (Hupp & Osterkamp 1985, 1996). We often found that the three pioneer species growing near river channels had adapted to frequent burial by sprouting adventitious roots (Fig. 4). They favored sites with coarse fragments, where disturbances were frequent and severe (Niiyama 1987). However, direct gradient analysis revealed that the modes of distribution clearly differed among the three pioneer species (Fig. 3).

Alnus hirsuta preferred low sand or gravel bars, which develop inside channel bends and are referred to as "point bars," and the riverbed substratum of abandoned channels, which contained not only coarse fragments but also substantial fine sediment. Such sites are not subject to severe disturbances, although they are easily submerged. *Alnus hirsuta* has higher submergence tolerance than other dominant species that form riparian forests in Hokkaido because it can survive for longer periods under submerged conditions (Takahashi et al. 1988). The site conditions favoring the growth of *Salix cardiophylla* var. *urbaniana* were characterized by

Fig. 4. Adventitious roots sprouting from a buried *Populus maximowiczii* trunk

frequent flooding, although this species was established on higher deposits than *A. hirsuta*. Only this species is able to survive on mid-channel islands and erosive floodplains subject to strong currents. Moreover, the underlying substratum was largely composed of boulders. *Populus maximowiczii* grew on higher surfaces than the other pioneer broad-leaved trees. Such high terraced deposits are generally created by high-magnitude floods. *Populus maximowiczii* established on a substratum that consisted largely of coarse fragments. In a mature stand, this species was sometimes mixed with *A. sachalinensis* in the understory (Fig. 5). The trees of *A. sachalinensis* were 30 to 50 years younger than those of *P. maximowiczii*, even though the trees seemed to be established on the same substratum. Thus, *P. maximowiczii* is a pioneer tree species prevailing on high terraced deposits. Partial disturbances caused by medium-magnitude floods can provide an opportunity for *A. sachalinensis* to succeed. Fine sediment brought by such floods on a cobble substratum may create the preferred sites for the establishment of conifers. Some stands reached a mature stage, reflecting high site stability. In contrast, *Picea jezoensis* and *Abies sachalinensis* tended to occur sporadically. These species can regenerate on higher floodplains where flooding rarely occurs and can survive under low-light conditions (Shibakusa et al. 1968). These traits enable conifers to become successors on high, stable deposits after partial disturbances of overstory trees and/or understory vegetation.

Fig. 5. A mature *Populus maximowiczii* stand mixed with *Abies sachalinensis* in the understory

Elevation both directly and indirectly influenced the physical characteristics of the establishing soil and possibly the flood-disturbance regime. The relative dominance of each of the three pioneer species and the two conifers overlapped substantially across the elevation gradient, rather than being separately distributed. The high amplitude and narrow dispersion of the Gaussian curves of *A. hirsuta* and *S. cardiophylla* var. *urbaniana* suggested that a high-stress environment, such as a high water level and frequent flooding, enabled only these species to dominate this habitat, without competition from other species. In contrast, *P. maximowiczii* and the two conifers showed lower and more scattered distribution patterns, which suggested that a low-stress environment with higher elevation provided more opportunities for their establishment.

9.5 Temporal and spatial variation among tree species

To examine temporal variation in the structure and composition of riparian forests, the stands were aged. We classified the stability of sites into three categories, i.e., stable, semi-active, and active, on the basis of elevation and distance. Active sites were defined as being located on bars or floodplains at a height ≤1.5 m and at a distance of ≤10 m from the stream channel. Floodplains with heights >1.5 m and

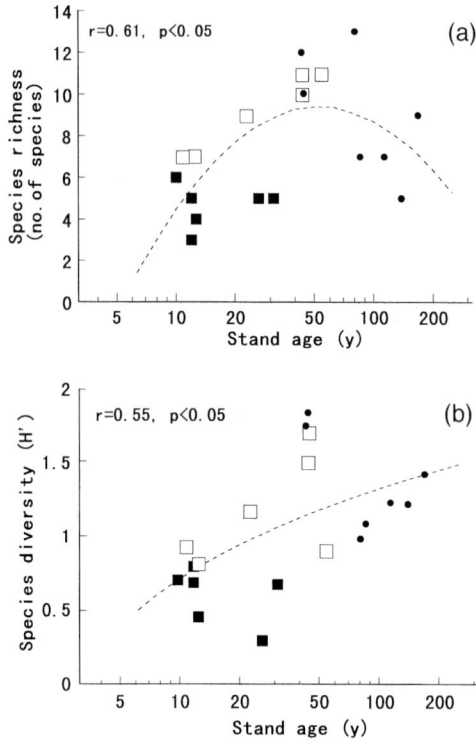

Fig. 6. Relationship between (a) species richness and stand age, and (b) species diversity and stand age (Nakamura et al. 1997). Black squares, white squares, and black circles indicate the degree of site stability, classified as active, semi-active, and stable, respectively

distances >10 m were grouped as stable sites. Floodplains with heights >1.5 m and distances ≤10 m or with heights ≤1.5 m and distances >10 m were grouped as semi-active sites.

The age of each stand should indicate the length of the stable period during which the stand has not suffered fatal disturbances. Therefore, all stands were classified into four groups of different stability: 10-year, 25-year, 50-year, and >80-year survival groups (Fig. 6). Species richness and diversity increased with the stability of the stand, although two stands in the 25-year group, which were pure stands of *A. hirsuta* or *S. cardiophylla* var. *urbaniana*, showed low richness and diversity.

The three pioneer species (*A. hirsuta*, *S. cardiophylla* var. *urbaniana*, and *P. maximowiczii*) and two conifers (*P. jezoensis* and *A. sachalinensis*) were considered together for an examination of changes in their relative dominance with stand age (Fig. 7). The dominance of the pioneers declined with age and fell almost to zero when stands were >80 years old. In contrast, the conifers became more dominant

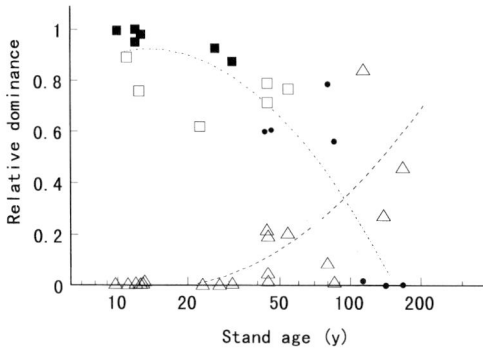

Fig. 7. Changes in relative dominance of pioneer species and conifers along the gradient of stand age (Nakamura et al. 1997). Black squares, white squares, and black circles indicate the sum of the relative dominance (calculated in terms of basal areas) of the three pioneer species that were established on active, semi-active, and stable sites, respectively. The triangles indicate the sum of the relative dominance of the two conifer species

over time, with dominance increasing sharply in stands >80 years old. Consequently, the species richness of stands between 50 and 80 years old was high. The three pioneers primarily occupied the active sites, rather than the semi-active sites, when stand age was similar. The two conifers were not found in active sites, and they were very rarely found in semi-active sites. The mature stands dominated by broad-leaved trees included upland species such as *T. japonica* and *Prunus ssiori* F. Schmidt. Although the relative dominance of the pioneers and the conifers displayed contrasting patterns with stand age, the basal areas of *P. maximowiczii* and *A. sachalinensis* were positively correlated (r^2=0.90, $p<0.001$). Mature stands dominated by *P. maximowiczii* were generally mixed with young *A. sachalinensis* in the understory, forming two-storied forests (Fig. 5).

Species richness and diversity increased with age until trees were 50 to 60 years old; thereafter, they decreased or reached a plateau. Johnson et al. (1976) and Bell (1980) also observed this temporal pattern, attributing the distribution to gradients of soil moisture and flooding frequency. These trends may be explained not only by the theory of succession but also by the stability index, which is defined in terms of the elevation and distance gradients. The stands on active sites suffered frequent, severe floods and failed to mature. They remained at an initial stage of development, while species that are intolerant of sediment burial and water submergence grew only in stable sites, with enhanced species diversity. The dominance of the three pioneers at active sites diminished species richness and diversity, whereas these species did not fully occupy semi-active sites, resulting in relatively high species richness and diversity.

The two conifers were frequently observed at sites where dominance of the three pioneers had declined. The high, stable terraced deposits created by high-magnitude floods consisted of fresh, fully exposed alluvium. These sites must initially have provided a primary habitat for the establishment of *P. maximowiczii* seedlings. The subsequent medium-magnitude floods may have partially disturbed the original stands and provided an opportunity for conifers to become established, with the resultant formation of conspicuous, two-storied forests. Conifers may have replaced the pioneer species over time, or *U. davidiana* var. *japonica*, *A. mono*, and *F. mandshurica* var. *japonica* may have succeeded pioneers on very stable deposits. With respect to available seed sources, the stable deposits were far from the river channel and close to hillslopes. They were therefore seeded by many upland species. This phenomenon may have introduced several dominant species that are common to hillslopes, such as *T. japonica* and *P. ssiori*, to floodplains, resulting in enhanced species richness and diversity.

Decreasing or steady trends in species richness and diversity after 50 years can be attributed to a thick cover of dwarf bamboo (*Sasa* spp.) in mature broad-leaved stands. The overbank deposition of sandy soil on high-terraced surfaces and the closeness to hillslopes likely provided favorable conditions for the extension of dwarf bamboo rhizomes into the substratum. Dwarf bamboo is the dominant understory species in mature floodplain forests and impedes seedling regeneration of other tree species by covering the forest floor with its thick leaves (Kudo 1980). Decreased species richness on high, stable terraced deposits was also caused by the replacement of shade-tolerant species in conifer-dominated stands (Bell 1980).

Figure 7 shows the contrasting patterns in relative dominance of pioneer species and conifers along the gradient of stand age. Transitional stands, consisting of a mixture of pioneer and late successional species, were not found in the study sites. Johnson et al. (1976) also pointed out this compositional dichotomy between pioneer and late successional species. In general, floods scour sediment that has been deposited at erosion-prone sites, such as areas adjacent to active channels, and may leave sediment at the margins of the valley floor (Nakamura & Kikuchi 1996). On the other hand, stable sites for late successional species are located far from river channels, form a high geomorphic surface in cross-sectional profiles, and are not frequently disturbed by floods. Consequently, floodplain areas are differentiated into high- and low-disturbance areas until the floodplain geomorphology is entirely reset by an extraordinary event.

9.6 Conservation of riparian forest dynamics

The Japanese archipelago is dominated by tectonically fragile, folded mountain ranges that are steep and particularly prone to earthquakes and volcanic activity. Additionally, the Asian monsoon climate tends to yield heavy rainfall, most often associated with typhoon-induced storms that move along the Japanese mountain ranges. Heavy snowfall also occurs in areas along the Sea of Japan and in the northern part of Japan, which results in periodic snow floods in spring. Furthermore, intense land utilization is present throughout these small Japanese islands owing to high population density and decreased land availability. Thus, the environmental and social settings of Japan predispose it to being a disaster-prone country, in which dam structures are common.

The mosaic patterns of riparian forests reflect the heterogeneity of site conditions created by frequent flood disturbances. Various geological and meteorological factors in Japan result in variation in frequency and intensity of flood disturbances. Therefore, various types of riparian forests are found in Japan. Dam structures, such as Sabo and reservoir dams, drastically change the disturbance regime and formation of geomorphic surfaces (Takagi & Nakamura 2003). Therefore, geomorphic adjustment caused by an altered flow regime can cause a substantial shift in riparian vegetation composition (Merritt & Cooper 2000).

Alteration of riparian forest ecosystems by regulation of river flow can also have a significant impact on aquatic ecosystems. Riparian forests are a source of organic matter, such as coarse woody debris (CWD) and litter. CWD creates variation in riverbed morphology and provides habitats for various aquatic organisms, while organic matter provides a food source. Following dam regulation, the frequency and quantity of organic matter supply will change, suggesting that structure and composition of aquatic ecosystems will also change.

Conservation of indigenous plant communities in riparian zones and protection of river health are strongly dependent on how we preserve the dynamic nature and processes of rivers. Variation of flood discharge and unconstrained valley floors should be maintained or restored. In recent years, two projects have been initiated to restore the original meandering channels and floodplains in Hokkaido (Nakamura 2003). The replacement or removal of dams has also been considered. Determining when and how much water and materials should be discharged to allow riparian plant species to become established and regenerate remains an important challenge.

References

Adams DE, Anderson RC (1980) Species response to a moisture gradient in central Illinois forests. Am J Bot 67:381-392

Araya T (1971) Morphological study of bed load movement in torrential rivers. Res Bull Coll Exp For Agri Hokkaido Univ 28:193-258 (in Japanese with English abstract)

Aruga M, Nakamura F, Kikuchi S, Yajima T (1996) Characteristics of floodplain forests and their site conditions in comparison to toeslope forests in the Tokachi River. J Jpn For Soc 78:354-362 (in Japanese with English abstract)

Bell DT (1980) Gradient trends in the streamside forest of central Illinois. Bull Torrey Bot Club 107:172-180

Bell DT, del Moral R (1977) Vegetation gradients in the streamside forest of Hickory Creek, Will County, Illinois. Bull Torrey Bot Club 104:127-135

Cappers RTJ (1993) Seed dispersal by water: a contribution to the interpretation of seed assemblages. Veg hist Archaeobot (2)3:173-186

Friedman JM, Osterkamp WR, Lewis Jr WM (1996) The role of vegetation and bed-level fluctuations in the process of channel narrowing. Geomorphology 14:341-351

Hupp CR, Osterkamp WR (1985) Bottomland vegetation distribution along Passage Creek, Virginia, in relation to fluvial landforms. Ecology 66:670-681

Hupp CR, Osterkamp WR (1996) Riparian vegetation and fluvial geomorphic processes. Geomorphology 14:277-295

Ishikawa S (1988) Floodplain vegetation of the Ibi River in central Japan I. Distribution behavior and habitat conditions of the main species of the river bed vegetation developing on the alluvial fan. Jpn J Ecol 38:73-84 (in Japanese with English abstract)

Johnson WC, Burgess RL, Keammerer WR (1976) Forest overstory vegetation and environment on the Missouri River floodplain in North Dakota. Ecol Monogr 46:59-84

Kikuchi T (2001) Vegetation and Landforms. Univ Tokyo Press, 220p (in Japanese)

Kudo H (1980) The variation of floor plants after withering of *Sasa Kurilensis* by unusual mass flowering. J Jpn For Soc 62:1-8 (in Japanese with English abstract)

Merritt DM, Cooper DJ (2000) Riparian vegetation and channel change in response to river regulation: a comparative study of regulated and unregulated streams in the Green River Basin, USA. Regul Rivers: Res Mgmt 16:543-564

Nakamura F (1995) Structure and function of riparian zone and implications for Japanese river management. Trans Jpn Geomorph Union 16:237-256

Nakamura F, Kikuchi S (1996) Some methodological developments in the analysis of sediment transport processes using age distribution of floodplain deposits. Geomorphology 16:139-145

Nakamura F, Yajima T, Kikuchi S (1997) Structure and composition of riparian forests with special reference to geomorphic site conditions along the Tokachi River, northern Japan. Plant Ecol 133:209-219

Nakamura F (2003) Restoration strategies for rivers, floodplains and wetlands in Kushiro Mire and Shibetsu River, northern Japan. Ecol Civil Eng 5:217-232 (in Japanese with English abstract)

Niiyama K (1987) Distribution of *Salicaceous* species and soil texture of habitats along the Ishikari River. Jpn J Ecol 37:163-174 (in Japanese with English abstract)

Niiyama K (1989) Distribution of *Chosenia Arbutifolia* and soil texture of habitats along the Satsunai River. Jpn J Ecol 39:173-182 (in Japanese with English abstract)

Shibakusa Y, Takahashi K, Saito Y (1968) Natural regeneration of *Abies sachalinensis* and its environmental factors with special reference to relative light intensity. Proceeding of the 79th Annual Conference of Japanese Forestry Society, pp 293-295 (in Japanese)

Sigafoos RS (1964) Botanical evidence of floods and flood-plain deposition. United State Geological Survey Professional Paper 485-A:A1-A35

Scott ML, Friedman JM, Auble GT (1996) Fluvial process and the establishment of bottomland trees. Geomorphology 14:327-339

Takagi M, Nakamura F (2003) The downstream effects of water regulation by the dam on the riparian tree species in the Satsunai River. J Jpn For Soc 85:214-221

Takahashi K, Fujimura Y, Koike T (1988) Tolerance deciduous broad-leaved trees in Hokkaido – Seasonal change of the tolerance. Trans Mtg Hokkaido Br Jpn For Soc 36:99-101 (in Japanese)

The Japan Meteorological Agency (2007) http://www.jma.go.jp/

10   Mosaic structure of riparian forests on the riverbed and floodplain of a braided river: A case study in the Kamikouchi Valley of the Azusa River

Shingo Ishikawa

Faculty of Science, Kochi University, 2-5-1 Akebonocho, Kochi 780-8520, Japan

10.1 Introduction

Ecological studies of riparian forests in mountain regions in Japan have mostly been conducted in upper reaches where floodplains are relatively narrow with incised meandering channels (for example, Kikuchi 1968; Aruga et al. 1996; Nakamura et al. 1997). These reaches are subject to frequent disturbances of various type and magnitude, and large floods can sometimes entirely remove vegetation from a reach (Sakio 1997).

There have been so few studies on riparian vegetation on wide floodplains in mountain regions in Japan because there are very few such reaches with wide floodplains. In addition, structures such as dams and river revetments have been intensively installed along riverine areas all over Japan, eliminating natural riparian forests.

In this section, I introduce the mosaic pattern and dynamics of riparian vegetation with special reference to the regeneration of a population of *Salix arbutifolia* Pall., formerly *Chosenia arbutifolia* (Pall.) Skvorts. on a wide floodplain with a braided channel network in the Kamikouchi Valley of the Azusa River, central Japan. The information is based mainly on the study of Shin et al. (1999).

10.2 Outline of the Kamikouchi Valley

The Kamikouchi Valley is in the upper reaches of the Azusa River in the Chubu-Sangaku National Park, thus the riparian areas have been protected from

Sakio, Tamura (eds) Ecology of Riparian Forests in Japan : Disturbance, Life History, and Regeneration
© Springer 2008

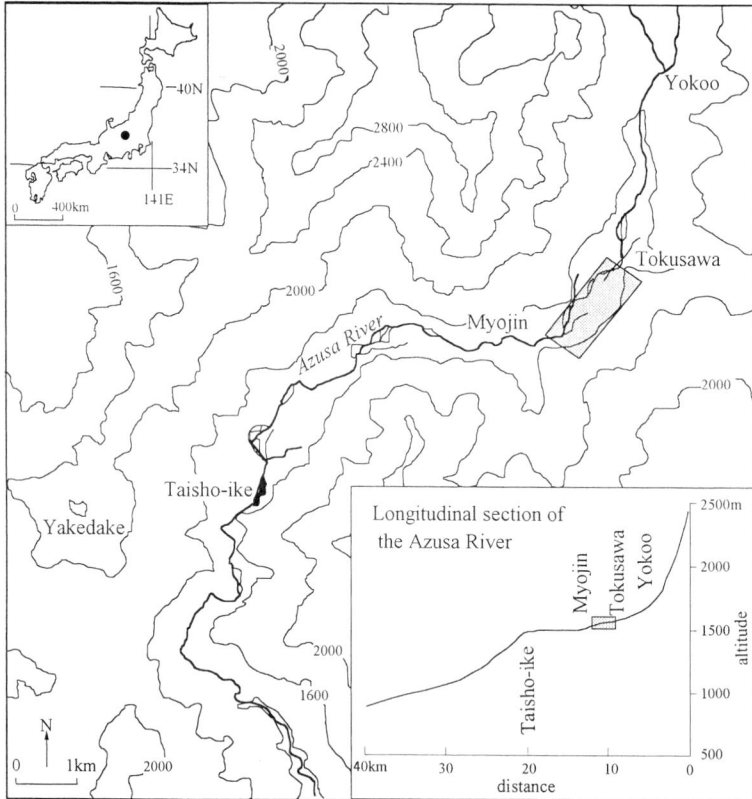

Fig. 1. Location of the Kamikouchi Valley with insert showing the longitudinal profile of the Azusa River. Hatching indicates the area corresponding to the vegetation map depicted in Fig. 2 (compiled from Shin et al. 1999)

destructive human activities. This valley has a considerably wider floodplain than those of other rivers flowing in mountainous regions. In the past, eruptions from Mt. Yakedake, the only volcano in the Northern Japan Alps, dammed the valley with enormous pyroclastic flow deposits and lava. Consequently, an abundance of rocks and gravels supplied from the mountains around the valley have raised the riverbed level (Iwata 1992; Shimazu 1995). The main study area is located between Tokusawa and Myojin, where the mean gradient of the riverbed is rather gentle (ca. 10–13‰) and the riparian vegetation is composed of diverse seral communities developing on a floodplain with a braided channel pattern.

The annual mean air temperature and annual mean precipitation from 1979 to 2000 were 4.6 °C and 2600 mm, respectively. The valley is at an altitude of about 1500 m and corresponds to the upper cool temperate zone. The mixed-forests found on many alluvial cones adjacent to the floodplain are composed of conifer-

Fig. 2. Riparian forest and active floodplain zone showing braided channel pattern in the Kamikouchi Valley

ous trees such as *Abies homolepis* Sieb. et Zucc. and deciduous trees such as *Ulmus davidiana* Planch. var. *japonica* (Rehder) Nakai, *Fraxinus mandshurica* Rupr. var. *japonica* Maxim. and *Pterocarya rhoifolia* Sieb. et Zucc. etc. These tree species have a close relationship with the riparian vegetation and play important roles in the successional process of the riparian forest. The mountain ridges and slopes around the valley support coniferous forests dominated by *Abies veitchii* Lindl., *Tsuga diversifolia* Masters, *Picea jezoensis* (Sieb. et Zucc.) Carr. var. *hondoensis* (Mayr) Rehder, *Larix kaempferi* Carr. and are often accompanied by *Betula ermanii* Cham., a subalpine deciduous broad-leaved tree.

The principal species of the riparian forests and scrubs developing along the river are *Salix arbutifolia*, *Salix cardiophylla* Trautv. & Mey. var. *urbaniana* (Seemen) Kudo, formerly *Toisusu urbaniana* (Seemen) Kimura, *Populus maximowiczii* A. Henry, *Salix rorida* Lackschewitz, *Alnus hirsuta* Turcz. var. *microphylla* (Nakai) Tatewaki., *L. kaempferi*, *U. davidiana* var. *japonica*, *F. mandshurica* var. *japonica*, and *A. homolepis*.

Among these species, *S. arbutifolia* is notable as a relic species from the last glacial maximum. This species occurs in several rivers in eastern Hokkaido (Tatewaki 1948; Ito 1986) and only along the Azusa River in Honshu (Kimura 1951), where scrub and forest dominated by this species can be observed in various age classes.

10.3 Mosaic structure of riparian vegetation

In the riparian vegetation of the floodplain from Myojin to Tokusawa, seven community types were recognized as follows: 1) Pioneer species scrub, 2) Young pioneer forest, 3) *S. rorida – A. hirsuta* var. *microphylla* forest, 4) *A. hirsuta* var. *microphylla* forest, 5) Old pioneer forest, 6) *U. davidiana* var. *japonica – A. homolepis* forest and 7) *L. kaempferi* forest (Fig. 3).

We can recognize two zones showing different vegetation patterns in Fig. 3: 1) braided channel with fragmented small pioneer species patches in the right bank, and 2) floodplain with well developed riparian forests spreading continuously on the left bank. The process forming this mosaic pattern of riparian vegetation is dis-

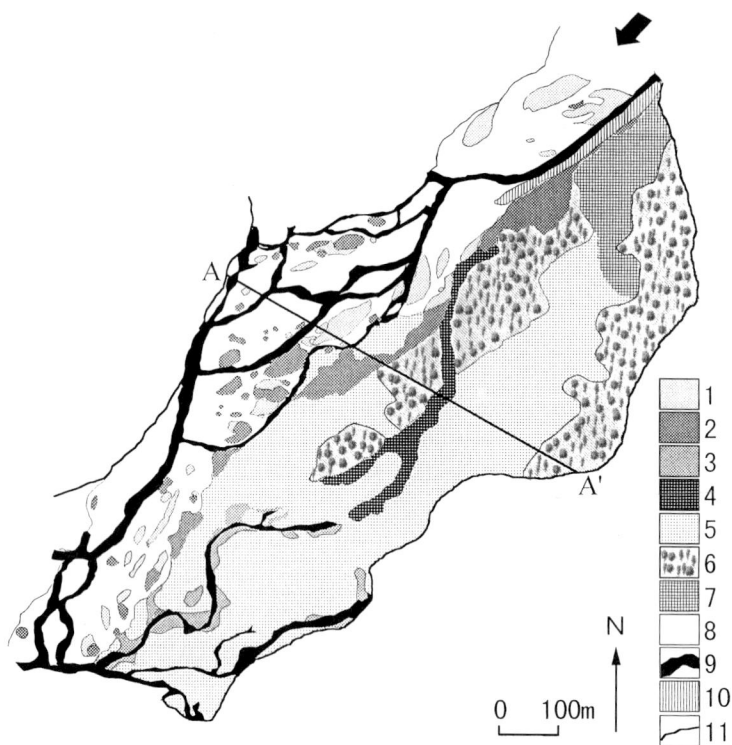

Fig. 3. Vegetation map of the floodplain from Myojin to Tokusawa in the Kamikouchi Valley as depicted in 1994. The line A- A' and arrow represent the belt transect and the direction of flow, respectively (compiled from Shin et al. 1999).

1, Pioneer species scrub; 2, Young pioneer forest; 3, *Salix rorida – Alnus hirsuta* var. *microphylla* forest; 4, *Alnus hirsuta* var. *microphylla* forest; 5, Old pioneer forest; 6, *Ulmus davidiana* var. *japonica – Abies homolepis* forest; 7, *Larix. kaempferi* forest; 8, river bed; 9, river channel; 10, artificial bank; 11, piedmont line

cussed in the sections below.

10.4 Geomorphic process and disturbance regime of the floodplain in Kamikouchi

The reach from Myojin to Tokusawa has a typical braided channel pattern (Fig. 2 and 3). Judging from the development and erosion processes of the alluvial cones adjacent to the floodplain and the tree ages of the forests in both the floodplain and the alluvial cone, drastic channel shifts from the left bank side to the right bank side occurred twice in the past 500 years (about 100 and 500 years ago) in this reach. That change is shown from B to C in Fig. 4 (Shimazu 1998).

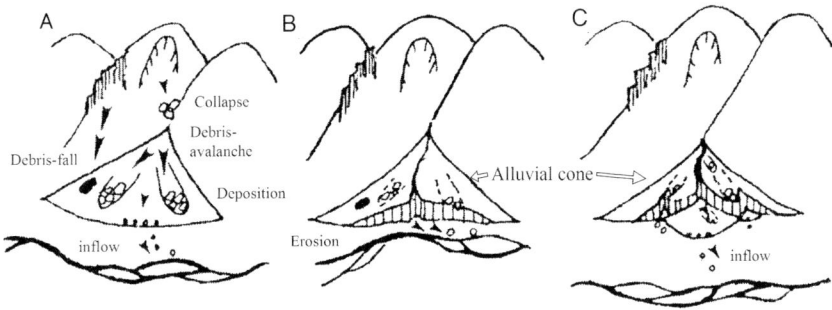

Fig. 4. Developmental process of alluvial cones and channel shift in the Azusa River in Kamikouchi (compiled from Shimazu 1998)

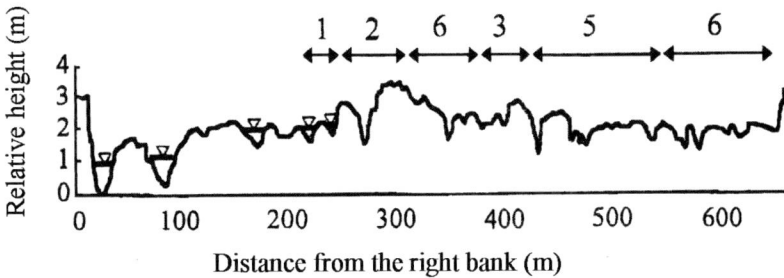

Fig. 5. Cross-sectional view of the distributions of typical plant communities along the belt transects (A-A'). Numerals above the profile indicate the community types shown in Fig.3 (compiled from Shin et al. 1999)

The cross-section of the belt transect A-A' (represented in Fig. 3) shows that the profile of the valley floor on the left bank, where diverse riparian forests have developed, is similar to that on the right bank where patches of young pioneer scrubs are scattered. It also shows that the relative height across the transect is within about 3 m (Fig. 5). In this reach, catastrophic disturbances, i.e., all vegetation removed, can not be observed to have occurred. The partial destruction and recovery of riparian forests have been investigated repeatedly using aerial photographs after 1947 (Fig. 6).

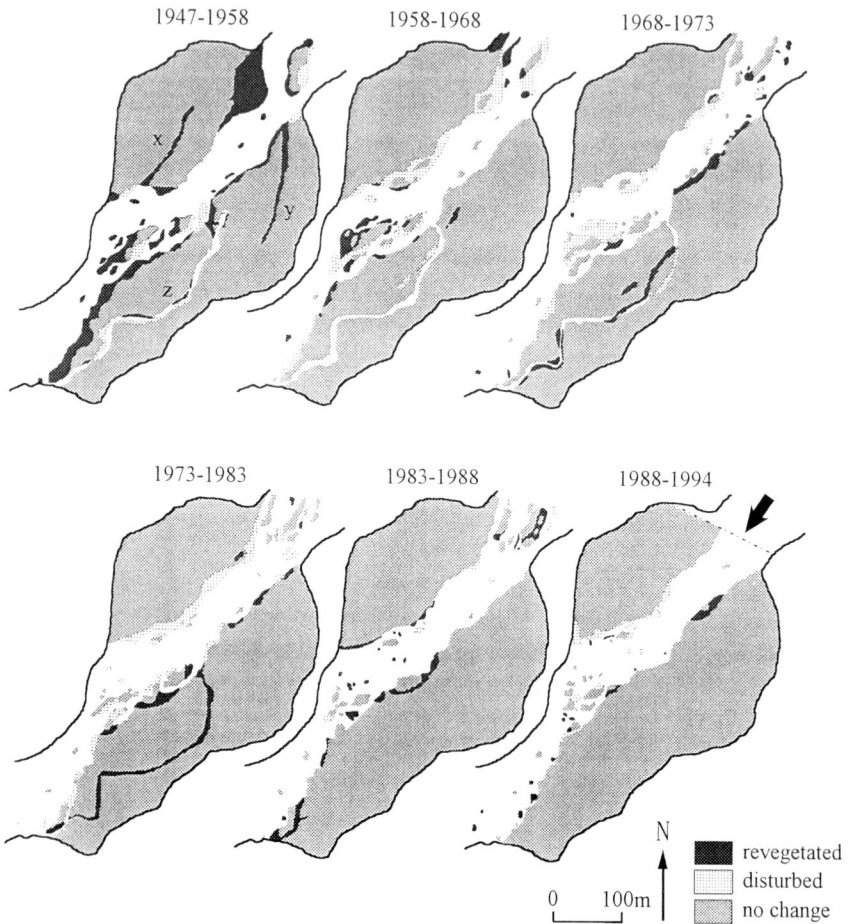

Fig. 6. Historical changes in riverbed landforms and riparian forest during the last 47 years. The arrow indicates the direction of flow. The wavy line shows the boundary of the aerial photograph (compiled from Shin et al. 1999)

10.5 Young pioneer scrubs and forests

Three of the community types shown in Fig. 3 occupy sites where no vegetation cover was found in aerial photographs taken before 1983: 1) Pioneer species scrub; 3) *S. rorida – A. hirsuta* var. *microphylla* comm.; and 4) *A. hirsuta* var. *microphylla* comm. However each of these three younger communities was found on different geomorphic surfaces of the floodplain. As described in some previous papers (Okuda 1978; Ishikawa 1983, 1987; Niiyama 1987, 1989), the distribution of diverse pioneer species is closely correlated with different soil water conditions, aeration, and nutrients of the habitats, which was also observed among the three communities.

The communities along abandoned channels with thin layers of fine sediments were dominated by *A. hirsuta* var. *microphylla*. On the other hand, *Salix rorida – A. hirsuta* var. *microphylla* community developed along channels originating from springs in the floodplain where finer sediments such as fine sand and silt were deposited along the channels.

On bars and islands, pioneer species such as *S. arbutifolia, S. cardiophylla* var. *urbaniana, P. maximowiczii,* and *A. hirsuta* var. *microphylla* invaded and grew into diverse communities with respective distinct dominants (Table 1). In these habitats, where plants are always subject to stress and disturbance through drought and erosion and deposition of sediments, patches dominated by *S. arbutifolia* occur most frequently (ca. 60% of all patches), suggesting that the fast taproot growth rate of *S. arbutifolia,* particularly in the early growth stage, is an effective ecological adaptation for survival in such a harsh environment. The seed dispersal season (from June to July) of *S. arbutifolia* occurs when large areas of the floodplain surface tend to stay moist from rain and the groundwater level is high, distinct advantages for seedlings to survive and become established. In contrast, the seed dispersal season of *P. maximowiczii* and *S. cardiophylla* var. *urbaniana* is from mid summer to early autumn, when the seedlings are usually subject to

Table 1. Mean percentage cover of common species in the pioneer species scrub established in open and active zones of the floodplain, categorized by the dominant species (compiled from Shin et al. 1999)

Dominant species	Number of Patches	Mean percentage cover					
		Sa	Sr	Ah	Pm	Sc	Be
Salix arbutifolia (Sa)	30	77	5	8	13	-	-
Salix rorida (Sr)	13	5	81	-	5	33	-
Alnus hirsuta var. *microphylla* (Ah)	3	4	11	56	5	-	22
Populus maximowiczii (Pm)	2	11	-	24	74	-	-
Salix cardiophylla var. *urbaniana* (Sc)	1	1	2	1	-	56	-
Betula ermanii (Be)	1	2	-	5	3	-	66

drought in sunny and warm weather.

Fluctuations in groundwater level cause shifts in the location of favorable positions on bars and islands for seedling establishment of pioneer species. The longevity of seed viability of salicaceous species is very short because of a very thin seed coat and a lack of dormancy (Nakajima, 1921). So the establishment of seedlings is possible only if the surface retains enough soil moisture and stays undisturbed for a sufficient period of time. Thus the phenological differences in fruiting and dispersal period of seeds among species are an important reason why each patch has a distinct dominant (Niiyama 1990; van Splunder et al 1995).

Furthermore, variation in the particle size of sediments has important ecophysiological implications for seedling establishment of pioneer species because it affects both groundwater level and aeration (Kozlowski 1982, Ishikawa 1994). Thus, texture of surface sediments also plays an important role in determining the most favorable establishment sites for pioneer species in riparian habitats, especially on bars and islands (Richards et al. 2002).

10.6 Seedling growth traits of salicaceous species

The difference in seedling growth traits among the salicaceous species found in the Kamikouchi Valley was investigated by means of a growth experiment that examined the effects of different sediment textures and humidity (Ishikawa 1994; Ishikawa & Asahina 1997) (Table 2).

Although all four species examined grow on bars and islands covered mainly with gravels, the root growth traits in the early stage of establishment are different among these species. The roots of *S. arbutifolia* seedlings grew faster than those of the other species under mesic conditions (water level of -15 cm in depth) in coarse sand, suggesting that this species is well adapted to fluctuations in groundwater level and has an advantage for survival on sites of low soil moisture availability. Conversely, in fine sand, all seedlings of *S. arbutifolia* lost their taproots and even some of the adventitious roots that emerged after the decay of

Table 2. Root length (mean ± SE ; mm) of the seedlings after 4 weeks of growth under four different conditions of particle size and humidity of sediments

Sediment size	Coarse Sand		Fine Sand	
Sediment water level	-2 cm	-15 cm	-2 cm	-15 cm
S. arbutifolia	67.8± 4.1	128.8± 3.3	9.0± 0.8	25.6 ± 1.0
S. rorida	80.4± 2.4	94.4± 6.0	68.5± 0.6	68.7 ± 6.8
S. cardiophylla var. *urbaniana*	50.5± 3.4	57.9± 2.2	51.3± 0.5	61.4 ± 5.0
P. maximowiczii	58.7± 4.6	34.8± 3.1	58.7± 4.6	37.0 ± 3.8

the taproots (Ishikawa 1994). In general, aeration is poor in saturated and/or fine-textured soils (Kozlowskii 1982), so the seedlings of *S. arbutifolia* are extremely intolerant of poorly aerated soil. This intolerance is thought to be one reason why this species has been restricted to limited areas and has become a relict species.

Although *S. cardiophylla* var. *urbaniana, P. maximowiczii, S. rorida* are also distributed in the upper reaches of rivers where sediments are composed mainly of gravel and coarse sand, their taproots grew well both in coarse and fine sands in the first 4 weeks, suggesting the relative tolerance of these three species to poorly aerated soil. The seedling growth traits of each species in the early establishment stage mentioned above are important determinants of suitable regeneration sites for each species. However, we often cannot observe any difference in sediment particle size among diverse scrubs and forests of pioneer species (Shin et al. 1999). This can be attributed to the fact that the surface sediment layers affecting seedling establishment of each species are thin and are thus frequently scoured and/or mixed with sediments deposited over the surface subsequent to establishment occurring.

The habitat mosaic resulting from fluvial processes shown in Fig. 6 provides a diverse range of surface conditions favorable for the establishment of an array of pioneer species. These processes also create a diverse age structure between patches.

10.7 Old pioneer and late successional forests

Following drastic channel shifts, separate patches of pioneer species on bars and islands are released from frequent disturbance by floods and become connected with one another to form larger patches with a diversity of pioneer species such as *S. arbutifolia, S. cardiophylla* var. *urbaniana, P. maximowiczii, S. rorida* coexisting in a continuous forest canopy (Fig. 3).

The resistance of saplings of salicaceous species to uprooting is higher than for those of *Alnus* (Karrenberg et al. 2003), indicating that salicaceous species have root systems that are well-developed and penetrate to depth (Honma et al. 2002). As pioneer trees grow taller, deep root systems enable trees to use water in deeper layers. However, the deeper depositional layers throughout the floodplain of the study area are composed mainly of cobbles and boulders within a matrix of coarse sands. Thus the micro scale heterogeneity of surface sediments that leads to colonization by a diversity of pioneer species becomes progressively insignificant over time for the development of pioneer species forests.

Several species of Betulaceae, such as *A. hirsuta* var. *microphylla* and *B. ermanii,* are also found on the floodplain. The forests dominated by these species are restricted to moist habitats, namely abandoned channels and secondary channels, because of their shallow root systems composed mainly of horizontal roots, although their seedlings occur widely on the floodplain in open areas.

Ulmus davidiana var. *japonica, F. mandshurica* var. *japonica* and *A. homolepis,* which are typical late successional species on alluvium in Japan

(Miyawaki 1988; Sakai et al. 1999), have invaded the undergrowth of old pioneer forests in the Kamikouchi Valley.

Ulmus seedlings are found in various habitats across a wide range of light and soil moisture conditions and disturbance regimes. However, they can survive only at sites with a thin herb layer, such as in newly abandoned channels in well-developed *Ulmus* forests and young forests of pioneer species, suggesting that this late successional species can invade only forest floors disturbed by floods (Wada & Kikuchi 2004). On the other hand, *Fraxinus* seedlings occur frequently in abandoned channels in *Alnus* and *Ulmus* forests that have wet conditions in addition to a disturbed forest floor.

Understory saplings and sub-trees of these late successional species in old pioneer species forests and well-developed forests dominated by the same species are 10 to 50 years younger than canopy trees and have undergone vigorous growth, indicating that these species can not only succeed the pioneer forests, but also regenerate at sites dominated by the same species (Shin et al. 1996; Wada & Kikuchi 2004).

Shin and Nakamura (2005) reported on the Rekifune River in eastern Hokkaido, where riparian forests dominated by *S. arbutifolia* have developed extensively. They pointed out that the dominant species varied among five geomorphic surfaces: gravel bar, lower and upper floodplains, secondary channel, and terrace. These five surfaces were classified based on flooding frequency. Moreover, *S. arbutifolia* dominated in the bar-braided section where gravel bars and lower and upper floodplains occupied wide areas of the entire floodplain, whereas late successional species such as *U. davidiana* var. *japonica* and *F. mandshurica* var. *japonica* were found at higher densities in the incised meandering section constrained in a narrow valley floor with wide terraces. We could not recognize clearly such diverse geomorphic surfaces as lower and upper floodplains and terrace in the study area of the Kamikouchi Valley. However, similar riparian forest types including late successional forests occurred in the floodplain, forming a complicated mosaic pattern of riparian vegetation. The floodplain in the Kamikouchi Valley comprises both an active zone with braided channel and a stable zone in the same section of river, suggesting that the disturbance regime in the floodplain along the Kamikouchi Valley is more complicated and diverse than that in either the bar-braided section or incised meandering section of the Rekifune River. In addition, these two zones are interchangeable over longer time periods across the wide floodplain (over 500 m average width). These overall fluvial disturbances thus create a mosaic of diverse vegetation stands in different successional stages.

10.8 Conclusion

In the Kamikouchi Valley, diverse communities of salicaceous species with *Salix arbutifolia* as the dominant were allowed to coexist through different species having different seed dispersal periods and seedling growth traits related to water availability and particle size distribution of sediment. Riparian forests representing various successional stages formed a mosaic vegetation structure on the floodplain. This mosaic was created not only by diverse floodplain sediment conditions, but also by diverse disturbance regimes associated with drastic channel shifts occurring over a longer time period (100–500 years) across the wide floodplain.

References

Aruga M, Nakamura F, Kikuchi S, Yajima T (1996) Characteristics of floodplain forests and their site conditions in comparison to toeslope forests in the Tokachi River. J Jpn For Soc 78:354-362 (in Japanese with English abstract)

Honma M, Yajima T, Kikuchi S (2002) Allometry and the root structure of *Chosenia arbutifolia, Toisusu urbaniana*, and *Populus maximowiczii* saplings. J Jpn For Soc 84:41-44 (in Japanese with English abstract)

Ishikawa S (1983) Ecological studies on the floodplain vegetation in the Tohoku and Hokkaido Districts, Japan. Ecol Rev 20:73-114

Ishikawa S (1987) Ecological studies on willow communities on the Satsunai River floodplain, Hokkaido, with special reference to the development of the Chosenia arbutifolia forest. Mem Fac Sci Kochi Univ Ser D (Biol) 8:57-67

Ishikawa S (1994) Seedling growth traits of three salicaceous species under different conditions of soil and water level. Ecol Rev 23:1-6

Ishikawa S, Asahina M (1997) Seedling growth traits of salicaceous species growing on the floodplain in the Azusa River, Kamikouchi. Research on the Dynamics of Riparian Plant Communities in the Azusa River, Kamikouchi, Research Group for Natural History in Kamikouchi, Matsumoto, pp 32-36 (in Japanese)

Ito K (1986) New locality of Hokkaido plants (3). J Jpn Bot 61:375-376 (in Japanese)

Iwata S (1992) Landform evolution and environmental conservation in tha Kamikouchi Valley, the Japanese Alps. Trans Jpn Geomorph Union 13:283-296 (in Japanese with English abstract)

Karrenberg S, Blaser S, Kollmann J, Speck T, Edwards PJ (2003) Root anchorage of saplings and cuttings of woody pioneer species in a riparian environment. Funct Ecol 17:170-177

Kıkuchi T 1968. Forest communities along the Oirase Valley, Aomori Prefecture. Ecol Rev 17:87-94

Kimura A (1951) A new locality of *Chosenia bracteosa* Nakai in Honshu. Ecol Rev 13:35-36

Kozlowski TT (1982) Response of woody plants to flooding. In: Kozlowski TT (ed) Flooding and plant growth. Academic Press, San Diego, pp 129-163

Miyawaki A (1988) Vegetation of Japan. (Vol. 9: Hokkaido). Shibundo, Tokyo (in Japanese)

Nakajima Y (1921) On the life duration of seed of Salix. Bot Mag Tokyo 35:17-42

Nakamura F, Yajima T, Kikuchi S (1997) Structure and composition of riparian forests with special reference to geomorphic site conditions along the Tokachi River, northern Japan. Plant Ecol 133:209-219

Niiyama K (1987) Distribution of salicaceous species and soil texture of habitats along the Ishikari River. Jpn J Ecol 37:163-174 (in Japanese with English summary)

Niiyama K (1989) Distribution of *Chosenia arbutifolia* and soil texture of habitatas along the Satsunai River. Jpn J Ecol, 39:173-182 (in Japanese with English summary)

Niiyama K (1990) The role of seed dispersal and seedling traits in colonization and coexistence of *Salix* spp. in a seasonally flooded habitat. Ecol Res 5:317-332

Okuda S (1978) Pflanzensoziologische Untersuchungen uber die Auenvegetation der Kanto-Eben. Bull Inst Environ Sci Tech Yokohama Nat Univ 4:43-112 (in Japanese with German summary)

Richards K, Brasington J, Hughes F (2002) Geomorphic dynamics of floodplains: ecological implications and a potential modelling strategy. Freshwater Biol 47:557-579

Sakai T, Tanaka H, Shibata M, Suzuki W, Nomiya H, Kanazashi T, Iida S, Nakashizuka T (1999) Riparian disturbance and community structure of a *Quercus-Ulmus* forest in central Japan. Plant Ecol 140: 99-109

Sakio H (1997) Effects of natural disturbance on the regeneration of riparian forests in a Chichibu Mountains, central Japan. Plant Ecol 132:181-195

Shimazu H (1995) Sediment transport along the Azusa River, central Japan. Geogr Rep Kanazawa Univ 7:53-60 (in Japanese with English abstract)

Shimazu H (1998) Landforms and sediment transport processes on the Furuikezawa alluvial cone. Research on landform changes, sediment transport processes, hydrological environment, and vegetation dynamics in the Azusa River, Kamikouchi. Research Group for Natural History in Kamikouchi, Matsumoto, pp 12-21 (in Japanese)

Shin N, Ishikawa S, Iwata S (1999) The mosaic structure of riparian forest and its formation pattern along the Azusa River, Kamikochi, central Japan. Jpn J Ecol 49:71-81(in Japanese with English abstract)

Shin N, Nakamura F (2005) Effects of fluvial geomorphology on riparian tree species in Rekifune River, northern Japan. Plant Ecol 178:15-28

Tatewaki M (1948) Distribution and community of *Chosenia arbutifolia*. Ecol Rev 11:77-86 (in Japanese)

van Splunder I, Coops H, Voesenik LACJ, Blom CWPM (1995) Establishment of alluvial forest species in floodplains: the role dispersal timing, germination characteristics and water level fluctuations. Act Bot Neerl 44:269-278

Wada M, Kikuchi T (2004) Emergence and establishment of seedlings of Japanese elm (*Ulmus davidiana* Planch. var. *japonica* (Rehder) Nakai) on a flood plain along the Azusa River in Kamikochi, central Japan. Veg Sci 21:27-38 (in Japanese with English abstract)

11 Coexistence of *Salix* species in a seasonally flooded habitat

Kaoru NIIYAMA

Forestry and Forest Products Research Institute (FFPRI), Department of Forest Vegetation, 1 Matsunosato, Tsukuba, Ibaraki 305-8687, Japan

11.1 Introduction

Riparian willows (genus *Salix*) are widely distributed in the northern hemisphere and dominate the vegetation along most rivers. Several riparian *Salix* species coexist on sand bars or floodplains (Cottrell 1996; Dionigi *et al.* 1985; Niiyama 1990; van Splunder *et al.* 1995). In Japan, *Salix* species dominate riparian forests but monodominant forests are very rare (Ishikawa 1979, 1980, 1982, 1983, 1987, 1988; Niiyama 1987, 1989; Yoshikawa & Hukusima 1999). These riparian *Salix* species have similar life-history traits; dioecy, flowering and seed dispersal in spring, anemochorous small hairy seeds, rapid germination, short seed longevity, etc., but there are subtle differences in regeneration traits, such as seed dispersal period, seed size, and root morphology of seedlings (Niiyama 1990). These subtle differences in regeneration traits would explain the dominance and coexistence of riparian *Salix* species in a seasonally flooded habitat.

Riparian *Salix* species largely depend on regeneration on new open sites after floods when erosion and sedimentation simultaneously occur along rivers. The seasonality of floods is affected by the rainfall regime. Monsoons and typhoons, which are major factors of rainfall in East Asia, determine the seasonal pattern of flooding in most rivers in Japan. Generally, the magnitude of rainfall during the monsoon and the frequency of typhoons decrease in the higher latitudes but snowfall is heavier there. The seasonality of flooding shows a gradient of large, distinctive snowmelt floods in northern rivers and unclear snowmelt floods in southern ones. Riparian *Salix* species are particularly abundant along northern rivers in Hokkaido and in the Tohoku district of northern Honshu (Ishikawa 1983).

This chapter describes the reason for the abundance of *Salix* species in northern Japan and the coexistence mechanisms of several riparian *Salix* species in the northernmost island of Hokkaido. The coexistence of closely related species is one of the major issues in plant ecology. The coexistence mechanisms of *Salix* species

Sakio, Tamura (eds) Ecology of Riparian Forests in Japan : Disturbance, Life History, and Regeneration
© Springer 2008

depend on several different spatial scales. The micro topographic scale (10^0~10^1 m), local heterogeneity of substrata (10^1~10^2 m) and regional scale distribution along a large river (10^3~10^5 m) are discussed here. There are 28 native *Salix* species in Japan, of which 13 species grow in riparian habitats. This chapter examines six common willows, *Salix rorida* Lackschewitz, *Salix schwerinii* E. Wolf (formerly *Salix pet-susu* Kimura), *Salix udensis* Trautv. & Mey. (formerly

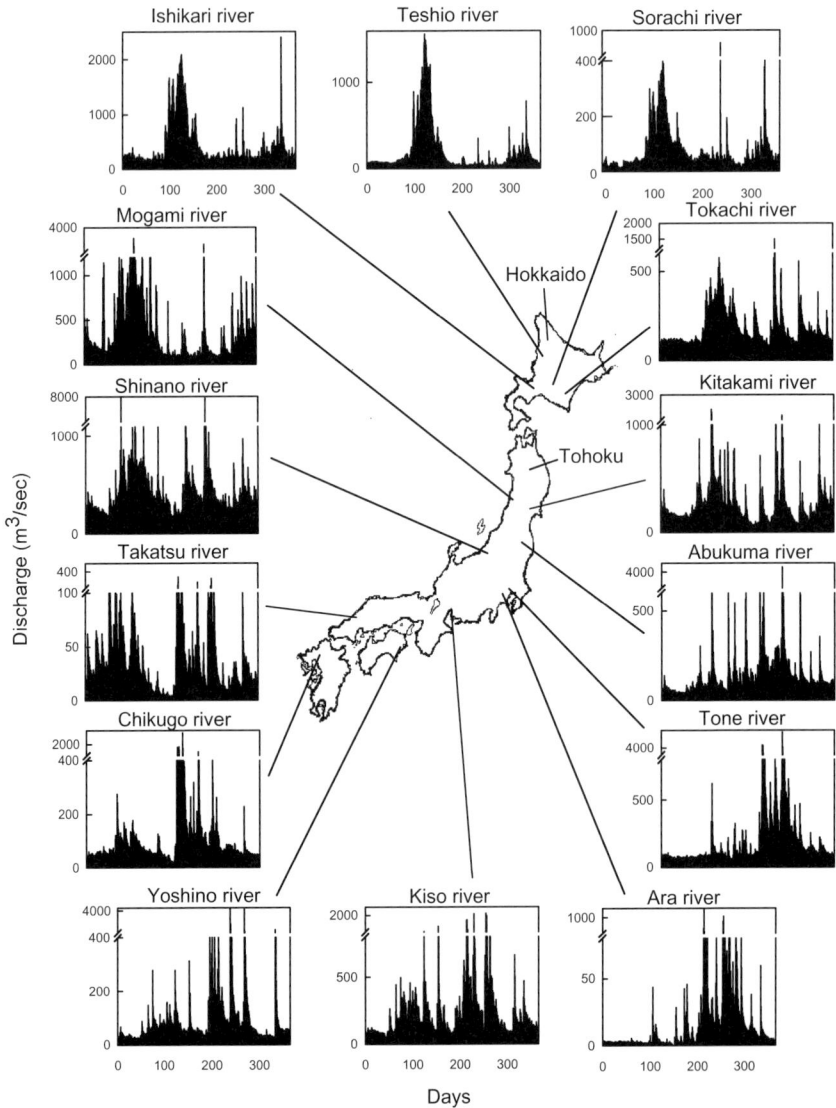

Fig. 1. Seasonal patterns of river discharge in Japan in1982 (From Niiyama 1995)

Salix sachalinensis Fr. Schm.), *Salix miyabeana* Seemen subsp. *miyabeana, Salix dolichostyla* Seemen (formerly *Salix jessoensis* Seemen) and *Salix triandra* L. (formerly *Salix subfragilis* Anders.), that occur in Hokkaido. Nomenclature follows (Ohashi 2000; Ohashi & Nakai 2006; Ohashi & Yonekura 2006). The main study sites are floodplains along the Sorachi River, Satsunai River and Ishikari River in the island of Hokkaido, Japan.

11.2 Seed dispersal and snowmelt floods

Seasonal changes in river discharge in northern Japan are characterized by snowmelt and unpredictable floods coinciding with rainfall from monsoons or typhoons (Fig. 1). The high water level of snowmelt floods gradually decreases in spring and the water level is usually stable in summer. On the other hand, in south-

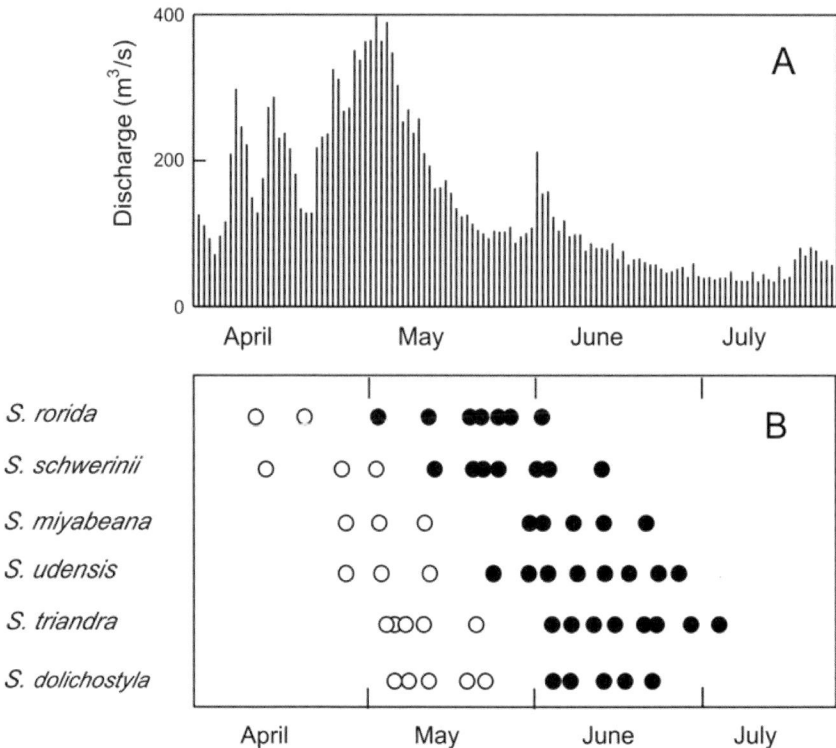

Fig. 2. Seasonality of river discharge and phenology of *Salix* species. A: discharge of Sorachi River in 1982. B: phenology of flowering (white circles) and seed dispersal (black circles) in 1982

Fig. 3. Zonation of *Salix* seedlings after the decline of high water level by snowmelting

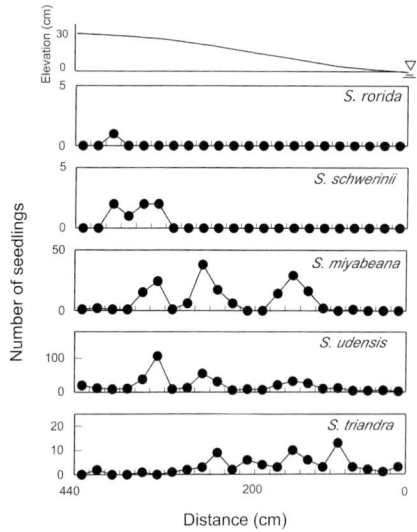

Fig. 4. Distributions of *Salix* seedlings along a 20 cm × 440 cm transect

western Japan floods by snowmelt are relatively unknown but monsoon and typhoon floods are fairly common. Riparian *Salix* species disperse their anemochorous small seeds in spring. For example in Hokkaido, most *Salix* species disperse their seeds from late May to early July (Fig. 2) when the high water level by snowmelt is gradually decreasing and many open sites are appearing along rivers. Many *Salix* species can use these open sites as safe sites for regeneration. The water level is usually stable during the summer.

Salix species can germinate on open sites where the soil surface is wet enough for germination and full sunlight is available for seedling growth. Riparian *Salix* species can grow one meter or more during one growing season. This is one of reasons that *Salix* species are more abundant in northern Japan. In contrast, monsoon and typhoon floods frequently destroy newly established *Salix* seedlings during the summer in southern Japan.

Established seedlings often suffer from repeated flooding. Catastrophic floods sometimes destroy the regeneration sites, but sedimentation is a more common type of flood disturbance to *Salix* seedlings or saplings. Riparian *Salix* species can develop adventitious roots from the stem (Carlson 1938; Houle & Babeux 1993). We have observed repeated adventitious root development resulting from the multi-layered sedimentation brought about by several flood events. This rooting ability helps to explain the dominance of *Salix* species in riparian areas in northern Japan.

11.3 Micro topographic scale distribution of *Salix* species

There is a microtopographic scale mechanism of coexistence on small slopes of riverbanks or sand bars. *Salix* species show staggered seed dispersal phenology (Fig. 2) when the water level decreases gradually. *Salix* seeds are trapped and germinate on the banks of rivers if soil moisture is sufficient for germination. The seeds may germinate within 24 hours under moist conditions but they lose viability within a few days (Niiyama 1990). As a result, their seedlings concentrate along the narrow shore line (Fig. 3).

There is a sequence of seed dispersal timings. Early seed dispersers, *S. rorida* and *S. schwerinii*, germinate at a relatively high elevation above the river level while other species germinate at a middle or lower elevation above the river level (Fig. 4). When *S. rorida* disperses seeds, the lower slope is under water. The relative elevation above the river water level affects the stability and moisture of the site. The higher elevation sites suffer drought but are stable against small floods, whereas the lower elevation sites can avoid drought but suffer frequent flood disturbances. Although this micro scale habitat segregation is not completely understood, it is one of the coexistence mechanisms of congeneric *Salix* species depending on the seasonality of water level. The same phenomenon, zonation of co-occurring *Salix* species resulting from dispersal phenology and seasonality of water level, was also reported in the Netherlands (van Splunder *et al.* 1995).

11.4 Soil texture and seedling establishment

Coexistence mechanisms at the local scale are the differentiation of regeneration traits among *Salix* species resulting from the heterogeneity of site substrata. Open sites that appear after floods show a wide variety of soil textures resulting from different sedimentation processes under different current velocities. The textures of open site substrata vary depending on the microtopography, vegetation types and flood magnitude.

The density and size of current-year seedlings were studied in nine quadrats in the floodplain of the Sorachi River. The abundance of *Salix* seedlings depended on the make-up of the substrata (Fig. 5). *Salix* seeds will encounter coarse, gravelly, sandy, or fine substrata.

S. rorida dominate in coarse substrata where safe sites for germination and seedling establishment are restricted to spaces between stones or partly disturbed fine sands because a large part of the surface is covered by gravel or stones. This species has the largest seeds (Fig. 6) and is the first of the co-occurring *Salix* species to disperse its seeds. The relatively large seeds and early dispersal are advantageous for preemptive competition, because early seed dispersal can preempt the restricted safe sites for germination and large seeds can develop a long tap root that can penetrate into the coarse substrata.

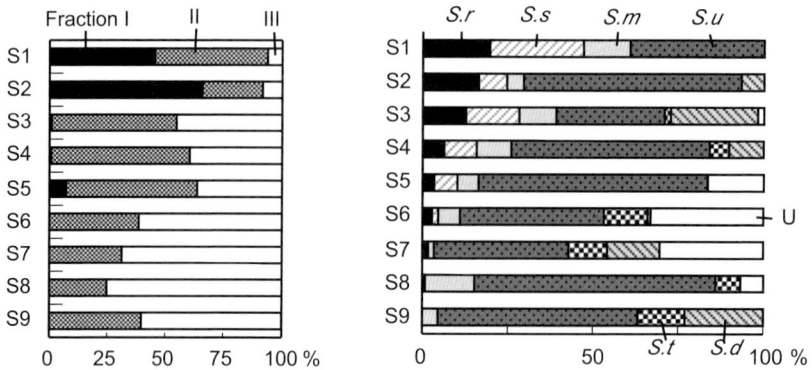

Fig. 5. Percentages of soil particle fractions and *Salix* seedlings in nine quadrats on a floodplain of the Sorachi River, Hokkaido, Japan. 2 cm > Fraction I >2mm, 2 mm > Fraction II > 0.25 mm, and Fraction III < 0.25 mm. *S.r: S. rorida, S.s: S. schwerinii, S.m: S. miyabeana, S.u: S. udensis, S.t: S. triandra, S.d: S. dolichostyla*, and U: unidentified

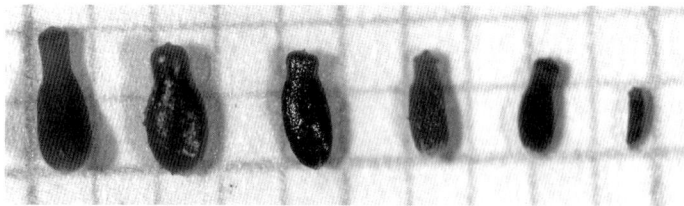

Fig. 6. Seeds of six *Salix* species; *S. roida, S. schwerinii, S. miyabeana, S. udensis, S. dolichostyla*, and *S. triandra* from left to right. Grid is 1 mm square

Surfaces of coarse substrata suffer more water stress than surfaces of fine-grained substrata. The experimental results also strongly indicate that substrata textures and moisture affect the regeneration success of salicaceous species (Ishikawa 1994; Yanai & Kikuzawa 1991).

On the other hand, all surfaces of fine textured substrata are safe sites for germination. Late seed dispersals, small seeds and shallow root systems are not crucial disadvantages for seedling establishment on fine textured substrata, which are wet and rich in nutrients. *S. triandra*, which is dominant in such fine substrata, disperses the smallest seeds among co-occurring *Salix* species. In this case, direct competition through seedling growth after germination is crucial for the dominance and coexistence of *Salix* species rather than preemptive competition before seedling growth. The advantages of early seed dispersal and relatively large seeds are not guaranteed in such fine substrata where sufficient water and nutrients usually provide for rapid growth.

The other four *Salix* species have intermediate traits between *S. rorida* and *S. triandra*. Among them, *S. schwerinii* has a similar seed size and dispersal phenology to *S. rorida*. The other three intermediate species, i.e., *S. udensis, S. dolichostyla* and *S. miyabeana,* do not show significant correlation with either coarse or fine fractions of substrata. However, these intermediate traits are not disadvantages for regeneration. The dominant species of this area, *S. udensis*, has intermediate seed size and intermediate dispersal phenology between *S. rorida* and *S. triandra*. In addition, this species shows a plasticity of seedling root morphology from taproot to shallow root types depending on the texture, which accelerates the dominance of this species under the mosaic structure of substrata at the local scale.

Extremely coarse or fine textured substrata are rare in the floodplains of middle-range rivers. Specializing in coarse or fine textures guarantees competitive advantages for species such as *S. rorida* and *S. triandra* on the extreme substrata, but they can not dominate in the intermediate substrata.

S. udensis and *S. miyabeana* have similar regeneration traits and co-occur in many places. Competitive exclusion between these two species would not happen, because this exclusion process is very slow and stochastic between similar species. Further, a flood usually brings competition back to the starting line of regeneration.

This type of coexistence has been considered as a limiting dissimilarity in competitive ability (Ågren & Fagerström 1984). Of course, its mechanism is considerable among other *Salix* species. The differences in regeneration traits are emphasized here, but it is subtle within congeners compared to the large differences between the species of other families or genera. For example, the alluvial fan species, *Ulmus davidiana* Planch. var. *japonica* (Rehder) Nakai, and delta zone species, *Fraxinus mandshurica* Rupr. var. *japonica* Maxim. do not co-occur with riparian *Salix* species. They can invade older *Salix* forests where flood disturbances are canceled by changes in river lines or the accumulation of thick sediments.

11.5 Habitat segregation along a river

This section examines regional scale habitat segregation. There is a gradient of substratum textures and/or differences in disturbance regimes caused by flooding along a large river such as the Ishikari River or Satsunai River in Hokkaido. In general, river morphology is divided into the valley zone, alluvial fan zone, meandering zone and delta zone (Fig. 7). Stony or gravelly coarse substrata are common in the valley zone, where floods are rare but destructive events. In the alluvial fan, sedimentation processes and changes in river lines are dominant and sedimentation of sandy and stony substrata is common. Sedimentation and erosion occur at the same time in the meandering zone of rivers where various types of textures are distributed continuously or mosaically. In the delta zone, flooding means waterlogging and anaerobic condition of the soil where physiological adaptation to lower oxygen concentration in the soil is a crucial ecological trait for the survival of *Salix* species. The six *Salix* species show their respective peaks of distribution along this riparian environmental gradient (Niiyama 1987, 1989, 1995).

Large salicaceous tree species, *Populus maximowiczii* A. Henry, *Salix cardiophylla* Trautv. & Mey. var. *urbaniana* (Seemen) Kudo (formerly *Toisusu urbaniana* (Seemen) Kimura) and *Salix arbutifolia* Pallas (formerly *Chosenia*

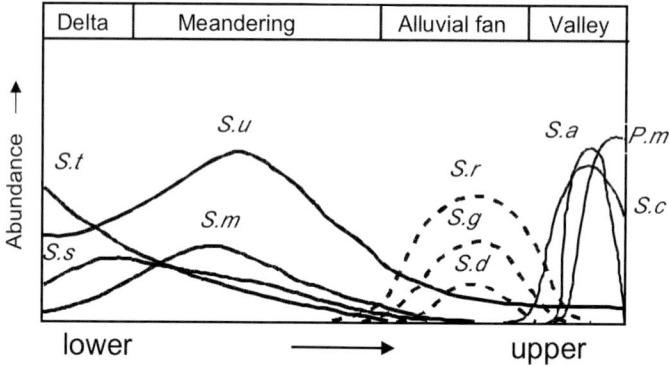

Fig. 7. Schematic distribution of salicaceous species along a large river in Hokkaido, Japan (From Niiyama 1995). *S.t: S. triandra, S.s: S. schwerinii, S.u: S. udensis, S.m: S. miyabeana, S.r: S. rorida, S.g: S. gracilistyla, S.d: S. dolichostyla, S.a: S. arbutifolia, P.m: Populus maximowiczii, and S.c: S. cardiophylla* var. *urbaniana*

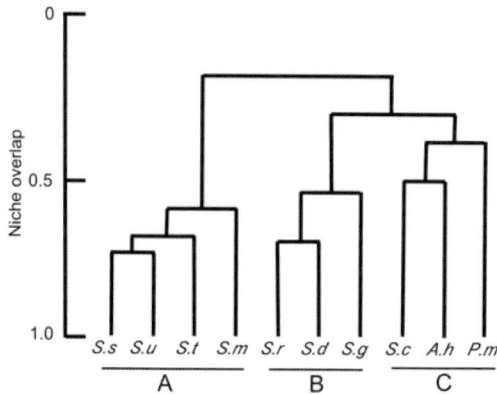

Fig. 8. Cluster analysis of niche overlaps between salicaceous species based on overlaps of soil textures of their habitats along the Ishikari River (From Niiyama 1987). *S.s: S. schwerinii, S.u: S. udensis, S.t: S. triandra, S.m: S. miyabeana, S.r: S. rorida, S.d: S.dolichostyla, S.g: S. gracilistyla, S.c: S. cardiophylla* var. *urbaniana, A.h: Alnus hirsuta* Turcz., and *P.m: Populus maximowiczii*

arbutifolia (Pall.) Skvorts.), are abundant in the upper reaches of rivers (Fig. 7) dominated by gravel and stony substrata. *S. dolichostyla*, *S. rorida* and *S. gracilistyla* Miq. are typical alluvial fan willows in Hokkaido. These species show relatively early seed disper-sal and large seeds from which long taproots grow. The other three species examined here, *S. udensis*, *S. miyabeana*, and *S. schwerinii*, are widely distributed in the middle and lower reaches of rivers. They have intermediate regeneration traits between the coarse-grained specialist, *S. rorida*, and the fine-grained special-ist, *S. triandra*, which is dominant under the fine-grained anaerobic conditions near the river mouths.

In the riparian area of the Ishikari River, salicaceous species can be classified into three groups by the extent of overlaps on the soil textures where they are growing (Fig. 8). Cluster analysis reveals that these three species groups can be identified based on the habitat niche overlaps between salicaceous species, where niche means species distribution along soil texture gradients. The niche overlap index is calculated based on the overlap of soil textures where salicaceous species are distributed. *S. schwerinii* and *S. udensis* show the largest niche overlap index.

These three species groups (A, B and C) well explain the distribution patterns along the rivers shown in Fig. 7. The three groups are arranged from river mouth to valley. The relationship between the regeneration traits of *Salix* species and the mosaic distribution of textures at the local scale also well explains the regional scale distribution along rivers.

11.6 Conclusion

The coexistence of riparian *Salix* species was examined based on flood seasonality and ecological traits at three spatial scales. A subtle differentiation of regeneration traits affects both microtopograpahic segregation and preemptive competition through the heterogeneity of substrata textures. Preferences for specific textures well explain the regional scale distribution of salicaceous species along large rivers. However, specialization and habitat segregation alone are not enough to explain the coexistence of closely-related *Salix* species. The similarity of regeneration traits slows the competitive exclusion process among them and flood disturbance usually resets their respective competitive ranks.

References

Ågren GI, Fagerström T (1984) Limiting dissimilarity in plants: randomness prevents exclusion of species with similar competitive abilities. Oikos 43:369-375

Carlson M. C. (1938) The formation of nodal adventitious roots in *Salix cordata*. Am J Bot 25:721-725

Cottrell TR (1996) Use of Plant Strategy Ordination, DCA and ANOVA to elucidate relationships among habitats of *Salix planifolia* and *Salix monticola*. J Veg Sci 7:237-346

Dionigi CP, Mendelssohn IA, Sullivan VI (1985) Effects of soil waterlogging on the energy status and distribution of *Salix nigra* and *S. exigua* (Salicaceae) in the Atchafalaya River Basin of Louisiana. Am J Bot 72:109-119

Houle G, Babeux P (1993) Temporal variations in the rooting ability of cuttings of *Populus balsamifera* and *Salix planifolia* from natural clones-populations of subarctic Quebec. Can J For Res 23:2603-2608

Ishikawa S (1979) Ecological studies of the plant communities on the Kitakami River floodplain. Ecol Rev 19:67-73

Ishikawa S (1980) Ecological studies of the floodplain willow forests in the Hokkaido district. Res Rep Kochi Univ 29:73-78

Ishikawa S (1982) Ecological studies of the floodplain willow forests in the Tohoku district. Res Rep Kochi Univ 31:95-104

Ishikawa S (1983) Ecological studies on the floodplain vegetation in the Tohoku and Hokkaido districts, Japan. Ecol Rev 20:73-114

Ishikawa S (1987) Ecological studies on the willow communities on the Satsunai River floodplain, Hokkaido, with special reference to the development of the *Chosenia arbutifolia* forest. Mem Fac Sci Kochi Univ Ser D 8:57-67

Ishikawa S (1988) Floodplain vegetation of the Ibi River in central Japan. I. Distribution behavior and habitat conditions of the main species of the river bed vegetation developing on the alluvial fan. Jpn J Ecol 38:73-84

Ishikawa S (1994) Seedling growth traits of three salicaceous species under different conditions of soil and water level. Ecol Rev 23:1-6

Niiyama K (1987) Distribution of salicaceous species and soil texture of habitats along the Ishikari River. Jpn J Ecol 37:163-174

Niiyama K (1989) Distribution of *Chosenia arbutifolia* and soil texture of habitats along the Satsunai River. Jpn J Ecol 39:173-182

Niiyama K (1990) The role of seed dispersal and seedling traits in colonization and coexistence of *Salix* species in a seasonally flooded habitat. Ecol Res 5:317-331

Niiyama K (1995) Life history traits of salicaceous species and riparian environment. Jpn J Ecol 45:301-306

Ohashi H (2000) A systematic enumeration of Japanese *Salix* (Salicaceae). J Jpn Bot 75:1-41

Ohashi H, Nakai H (2006) *Salix dolichostyla* Seemen, a correct name for *S. jessoensis* Seemen (Salicaceae). J Jpn Bot 81:75-90

Ohashi H, Yonekura K (2006) Additions and Corrections in Salicaceae of Japan 1. J Jpn Bot 81:35-46

van Splunder I, Coops H, Voesenek LACJ, Blom CWPM (1995) Establishment of alluvial forest species in floodplains: the role of dispersal timing, germination characteristics and water level fluctuations. Act Bot Neerl 44:269-278

Yanai S, Kikuzawa K (1991) Characteristics of seed germination and seedling establishment in three *Salix* species demonstrated by field germination experiment. Jpn J Ecol 41:145-148

Yoshikawa M, Hukusima T (1999) Distribution and developmental patterns of floodplain willow communities along the Kinu River, central Japan. Veg Sci 16:25-38

Part 6

Riparian forests in lowland regions

12 Process of willow community establishment and topographic change of riverbed in a warm-temperate region of Japan

Mahito KAMADA

Division of Ecosystem Design, Institute of Technology and Science, the University of Tokushima, 2-1 Minami-Josanjima, Tokushima 770-8506, Japan

12.1 Introduction

Distinct and separate arcuate bands formed by *Salix* species are frequently observed near low-water channels on bars (McKenny et al. 1995; Okabe et al. 2001). Each band of willow community seems to be evenly aged with gaps in ages between the bands, and tree age increases moving inland from the stream, as reported for cottonwoods (Everitt 1968; Noble 1979; Bradley & Smith 1986).

The occurrence pattern of *Salix* species has been explained in relation to flooding regime, and seed-dispersal timing (Nanson & Beash 1977; Noble 1979; Niiyama 1990; Johnson 1994). Sediment size, riverbed gradient and elevation from low-water channel also act as limiting factors for species distribution (Ishikawa 1983; van Splunder et al. 1995). However these alone are not sufficient to explain discontinuous establishment and formation of bands for willow communities.

At the start of riparian *Salix* species' life cycle, sufficient moisture is required for germination (Niiyama 1990). However, the surface inside a bar where older *Salix* bands occur is dry and considered to be inadequate for establishing a new community. Here, a question arises regarding how the older community colonized such a place. In addition, *Salix* species produce an abundance of wind-blown seeds every year, therefore recruitment can be expected to occur every year and not to form separate bands made of different cohorts. With this in mind, it remains unclear how the bands were formed.

The main focus of this paper is to clarify the process forming the bands of willow communities by studying the following: (i) Hydrological characteristics

Sakio, Tamura (eds) Ecology of Riparian Forests in Japan : Disturbance, Life History, and Regeneration
© Springer 2008

that are necessary for willow seedling establishment in relation to tolerance against submerged and drought conditions. (ii) Temporal change of riverbed topography caused by the establishment of willow communities. Finally, the influence of topographical change to riparian ecosystem is also discussed.

12.2 Study area and willow communities on the bar

12.2.1 Characteristics of the basin and the studied bar

The study area is located in the basin of the Yoshino River in Shikoku, western Japan (Fig. 1). Length, drainage area and design-flood discharge of the river are 198 km, 3750 km^2 and 19,000 m^3/sec, respectively. The area belongs to a warm-temperate zone, and no snowmelt floods occur. Mean annual temperature and precipitation in Tokushima City is 16.7 °C and 1337 mm, respectively. Most of the precipitation falls from May to September with monsoon rains and typhoons. The active floodplain is occasionally inundated.

The surveyed bar was selected in a lower reach, ranging 19 km to 21 km from the river mouth. It is 500 m wide and 2 km long with the average bed slope being approximately 1/1100. Average sediment size covering bar surface is about 30 mm.

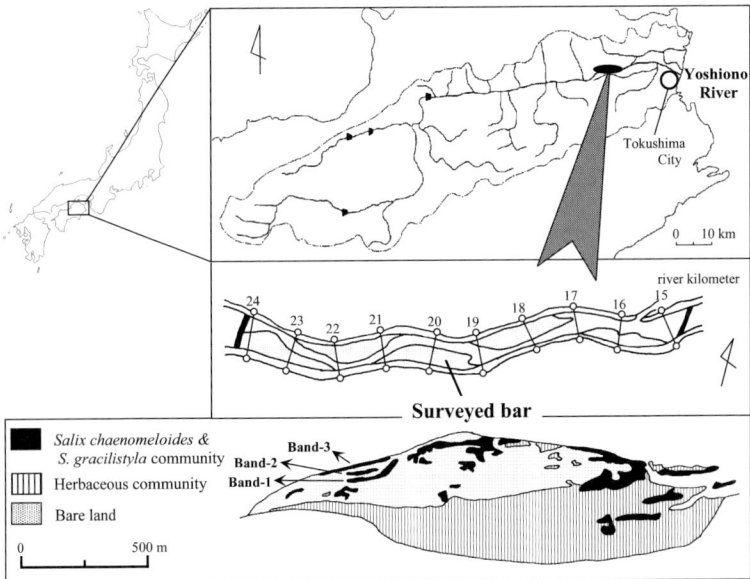

Fig. 1. Map showing surveyed basin and the bar

Willow communities dominated by *Salix chaenomeloides* Kimura are widely distributed along the low water channel (Kamada & Okabe 1998). Target communities in the study are located on the upstream part of the bar, where the distinct and separate bands-1, 2 and 3 are formed (Fig. 1). All bands are composed of *Salix chaenomeloides* and *Salix gracilistyla* Miq., although bands-1 and 2 are dominated by *S. chaenomeloides*, with the exception of band-2 which is dominated by *S. gracilistyla*.

12.2.2 Age of willow bands

Sliced samples of willow stems were collected from each community forming the band, in order to ascertain the established year of the cohort. Trees were excavated in 1996 and samples were obtained from the lowest part of the stems because the bottom most part of the stems of all trees was buried. Sediment depth was measured at the same time.

Table 1 shows the year when each willow band became established, elevation of the bar surface where sampled trees were standing, and elevation of the lowest

Table 1. Year of establishment of willow bands and elevation of original and present bar surface

	Estimated year of colonization	Elevation of bar surface (m)	Elevation of the lowest-part of stem (m)
Band-1	1977	8.29	5.39
Band-2	1986	7.43	5.53
Band-3	1994	5.38	5.38

Fig. 2. *Salix chaenomeloides* seedlings along the fringe of the bar. The cohort occurred in 1994

part of stem of sampled trees. Tree rings showed that individuals in band-1 and 2 established in 1977 and 1986. The sudden occurrence of band-3 was directly observed at the bar in 1994 (Fig. 2). Age and bar-surface elevation of each band became older and higher with increase in distance from the fringe of low-water channel; from band-3 to band-1. However, elevation of the lowest part of the stems, in terms of the original surface, was very similar to each other among the bands. It can be said that elevation of colonized bar-surface is approximately equal to water surface of low-water channel and groundwater.

This result indicates that bar bed around band-1 has aggraded by about 3m from 1977 to 1996, and that around band-2 has aggraded by about 2m from 1986 to 1996.

12.3 Tolerance of willow seedlings against submerged and drought conditions

12.3.1 Experimental design to test tolerance

In order to understand the process of band formation, it is important to know response of seeds and seedlings of willow to fluctuating water level. Regarding germination traits, Niiyama (1990) showed that seeds of riparian *Salix* species are less tolerant to dry conditions and potential for germination is lost within 10 days after seed-dispersal. Therefore, the aim of experiments was to clarify tolerance of the seedlings of *S. chaenomeloides* to drought and submerged conditions, referring to research carried out by Ishikawa (1994). This species is the dominant component in the bands, as well as the other communities on the bar.

Prior to the experiments, seeds of *S. chaenomeloides* were collected from the communities and immediately sown in 15 × 20 × 3 cm containers filled with fine soil and sufficient moisture. Then seeds were allowed to germinate and grow over

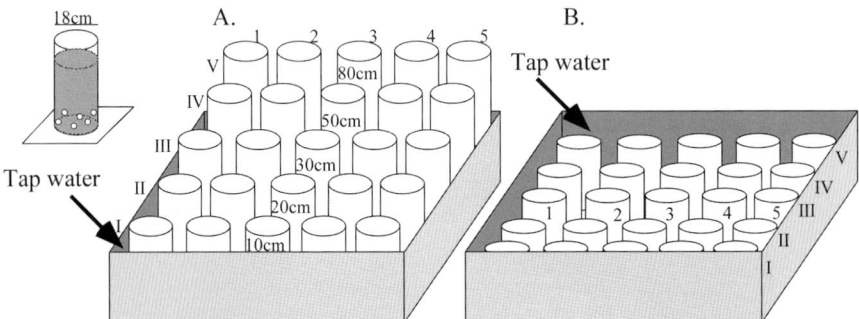

Fig. 3. Experimental design for testing tolerance of *Salix chaenomeloides* seedlings against drought (A) and submergence (B). Tap water was continuously supplied to tanks to keep the water level constant

a period of 7 days.

Growth trait of *S. chaenomeloides* seedlings was examined under different conditions of water-level and submerged duration (Fig. 3). Sediment (ø < 2 mm) that was collected from under the young willow community on the bar was put into pots measuring 18 cm in diameter, and 10 seedlings were transplanted into each pot from the germination container. Holes were made at the lower part of the pots to allow water intake.

Experiment A: Twenty-five pots were set in a tank to be 5 cases with different ground-water levels (I - V); water surface was kept at -10cm, -20cm, -30cm, -50 cm and -80cm from soil surfaces in the pots, respectively (Fig. 3-A). Every 7 days, the number of surviving seedlings was counted. Following this all seedlings were removed from each pot in every case in order to measure shoot length. The experiment continued for 35 days.

Experiment B: (i) Twenty-five pots were submerged in a tank under 5 cases (I - V) of different durations; 7, 14, 28, 56, 84 days, respectively. During the experiment, pots were submerged in a water tank and seedlings in the pots were kept at a level -10cm from the water surface (Fig. 3-B). On the scheduled end day for each case, number of surviving seedlings was counted. Maximal duration of submergence was 84 days. (ii) Then a pot from every case, 5 pots in total, was lifted and put on the block in order to adjust the level of soil surface to 15 cm above the water surface. Next, every 7 days, the number of surviving seedlings was counted and the seedlings were removed to measure shoot length. Seedlings, which were submerged for 84 days, were unused in the experiment, because of limitation of experiment period.

Both experiments A and B were done outside with natural light and temperature conditions from June to August in 1996. The water tanks were protected by a roof to avoid rain entering them. Tap water was continuously supplied to the tanks during the experiments to maintain the water level.

12.3.2 Survivorship and growth pattern of *Salix chaenomeloides* seedlings under stressed conditions

Figure 4 shows survival rate of seedlings and shoot growth under experiment A (Fig. 3-A). When the water level from the soil surface was within -20cm almost all seedlings could survive and grow successfully. However when the water level fell to -30cm, all seedlings died within 7 days. These results show that the seedlings are less tolerant to drought condition.

Seedlings are highly tolerant to the submerged condition (Fig. 3-B), where 86 % of the seedlings survived even in the case of being submerged for a duration up to 84 days, although no seedlings could grow in the water. Figure 5 shows growth pattern of shoots after the seedlings were partially raised from the water. All seedlings grew regardless of submerged duration. Seedlings which were submerged for a longer duration grew better compared to seedlings which were submerged for a shorter duration. This result is probably due to the difference of outside temperature during experiment B-ii period which was in June and July for

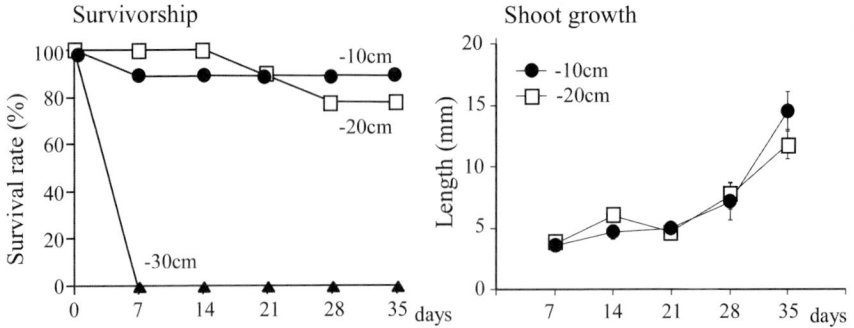

Fig. 4. Survivorship and growth pattern of *Salix chaenomeloides* seedlings under conditions of different underground water-level (Experiment A)

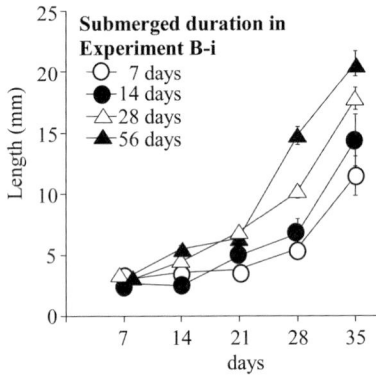

Fig. 5. Growth pattern of shoots of *Salix chaenomeloides* seedlings during experiment B-ii. Seedlings, which were submerged for 84 days, were not used, because of limitation of duration of the experiment.

the seedlings which were submerged for 7 days, and August and September for the seedlings which were submerged for 56 days. The later period had a higher temperature and was better for growing.

Hydrological conditions for successful colonization and establishment of *Salix chaenomeloides* seedlings can be summarized as follows. (i) If water level falls to -30cm from the level of colonized bar surface in initial growing season, no seedlings can survive. When water level becomes high, seedlings can survive but cannot grow in the water. (ii) Seedlings are highly tolerant to the submerged condition, but they are not able to grow in the water. Therefore, only when the level of bar-surface is kept at a level between 0 and 30 cm from water level during growing season, seedlings can establish on the bar.

The same has been observed for seedlings of *S. gracilistyla*, with the exception of seed-dispersal period. Seed-dispersal and germinating period is May for *S.*

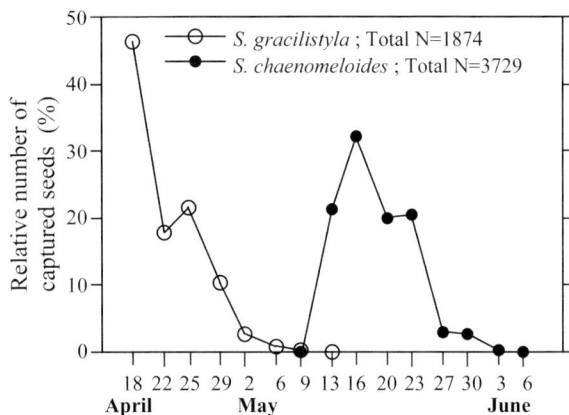

Fig. 6. Relative number of dispersed seeds of *Salix chaenomeloides* and *S. gracilistyla*. Six seed-traps (1m × 1m) were set along the bands and seeds flying into the traps were collected every 3 or 4 days during dispersal period in 1996

chaenomeloides and April for *S. gracilistyla* (Fig. 6).

12.4 Actual process of willow bands formation

12.4.1 Hydrological conditions in the year willow bands formed

Hydrological characteristics from 1970 to 1994 are presented in Fig. 7 according to hourly discharge data, which was officially recorded at an observation gauge 22 km from the river mouth by the Japanese Ministry of Land, Infrastructure and Transport.

In the years of 1977, 1986 and 1994 when the willow bands were formed, specific characteristics were evident. First, during the seed-dispersal and germinating period, April and May, the water-levels are relatively stable compared to other years. The standard deviation (S.D.) of water levels was less than 25 cm (Fig. 7-1). This indicates that fringe of low-water channel had been maintained at the same elevation during both seed-dispersal and germinating periods, and thus a wet site for germination had been constantly kept at the same elevation.

Second, when average water-level of seed-dispersal and germinating season (April and May) is subtracted from that of growing season (from June to October), values fit in a range from 0cm to 25cm in 1977, 1986 and 1994 (Fig. 7-2). In other words, the seedlings were not submerged during the growing season, and the bar surface where the seedlings colonized was kept in a wet condition during those years. These hydrological conditions perfectly satisfy conditions clarified by the experiments, and thus there was a chance for *S. chaenomeloides* and probably for *S. gracilistyla* to colonize on the bar.

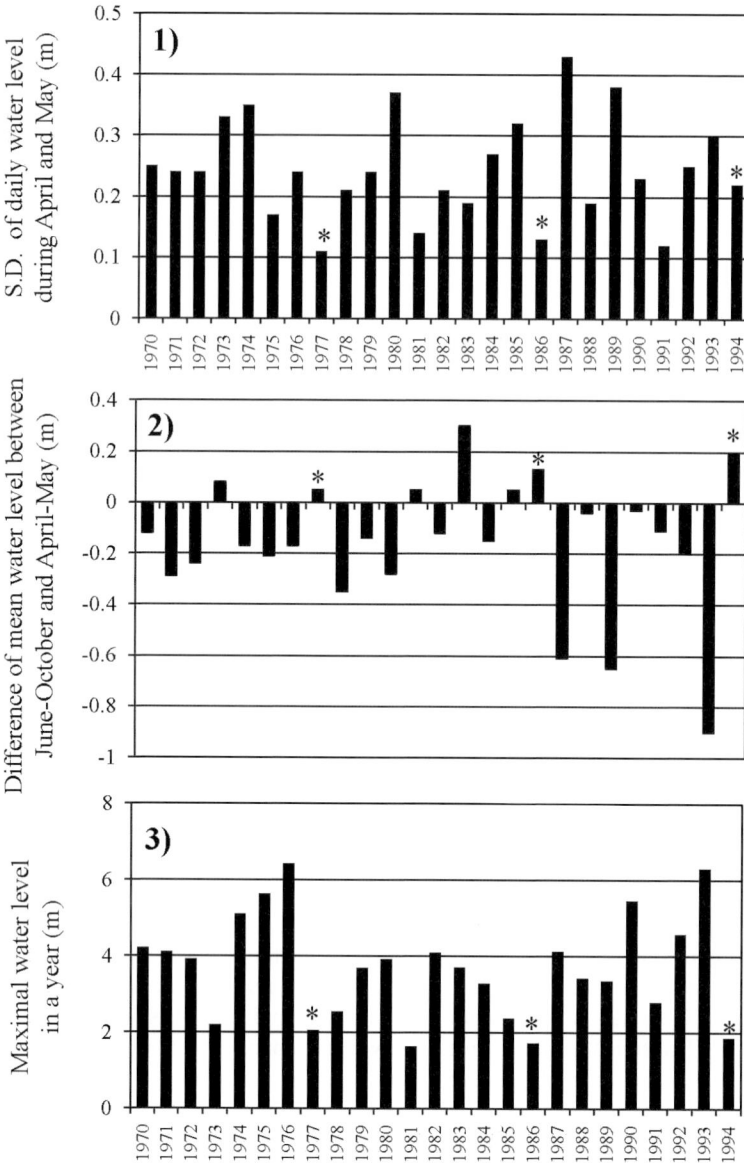

Fig. 7. Hydrological characteristics from 1970 to 1994. Asterisks indicate the year, 1977, 1986 and 1986 when willow-bands were formed. 1) standard deviation (S.D.) of water levels during the seed-dispersal and germinating period (April and June), 2) difference of water level (cm), which is obtained by subtracting average water-level of seed-dispersal and germinating season (April and May) from that of growing season (from June to October), and 3) maximal water level (m) in each year

Third, no large floods occurred in those years (Fig. 7-3). Newly emerged seedlings grow only to approximately 10 cm in the first year and such small seedlings are easily washed away by floods. A hydrological condition without large flood allowed small seedlings to grow.

Such a hydrological condition with stable water-level in seed-dispersal and germinating periods, low water-level during growing season and no large floods as described above is uncommon, and as a result, willow communities established intermittently.

12.4.2 Topographical change of riverbed

Cross-sectional profiles of riverbed in 1977, 1986 and 1995, when each willow band was established, are shown in Fig. 8. The data was provided by the Japanese Ministry of Land, Infrastructure and Transport who measure cross-sections at 200 m intervals almost every year. Data of 1995 was used instead of 1994, because measurement was not conducted in that year. The topography of bar bed has drastically changed. A flat bar bed in 1977 changed to sloping one in 1986, and became a steep slope in 1995.

Average water levels during seed-dispersal period (April - May) at the cross-section were estimated for 1977, 1986 and 1994 (1995), as shown in Fig. 8. The estimation was carried by non-uniform flow calculation using the data set of topography of the above mentioned years and discharge data from 1977, 1986 and

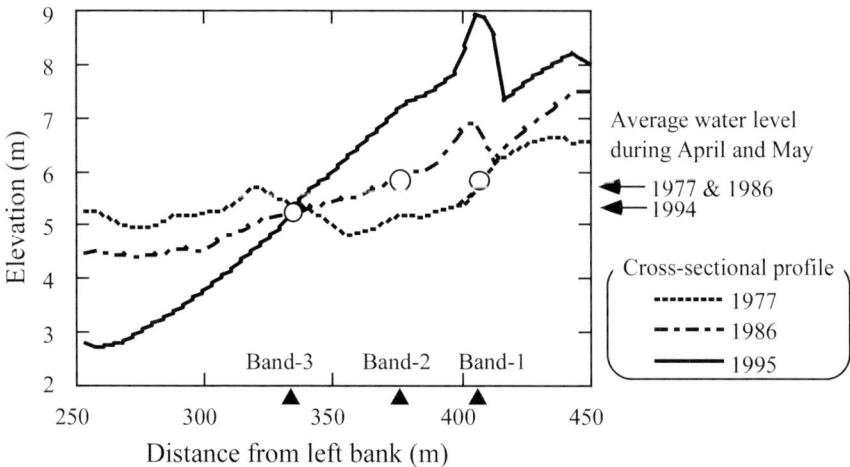

Fig. 8. Cross-sectional profiles of riverbed in 1977, 1986 and 1995. Water-levels in seed-dispersal period (April - May) were estimated for the years when each willow band was established. The elevation of each willow band in established year corresponds well with the water-level during seed-dispersal period. After establishment of willow communities, the bar bed around the willow bands has raised and the bar bed area has extended to the low-water channel side, while the area in low-water channel has become deeper

1994.

When bands-1 and 2 were established on the bar, elevation of the bar-surface around the bands was lower than that of the present; below 2.9m for band-1, below 1.9m for band-2 (Table 1). For every band, bar-surface elevation in colonized year was matched with the water-level of the same year (Fig. 8). This indicates that each willow community occurred along former fringe of low-water channel. In other words, it can be said that an arcuate band of willow community represents the fringe of low-water channel in the year when the community established.

Drastic change of topography of the bar was caused by establishment and development of willow communities; the bar bed around the willow bands has risen and bar area has extended to low-water channel, while the area of low-water channel has become deeper (Fig. 8). This is a result of effective trapping of sediment by willow trees developed on the bar. Submerged trees increase flow resistance and promote deposit of sediments around stems during floods (Bradley & Smith 1986; Odum 1990; Watts & Watts 1990; McKenney et al. 1995; Tsujimoto & Kitamura 1996; Kamada & Okabe 1998). Trapped sediment tends to form itself into a long mound resembling an embankment. Willow trees and the mound significantly restrict stream width in the event of small or medium-sized floods, and the stream-bed has been continuously scoured (Kamada et al. 2004).

The area of the bar expanding toward low-water channel becomes a habitat for the next generation of willow trees. When hydrological conditions are satisfied, willows, which seeds are dispersed along the fringe of the expanded bar, can germinate and grow there (Fig. 2).

12.5 Stabilization of river-system and its influence on riparian ecosystem

It has been known that willow communities drastically expanded on bars after the 1970s in the Yoshino River as well as other rivers in Japan (Kamada et al. 1997; Kamada et al. 2004). Willow bands, which process has been mentioned above, occurred in that period through stabilization of river-system as a result of human activities in and along rivers. Stabilization of river system has seriously altered the distribution of vegetation and their dynamics (Kamada et al. 2004).

In lower reaches of the Yoshino River, a large amount of gravel was dredged to obtain materials for building construction during a rapid economic growth period of Japan during the 1960s in particular. Bar surface, in terms of habitat for willow trees, had been seriously disturbed and no plant communities had established on bars. Gravel dredging caused degradation of riverbed (Takebayashi et al. 2003), and increased risk of bridge destruction. The Japanese government, therefore, started to regulate gravel dredging in the early 1970s. After cessation of gravel dredging due to the regulation, plants became free from disturbance and were able to grow and expand their area (Kamada et al. 2004).

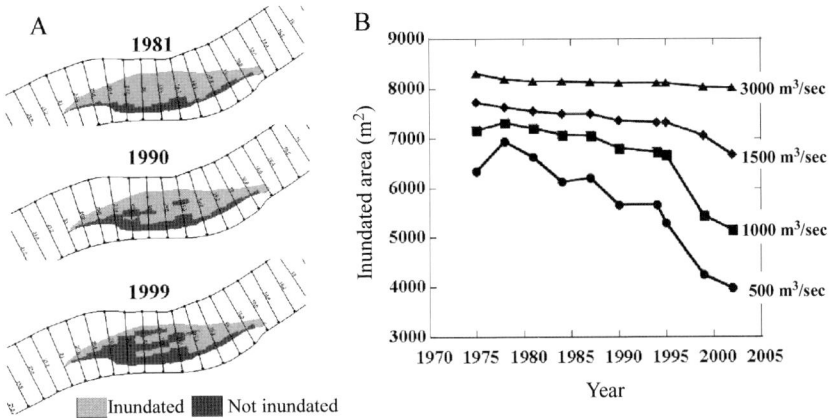

Fig. 9. Temporal change of bar area inundated by floods. (A) Example in the case of 500 m³/sec discharge, and (B) 4 cases of different discharges

Degradation of the riverbed in low-water channel due to gravel dredging resulted in comparative aggregation of bar bed elevation. It changed intensity and frequency of flood disturbance, which influenced willow seedlings establishment.

Figure 9 shows temporal change of bar area inundated by floods for 4 cases of different discharges. The inundated area was estimated by non-uniform flow calculation using topographic data from 1975 to 2002 and given discharges of 500m³/sec, 1000m³/sec, 1500m³/sec, 3000m³/sec, respectively.

The inundated area has gradually reduced year by year, particularly when there is less than 1000 m³/sec discharge (Fig. 9-B). Floods of 500 m³/sec and 1000 m³/sec discharges occur about three and two times per year respectively in the Yoshino River. This suggests that the area which is frequently disturbed by small floods has been drastically reduced and, as a result, survival chance for plants on the bar has increased. That is to say, plants on bars become free from flood disturbance and are able to expand their distribution successfully.

The river system has also been altered through civil engineering works, such as dam construction and embankment for flood control and stable water supply. Such works have stabilized physical conditions of the river by several means, such as reducing flood discharge and straightening and fixing channel course (Brookes 1988; Kamada et al. 2004), and, as a result, have altered vegetation dynamics and caused woodland expansion (Décamps et al. 1988; Johnson 1994).

This is the background of rapid development of willow communities in recent years with expansion of willow communities becoming obvious from the 1970s in the Yoshino River (Kamada et al. 1997; Kamada et al. 2004). Willow communities established on bars have enhanced stabilization of physical condition of the bar through aggrading bar bed by trapping sediment.

At bars which are stabilized by both human activities and willow communities, invasive alien plant species, *Eragrostis curvula* Nees, has been intruding gravel areas and expanding rapidly (Kotera et al. 1998). Expansion of *E. curvula* over bare areas threatens species which require bare area covered with gravel as a part of their life cycle, such as *Sterna albifrons* Pallas and *Charadrius dubius* Scopoli. Invasion of *E. curvula* over bare areas of the gravel bar is considered to be enhanced by willow shrubs, because the shrubs protect seedlings of *E. curvula* from flood disturbance (Kamada 2004).

12.6 Conclusion

Required hydrological conditions for successful colonization and establishment of *Salix chaenomeloides* are clarified as follows. (i) Water level during seed-dispersal and germination season should be stable. (ii) Average water level should be in a range between 0cm and -30cm from the level of colonized bar surface during growing season. (iii) No large floods occur. Such conditions are also probably required for *S. gracilistyla*. Hydrological conditions in years when willow communities were established on the bar, 1977, 1986 and 1994, completely satisfied these conditions.

Stabilization of river system caused by civil engineering works enhances establishment of willow communities, and then the willow communities developed on the bar trap sediments during floods and form mounds. The bar expands toward low-water channel and provides a colonization site for the next generation of willows along the fringe of low-water channel, when the hydrological condition is suitable. Bands of willow communities are formed in the process.

The developed willow community and mound act as a bank and restrict stream width in the event of small or medium-sized floods and, as a result, the riverbed at the low-water channel is continuously scoured and deepened. Aggrading of bar bed and deepening of low-water channel reduces inundated area and increases the area free from flood disturbance. This enhances physical stabilization of the bar and hence assists alien plant species to establish and develop, as well as willows.

The process is highly interactive and accelerates the change of river system, from dynamic to stable system.

Acknowledgements

I am grateful to Prof. T. Okabe of the University of Tokushima for his collaborative works. I am also indebted to the Tokushima Office of River and National Highway, Shikoku Regional Development Bureau, Ministry of Land, Infrastructure and Transport of Japan, for providing hydrological and topographic data. This study was partly funded by JSPS's Grant-in-Aid for Scientific Research (nos. 18201008, 19208013)

References

Bradley CE, Smith DG (1986) Plains cottonwood recruitment and survival on a prairie meandering river floodplain, Milk River, southern Alberta and northern Montana. Can J Bot 64:1433-1442

Brookes A (1988) Channelized Rivers, Perspectives for Environmental Management. Wiley, Chichester

Décamps H, Fortuné M, Gazelle F, Pautou G (1988) Historical influence of man on the riparian dynamics of a fluvial landscape. Landscape Ecol 1:163-173

Everitt BL (1968) Use of the cottonwood in an investigation of the history of a floodplain. Am J Sci 266:417-459

Ishikawa S (1983) Ecological studies on the floodplain vegetation in the Tohoku and Hokkaido districts, Japan. Ecol Rev 20:73-114

Ishikawa S (1994) Seedling growth traits of three Salicaceous species under different conditions of soil and water level. Ecol Rev 23:1-6.

Johnson WC (1994) Woodland expansion in the Platte River, Nebraska: patterns and causes. Ecol Monogr 64:45-84

Kamada M (2004) Ecological and hydraulic factors influencing expansion of invasive plant species, *Eragrostis curvula*, at bar in the Yoshino River, Shikoku, Japan. Proceedings of the Second Annual Joint Seminar between Korea and Japan on Ecology and Civil Engineering, "Ecohydraulics and Ecological Process – Principle, Practice and Evaluation":35-40

Kamada M, Okabe T, Kotera I (1997) Influencing factors on distribution changes in trees and land-use types in the Yoshino River, Shikoku, Japan. Environmental System Research 25:287-294 (in Japanese with English abstract)

Kamada M, Okabe T (1998) Vegetation mapping with the aid of low-altitude aerial photography. Appl Veg Sci 1:211-218

Kamada M, Woo H, Takemon Y (2004) Ecological engineering for restoring river ecosystems in Japan and Korea. In: Hong SK, Lee JA, Ihm BS, Farina A, Son Y, Kim ES, Choe JC (eds) Ecological Issues in a Changing World - Status, Response and Strategy, Kluwer Academic Publishers, Dordrecht, The Netherlands, pp 337-354

Kotera I, Okabe T, Kamada M (1998) Physical factors affecting to distribution of plant communities on a sand bar. Environmental System Research 26:231-237 (in Japanese with English abstract)

McKenny R, Jacobson RB, Wertheimer RC (1995) Woody vegetation and channel morphogenesis in low-gradient, gravel-bed streams in the Ozark Plateaus, Missouri and Arkansas. Geomorphology 13:175-198

Niiyama K (1990) The role of seed dispersal and seedling traits in colonization and coexistence of *Salix* species in a seasonally flooded habitat. Ecol Res 5:317-331

Noble MG (1979) The origin of *Populus deltoids* and *Salix interior* zones on point bars along the Minnesota River. Am Midl Nat 102:69-102

Okabe T, Anase Y, Kamada M (2001) Relationship between willow community establishment and hydrogeomorphic process in a reach of alternative bars. Proc. of XXIX IAHR Congress, B: 340-345

Takebayashi H, Egashira S, Okabe T (2003) Numerical analysis of braided streams formed on beds with non-uniform sediment. Ann J Hydraul Eng JSCE 47: 631-636 (in Japanese with English abstract)

Tsujimoto T, Kitamura T (1996) Deposition of suspended sediment around vegetated area and expansion process of vegetation. Ann J Hydraul Eng JSCE 40:1003-1008 (in Japanese with English abstract)

van Splunder I, Coops H, Voesenek LACJ, Blom CWPM (1995) Establishment of alluvial forest species in floodplains: the role of dispersal timing, germination characteristics and water level fluctuations. Act Bot Neerl 44:269-278

Watts JF, Watts GD (1990) Seasonal change in aquatic vegetation and its effect on river channel flow. In: Thornes JB (ed) Vegetation and Erosion. Wiley, New York, pp 257-267

13 Growth and nutrient economy of riparian *Salix gracilistyla*

Akiko SASAKI[1], Takayuki NAKATSUBO[2]

[1]Coastal Environment and Monitoring Research Group, Institute of Geology and Geoinformation, National Institute of Advanced Industrial Science and Technology (AIST), 2-2-2 Hiro-Suehiro, Kure, Hiroshima 073-0133, Japan

[2]Department of Environmental Dynamics and Management, Graduate School of Biosphere Science, Hiroshima University, 1-7-1 Kagamiyama, Higashi-Hiroshima, Hiroshima 739-8521, Japan

13.1 Introduction

Bars are characteristic fluvial landforms created by the deposition of sediment resulting from fluvial processes. In mountainous regions, such as those in Japan, rapid stream flow and occasional flooding often create sandy to gravelly bars, especially in the middle reaches of rivers (Fig. 1), and these bars are important habitats for riparian plants. Drought of the surface soil is important in determining seedling establishment on bars (McLeod & McPherson 1973). Plants that do become established in these habitats are exposed to severe environmental conditions, including occasional submergence and sand burial (Viereck 1970; Walker & Chapin 1986). Plants colonizing these habitats have some ecological traits that help them to tolerate these stresses (White 1979).

Another important environmental factor in sandy to gravelly bars, which has received less attention, is low nutrient availability. Although floodplain alluvium is generally considered to be fertile (Brinson 1990; Schnitzler 1997), the soil nutrient content in bars, where high current velocity removes accumulated organic matter and prevents fine particle deposition, is often very low (Naohara 1936–1937; Pinay et al. 1992; Nakatsubo 1997). In these nutrient-poor habitats, symbiotic nitrogen fixation can be important for pioneer trees including *Alnus* spp. and *Elaeagnus* spp. (Larcher 2001). In addition, most herbaceous species growing on bars are colonized with mycorrhizal fungi (Nakatsubo et al. 1994), which may also contribute to nutrient absorption by the host plants.

Salix are dominant tree species widely seen on bars in the Northern Hemisphere

Sakio, Tamura (eds) Ecology of Riparian Forests in Japan : Disturbance, Life History, and Regeneration
© Springer 2008

(Viereck 1970; Ishikawa 1988; Cooper & Van Haveren 1994). Although the nutrient economies of some *Salix* species were studied under experimental conditions (Rytter 2001; Von Fricks et al. 2001), there is little information on the nutrient economy of *Salix* stands in riparian habitats. In this chapter, we introduce the nutrient economy and sources of *Salix gracilistyla* Miq., a shrubby willow that dominates the upper to middle reaches of rivers in western Japan. *S. gracilistyla* grows mainly on sandy to gravelly bars (Niiyama 1987, 1989; Ishikawa 1988), and growth of the species is likely to be limited by nutrients. Because a plant's nutrient requirements depend on its growth rate (Rodgers & Barneix 1988; Schenk 1996), we first describe the growth pattern and production of *S. gracilistyla*. Next we discuss the nutrient economy of *S. gracilistyla* stands and the possible nutrient sources for this species.

13.2 Growth pattern of *Salix gracilistyla*

Like other pioneer plants, *Salix gracilistyla* has several ecological traits that enable individuals to be the first colonizers in highly unstable habitats, such as high fecundity, high germination capacity, and tolerance of submergence (White 1979; Niiyama 1990; Ishikawa 1994). *S. gracilistyla* also has a strong ability to sprout and to form adventitious roots. After a severe disturbance by flooding, *S. gracilistyla* restores its plant organs rapidly by sprouting new shoots and forming adventitious roots from stems buried in the sand (Sasaki & Nakatsubo 2003). Thus, *S. gracilistyla* often shows a characteristic shrub form, and the distinction

Fig. 1. Meandering stream and bars created by deposition of sediment in the middle reaches of the Ohtagawa River in Hiroshima, Japan. Arrow (lower right) shows the flow direction (© Geographical Survey Institute)

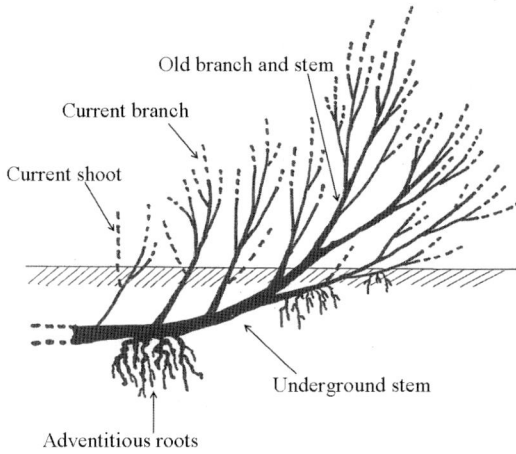

Fig. 2. Growth pattern of *Salix gracilistyla* (reproduced with permission from Sasaki and Nakatsubo 2003)

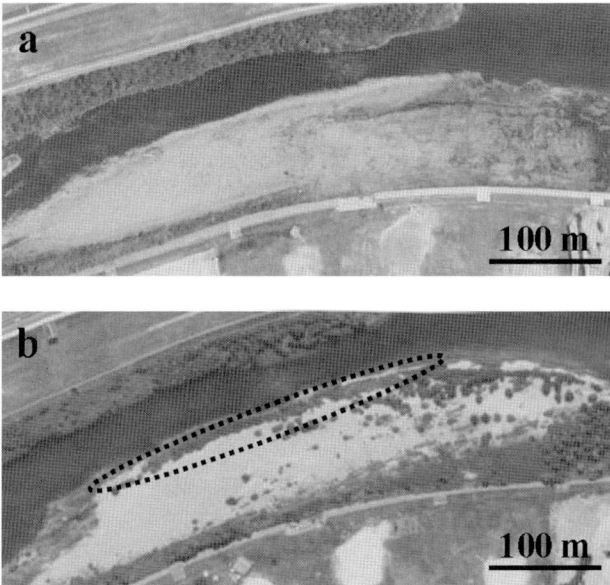

Fig. 3. Vegetation changes at a bar in the middle reaches of the Ohtagawa River in Hiroshima, Japan. Aerial photographs were taken in 1990 (a; © Geographical Survey Institute) and 2001 (b; © Digital Earth Technology Corporation). The area of the *Salix gracilistyla* stand (enclosed by dotted line) greatly increased over this decade. The distal part of the bar was dominated by other *Salix* species, including *Salix. chaenomeloides* Kimura and *Salix triandra* L. (*S. subfragilis* Anders.)

between aboveground and belowground parts is not definite (Fig. 2). As a result of repeated sand burial and subsequent vegetative growth (i.e., production of new shoots and adventitious roots), the stand area of *S. gracilistyla* expands vegetatively. The aerial photographs in Figure 3 illustrate a *S. gracilistyla* stand that developed rapidly during a single decade. In this area, most *S. gracilistyla* stands consist of shoots that sprouted from stems buried in sand. Vegetative growth is an extremely important developmental pattern for *S. gracilistyla* stands establishing on bars that are exposed to frequent disturbances by flooding.

13.3 Biomass and production of *Salix gracilistyla*

Salix gracilistyla has a strong ability to grow and rapidly increases its stand area, implying that this species has a high productivity. Sasaki and Nakatsubo (2003) reported that the aboveground biomass of a *S. gracilistyla* stand increased considerably during the growth period (May to September) and reached its maximum at the end of the period (Fig. 4). The aboveground biomass of a *S. gracilistyla* stand at the end of the growth period (2.2 kg m^{-2}; in September) is small compared with those of climax forests (Table 1) and riparian forests (10–30 kg m^{-2}; Brinson 1990), although the value is within the range of values reported for young pioneer trees (Table 1). This reflects the shrubby growth form of *S. gracilistyla* and is mainly due to the large numbers of stems buried in the sand, as described above (Fig. 2). The aboveground net production of a *S. gracilistyla* stand (1.3 kg m^{-2} yr^{-1}; Sasaki & Nakatsubo 2003) is within the range of values reported for riparian

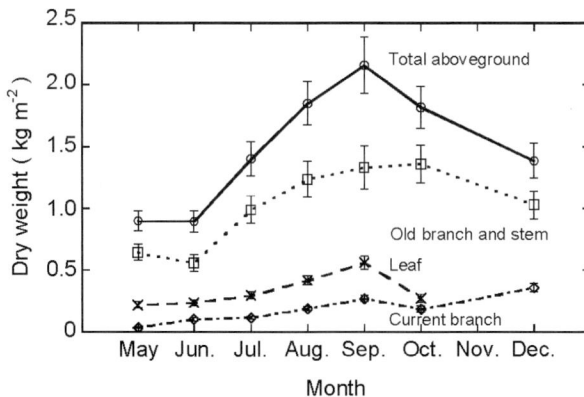

Fig. 4. Seasonal change in the aboveground biomass of a *Salix gracilistyla* stand. Each value represents the mean of three quadrats, and vertical bars represent standard errors (reproduced with permission from Sasaki and Nakatsubo 2003)

Table 1. Aboveground biomass and net primary production in several forest stands (modified from Sasaki and Nakatsubo 2003)

Species	Habitat	Location	Stand age (yr)	Biomass (kg m^{-2})	Aboveground net production (kg m^{-2} yr^{-1})	Reference
Climax forests						
Evergreen oak forest[1]	Hillslope	Minamata, Japan	50-60	34-37	-	Nagano and Kira (1978)
Evergreen oak forest[1]	Hillslope	Minamata, Japan	50-60	-	1.3-1.4	Kira and Yabuki (1978)
Fagus crenata	Hillslope	Niigata, Japan	-	15-36[2]	0.5-0.9[2]	Maruyama (1977)
Fagus crenata	Plantation[3]	Niigata, Japan	35-50	20-24	1.4-1.6	Tadaki et al. (1969)
Pioneer trees						
Alnus japonica	Swamp	Nagano, Japan	15	8.5	1.6	Tadaki et al. (1987)
Alnus maximowiczii	Timber line	Mt. Fuji, Japan	-	1.9	0.6	Sakio and Masuzawa (1987)
Alnus rubra	Alluvial plain	Oregon, USA	33	24	-	Zavitkovski and Stevens (1972)
Alnus rubra	Alluvial plain	Oregon, USA	10-15	-	2.2	Zavitkovski and Stevens (1972)
Populus deltoides	Plantation[3]	Lucknow, India	10	1.2-4	0.2-0.6	Singh (1998)
Salix gracilistyla	Fluvial bar	Hiroshima, Japan	10	2.2	1.3	Sasaki and Nakatsubo (2003)
Salix triandra (*S. subfragilis*)	Abandoned paddy field	Mie, Japan	10	6	2	Kawaguchi et al. (2005)
Salix viminalis	Experimental condition[4]	Uppsala, Sweden	3	1.2-3.4	0.5-1.3	Rytter (2001)

[1] Dominated by *Castanopsis cuspidata* and *Cyclobalanopsis* spp.

[2] Diameter at breast height of trees ≥ 4.5 cm.

[3] Not fertilized.

[4] Clonal cuttings were planted in a lysimeter

forests (0.65–2.14 kg m^{-2} yr^{-1}; Naiman et al. 1998) and for young pioneer trees (Table 1). In addition, in spite of the species' shrubby growth form and small aboveground biomass, the aboveground net production of a *S. gracilistyla* stand is comparable to those of temperate climax forests having large amounts of above-ground biomass (Table 1). Young forests dominated by pioneer trees, including *Salix* sp., frequently have high productivity (Tadaki et al. 1987; Saito 1990; Kopp et al. 2001; Kawaguchi et al. 2005). Therefore, it is possible that the high productivity of the *S. gracilistyla* stand reported by Sasaki and Nakatsubo (2003) can be explained by the young stand age.

13.4 Nutrient economy of *Salix gracilistyla*

The potential for high net production of *Salix gracilistyla* stands suggests that this species has a large nutrient demand. Deciduous tree stands growing in nutrient-poor habitats often use nutrients conservatively, showing a high resorption efficiency during leaf senescence (Small 1972; Stachurski & Zimka 1975; Boerner 1984), and *S. gracilistyla* growing on bars may also use nutrients conservatively.

In western Japan, bud break of *S. gracilistyla* occurs in February. Flowering and seed maturation occur from March to April, and leaf emergence starts at the beginning of April. Leaves and shoots expand from May to September, and in September the aboveground biomass of *S. gracilistyla* stands reaches its maximum. At the end of October, leaves turn yellow and begin to shed, and leaf shedding ends by January (Sasaki & Nakatsubo 2007). The seasonal change in nitrogen and phosphorus concentrations in leaves of *S. gracilistyla* is similar to that of other deciduous trees that have been previously reported. Nitrogen and phosphorus concentrations decrease as leaves expand, show little fluctuation during the growth period, and decline considerably during the senescence period owing to nutrient resorption (Fig. 5; e.g., Chapin et al. 1980; Sakio & Masuzawa 1992). The nutrient concentrations in branches show slight seasonal changes, and there are significant differences in nitrogen and phosphorus concentrations between branch and stem age classes (Sasaki & Nakatsubo 2007).

Figure 6 illustrates the seasonal changes in nitrogen and phosphorus in a *S. gracilistyla* stand calculated based on biomass and nutrient concentrations in each aboveground plant organ. The total amounts of nitrogen and phosphorus in the aboveground organs increase during the growing period, in parallel with increases in the aboveground biomass. Total nitrogen and phosphorus in the aboveground

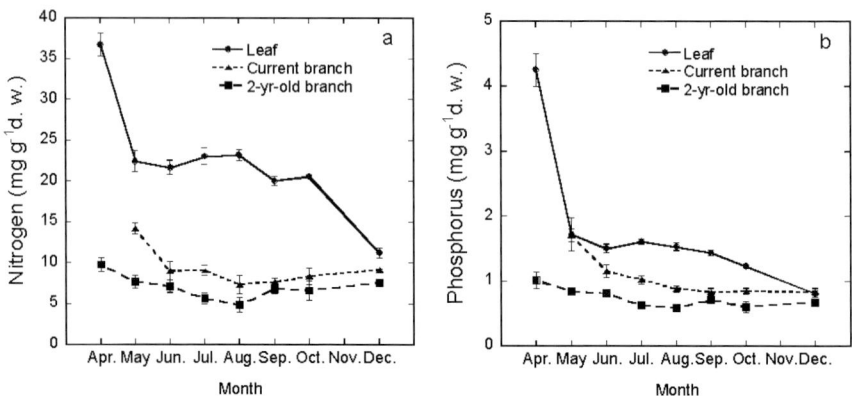

Fig. 5. Seasonal changes in nitrogen (a) and phosphorus (b) concentrations in leaves, current branches, and 2-yr-old braches of *Salix gracilistyla*. Vertical bars represent standard errors ($n = 3$) (reproduced from Sasaki and Nakatsubo 2007)

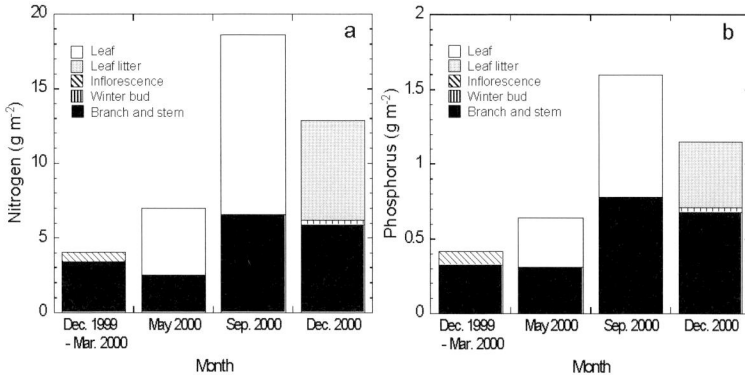

Fig. 6. Seasonal changes in nitrogen (a) and phosphorus (b) amounts in the aboveground parts of a *Salix gracilistyla* stand from December 1999 to December 2000 in the Ohtagawa River, Hiroshima, Japan (modified from Sasaki and Nakatsubo 2007)

organs reached their maximum values (18.6 g N m^{-2}, 1.6 g P m^{-2}) at the end of the growth period. The annual net increases of nitrogen and phosphorus in the aboveground organs of a *S. gracilistyla* stand were estimated to be 9.9 g N m^{-2} yr^{-1} and 0.83 g P m^{-2} yr^{-1}, respectively (Sasaki & Nakatsubo 2007). These values are comparable to those in temperate climax deciduous forests. For instance, the annual nitrogen and phosphorus increases in the aboveground parts of a *Fagus crenata* forest were reported to be 7.9 g N m^{-2} yr^{-1} and 0.45 g P m^{-2} yr^{-1} (Kawahara 1971), and those of a *Quercus* forest were 9.2 g N m^{-2} yr^{-1} and 0.97 g P m^{-2} yr^{-1} (Johnson & Risser 1974). These findings suggest that *S. gracilistyla* stands have a relatively high nutrient demand.

Many deciduous trees resorb nutrients from leaves into perennial organs during the abscission period and use the stored nutrients for the next bud break and growth (e.g., Chapin et al. 1980; Bollmark et al. 1999). Resorption efficiency refers to the percentage of the nutrient pool withdrawn from mature leaves before leaf abscission (Van Heerwaarden et al. 2003). The amount of nitrogen and phosphorus in mature leaves in a *S. gracilistyla* stand were 12 g N m^{-2} and 0.82 g P m^{-2}, respectively, and more than half of these amounts were released through leaf litter (6.7 g N m^{-2} and 0.44 g P m^{-2}) during autumn senescence (Fig. 6). Assuming that the decrease of foliar nutrient content during senescence is mostly owing to withdrawal of nutrients into perennial organs (nutrient resorption), the resorption efficiency of nitrogen and phosphorus from *S. gracilistyla* leaves are 44% and 46%, respectively (Sasaki & Nakatsubo 2007), as calculated by the following equation:

resorption efficiency (%) = 100 × [nutrient amount resorbed during senescence (g m^{-2})/nutrient amount in mature leaves (in September; g m^{-2})].   (13. 1)

Nutrient resorption efficiency of several pioneer trees, including other *Salix* species, are higher than 60% (Chapin & Kedrowski 1983; Escudero et al. 1992; Sakio & Masuzawa 1992). Aerts (1996) analyzed the data on nutrient resorption from leaves into perennial organs from a wide variety of studies, including data on pioneer trees and climax forests in the tropical, temperate, and boreal zones. Aerts reported that mean nutrient resorption efficiency of deciduous shrubs and trees was 54% for nitrogen and 50% for phosphorus. Thus, compared with other deciduous trees, the nutrient resorption efficiency from *S. gracilistyla* leaves is not high. This fact suggests that *S. gracilistyla* growing on sandy to gravelly bars do not use nitrogen and phosphorus conservatively.

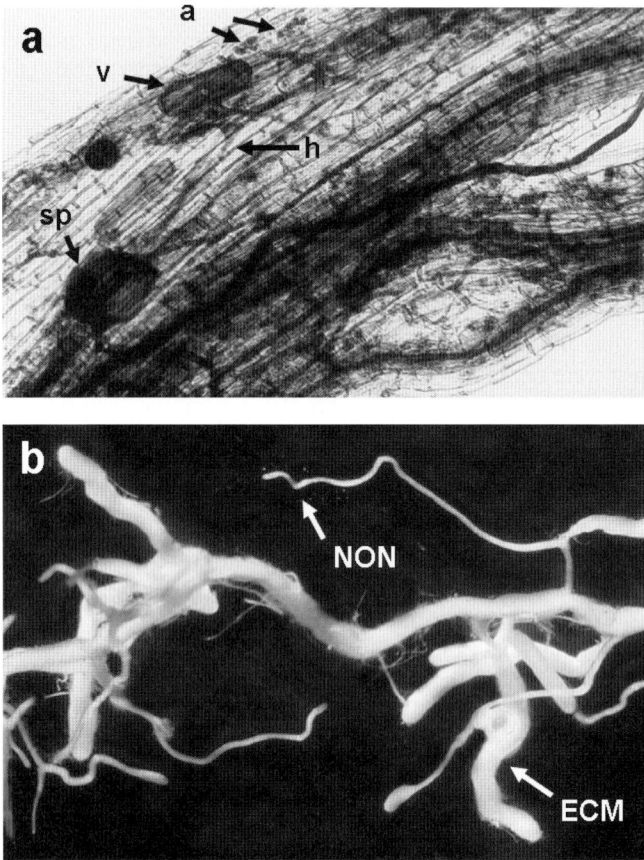

Fig. 7. Arbuscular and ectomycorrhizal colonization on *Salix gracilistyla* roots: (a) fine root colonized with arbuscular mycorrhizal fungi (400×, a: arbuscule; h: internal hypha; sp: spore; v: vesicle); (b) fine roots with (ECM) and without (NON) ectomycorrhizal colonization (25×)

13.5 Nutrient sources

As described above, *Salix gracilistyla* acquires large amounts of nutrients every year, in spite of the low-nutrient soil on bars. Therefore, it is likely that in addition to mineral nutrients in the soil there are other nutrient sources in sandy to gravelly bars. Hyporheic water may be one such nutrient source available for *S. gracilistyla* growth in riparian areas. The hyporheic zone is defined as subsurface areas where saturated sediments are hydrologically linked to the open stream channel and where stream water and subsurface water mix (Stanford & Gaufin 1974; Findlay 1995). Harner and Stanford (2003) indicated that the nutrient-rich upwelling of hyporheic water fertilizes riparian trees.

In addition, it is possible that mycorrhizal associations have an important role in *S. gracilistyla* growth on bars. Many mycorrhizal fungi promote nutrient absorption by host plants, and are dependent on the host for organic carbon as an energy source (Smith & Read 1997). Mycorrhizal associations are generally divided into seven distinct types, and the most widespread types among these are arbuscular mycorrhiza and ectomycorrhiza (Smith & Read 1997). Although most plants form only one type of mycorrhiza, some tree species (including *Salix* spp.) form both arbuscular mycorrhiza and ectomycorrhiza (Lodge 1989; Dhillion 1994; Van der Heijden & Vosatka 1999; Hashimoto & Higuchi 2003) and are called

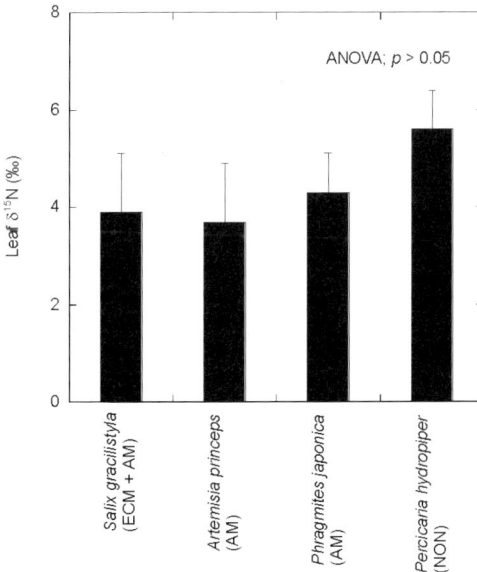

Fig. 8. $\delta^{15}N$ of plant leaves gathered at a bar in the Ohtagawa River, Hiroshima, Japan. ECM: colonized with ectomycorrhizal fungi; AM: colonized with arbuscular mycorrhizal fungi; NON: non-mycorrhizal plant. Vertical bars represent standard errors ($n = 3$) (Sasaki et al. unpublished data)

dual mycorrhizal plants. Figure 7 shows the roots of a *S. gracilistyla* growing in a bar that was colonized with arbuscular and ectomycorrhizal fungi. Sasaki et al. (2001) reported the maximum colonization levels of arbuscular and ectomy-corrhizal fungi in the roots of *S. gracilistyla* to be 30% and 80%, respectively.

Arbuscular mycorrhizal fungi stimulate the growth of host plants mainly by promoting phosphorus absorption (Smith & Read 1997). Van der Heijden (2001) conducted a culture experiment using *Salix* cuttings and reported positive short-term (< 12 weeks) effects of arbuscular mycorrhizal colonization on plant growth and phosphorus uptake. Thus, it is possible that arbuscular mycorrhizal coloni-zation stimulates *S. gracilistyla* growth on bars. However, the direct effect of arbuscular mycorrhizal colonization on the growth of *S. gracilistyla* seedlings was not observed in a pot culture experiment (Sasaki et al. 2001). Instead, nitrogen application significantly promoted plant growth, suggesting that the growth of *S. gracilistyla* seedlings is largely limited by the availability of nitrogen in the field (Sasaki et al. 2001).

Several culture experiments demonstrated that ectomycorrhizal fungi de-compose organic matter and transfer nitrogen to the host plants (Abuzinadah & Read 1988; Read et al. 1989; Read 1993). This ability may explain how *S. gracilistyla* growing in nutrient-poor habitats can acquire sufficient nitrogen for growth. Measurements of ^{15}N abundance (expressed as δ^{15}N) in plant tissue provide useful information concerning the source of nitrogen utilized by plants and mycorrhizal fungi (Nadelhoffer et al. 1996; Michelsen et al. 1998). Although δ^{15}N in plant tissues is determined by various factors (Högberg 1997), plants associated with ectomycorrhizal fungi tend to show lower δ^{15}N values than non-mycorrhizal plants or plants associated with arbuscular mycorrhizal fungi (Michelsen et al. 1996, 1998; Schmidt & Stewart 1997). We compared the δ^{15}N value of *S. gracilistyla* leaves with those of a non-mycorrhizal plant (*Polygonum hydropiper* L.) and two arbuscular mycorrhizal plants (*Artemisia princeps* Pampan. and *Phragmites japonica* Steud.) growing on a bar in the middle reaches of the Ohtagawa River in Hiroshima, Japan. Although *S. gracilistyla* associated with both arbuscular and ectomycorrhizal fungi showed slightly lower δ^{15}N values than the non-mycorrhizal plant, no significant difference among the four species was observed (Fig. 8).

13.6 Conclusions

This chapter introduced the developmental patterns and nutrient economy of *Salix gracilistyla* stands, which are dominant on sandy to gravelly bars in the upper to middle reaches of rivers in western Japan. In spite of the low-nutrient soil in these bars, this species has high productivity. After a severe disturbance by flooding, *S. gracilistyla* restores its plant organs rapidly, and spreads the stand area vegetatively by sprouting new shoots and forming adventitious roots. The annual net nitrogen and phosphorus increases in the aboveground organs of a *S. gracilistyla* stand are comparable to those in temperate deciduous forests, suggesting this species has a large nutrient demand. In comparison with other deciduous trees, however, the nutrient resorption efficiency from *S. gracilistyla* leaves is not high (44% for N, 46% for P), suggesting that this species absorbs sufficient nitrogen and phosphorus for its growth from soil and/or hyporheic water. In addition, ectomycorrhizal symbiosis may contribute to the nutrient absorption of this species. Additional research will be necessary to test these hypotheses.

Acknowledgements

We thank Dr. Y. Mishima, National Institute of Advanced Industrial Science and Technology (AIST), for stable isotope analysis of the plant samples.

References

Abuzinadah RA, Read DJ (1988) Amino acids as nitrogen sources for ectomycorrhizal fungi: utilization of individual amino acids. Trans Brit Mycol Soc 91:473-479

Aerts R (1996) Nutrient resorption from senescing leaves of perennials: are there general patterns? J Ecol 84:597-608

Boerner RE (1984) Foliar nutrient dynamics and nutrient use efficiency of four deciduous tree species in relation to site fertility. J Appl Ecol 21:1029-1040

Bollmark L, Sennerby-Forsse L, Ericsson T (1999) Seasonal dynamics and effects of nitrogen supply rate on nitrogen and carbohydrate reserves in cutting-derived *Salix viminalis* plants. Can J For Res 29:85-94

Brinson MM (1990) Riverine forests. In: Lugo AE, Brinson MM, Brown S (eds) Forested wetlands (Ecosystems of the world vol. 15). Elsevier Sci Pub BV, Amsterdam, pp 87-141

Chapin FS III, Kedrowski RA (1983) Seasonal changes in nitrogen and phosphorus fractions and autumn retranslocation in evergreen and deciduous taiga trees. Ecology 64:376-391

Chapin FS III, Johnson DA, McKendrick JD (1980) Seasonal movement of nutrients in plants of differing growth form in an Alaskan tundra ecosystem: implications for herbivory. J Ecol 68:189-209

Cooper DJ, Van Haveren BP (1994) Establishing felt-leaf willow from seed to restore Alaskan, U.S.A., floodplains. Arc Alp Res 26:42-45

Dhillion SS (1994) Ectomycorrhizae, arbuscular mycorrhizae, and *Rhizoctonia* sp. of alpine and boreal *Salix* spp. in Norway. Arc Alp Res 26:304-307

Escudero A, del Arco JM, Sanz IC, Ayala J (1992) Effects of leaf longevity and retranslocation efficiency on the retention time of nutrients in the leaf biomass of different woody species. Oecologia 90:80-87

Findlay S (1995) Importance of surface-subsurface exchange in stream ecosystems: the hyporheic zone. Limnol Oceanogr 40:159-164

Harner MJ, Stanford JA (2003) Differences in cottonwood growth between a losing and a gaining reach of an alluvial floodplain. Ecology 84:1453-1458

Hashimoto Y, Higuchi R (2003) Ectomycorrhizal and arbuscular mycorrhizal colonization of two species of floodplain willows. Mycoscience 44:339-343

Högberg P (1997) ^{15}N natural abundance in soil-plant systems. New Phytol 137:179-203

Ishikawa S (1988) Floodplain vegetation of the Ibi River in central Japan I. Distribution behavior and habitat conditions of the main species of the river bed vegetation developing on the alluvial fan. Jpn J Ecol 38:73-84 (in Japanese with English summary)

Ishikawa S (1994) Seedling growth traits of three salicaceous species under different conditions of soil and water level. Ecol Rev 23:1-6

Johnson FL, Risser PG (1974) Biomass, annual net primary production, and dynamics of six mineral elements in a post oak–blackjack oak forest. Ecology 55:1246-1258

Kawaguchi S, Saito H, Kasuya N, Ikeda T, Imamura Y (2005) Primary production of a young *Salix subfragilis* community in wet abandoned paddy field. J Jpn For Soc 87: 430-434 (in Japanese with English summary)

Kawahara T (1971) The return of nutrients with litter fall in the forest ecosystems. II. The amount of organic matter and nutrients. J Jpn For Soc 53:231-238 (in Japanese with English summary)

Kira T, Yabuki K (1978) Primary production rates in the Minamata forest. In: Kira T, Ono Y, Hosokawa T (eds) Biological production in a warm-temperate evergreen oak forest of Japan (JIBP synthesis vol. 18). Univ Tokyo Press, Tokyo, pp 131-138

Kopp RF, Abrahamson LP, White EH, Volk TA, Nowak CA, Fillhart RC (2001) Willow biomass production during ten successive annual harvests. Biomass Bioenerg 20:1-7

Larcher W (2001) The utilization of mineral elements. In: Larcher W (ed) Physiological plant ecology, 4th edn. Springer-Verlag, Berlin, pp 185-229

Lodge DJ (1989) The influence of soil moisture and flooding on formation of VA-endo- and ecto-mycorrhizae in *Populus* and *Salix*. Plant Soil 117:243-253

Maruyama K (1977) Beech forests in the Naeba Mountains. I. Comparison of forest structure, biomass and net productivity between the upper and lower parts of beech forest zone. In: Shidei T, Kira T (eds) Primary productivity of Japanese forests (JIBP synthesis vol. 16). Univ Tokyo Press, Tokyo, pp 186-201

McLeod KW, McPherson JK (1973) Factors limiting the distribution of *Salix nigra*. Bull Torrey Bot Club 100:102-110

Michelsen A, Schmidt IK, Jonasson S, Quarmby C, Sleep D (1996) Leaf ^{15}N abundance of subarctic plants provides field evidence that ericoid, ectomycorrhizal and non- and arbuscular mycorrhizal species access different sources of soil nitrogen. Oecologia 105:53-63

Michelsen A, Quarmby C, Sleep D, Jonasson S (1998) Vascular plant ^{15}N natural abundance in heath and forest tundra ecosystems is closely correlated with presence and type of mycorrhizal fungi in roots. Oecologia 115:406-418

Nadelhoffer K, Shaver G, Fry B, Giblin A, Jhonson L, Mckane R (1996) ^{15}N natural abundances and N use by tundra plants. Oecologia 107:386-394

Nagano M, Kira T (1978) Aboveground biomass. In: Kira T, Ono Y, Hosokawa T (eds) Biological production in a warm-temperate evergreen oak forest of Japan (JIBP synthesis vol. 18). Univ Tokyo Press, Tokyo, pp 69-82

Naiman RJ, Fetherston KL, McKay SJ, Chen J (1998) Riparian forests. In: Naiman RJ, Bilby RE (eds) River ecology and management. Springer-Verlag, Berlin, pp 289-323

Nakatsubo T (1997) Effects of arbuscular mycorrhizal infection on the growth and reproduction of the annual legume *Kummerowia striata* growing in a nutrient-poor alluvial soil. Ecol Res 12:231-237

Nakatsubo T, Kaniyu M, Nakagoshi N, Horikoshi T (1994) Distribution of vesicular-arbuscular mycorrhizae in plants growing in a river floodplain. Bull Jpn Soc Microb Ecol 9:109-117

Naohara K (1936–1937) Ecological studies on the vegetation of the basin of the River Abukuma. Ecol Rev 2:180-191, 306-318; 3:35-46 (in Japanese with English summary)

Niiyama K (1987) Distribution of salicaceous species and soil texture of habitats along the Ishikari River. Jpn J Ecol 37:163-174 (in Japanese with English summary)

Niiyama K (1989) Distribution of *Chosenia arbutifolia* and soil texture of habitats along the Satsunai River. Jpn J Ecol 39:173-182 (in Japanese with English summary)

Niiyama K (1990) The role of seed dispersal and seedling traits in colonization and coexistence of *Salix* species in a seasonally flooded habitat. Ecol Res 5:317-331

Pinay G, Fabre A, Vervier P, Gazelle F (1992) Control of C, N, P distribution in soils of riparian forests. Landscape Ecol 6:121-132

Read DJ (1993) Plant–microbe mutualisms and community structure. In: Schulze ED, Mooney HA (eds) Biodiversity and ecosystem function (Ecological studies vol. 99). Springer, Berlin, pp 181-203

Read DJ, Leake JR, Langdale AR (1989) The nitrogen nutrition of mycorrhizal fungi and their host plants. In: Boddy L, Marchant R, Read DJ (eds) Nitrogen, phosphorus and sulphur utilization by fungi. Cambridge Univ Press, Cambridge, pp 181-204

Rodgers CO, Barneix AJ (1988) Cultivar differences in the rate of nitrate uptake by intact wheat plants as related to growth rate. Physiol Plant 72:121-126

Rytter R-M (2001) Biomass production and allocation, including fine-root turnover, and annual N uptake in lysimeter-grown basket willows. For Ecol Manage 140:177-192

Saito H (1990) Some features of dry-matter production in a young *Alnus sieboldiana* community on dumped detritus with special reference to reproductive parts. J Jpn For Soc 72:208-215 (in Japanese with English summary)

Sakio H, Masuzawa T (1987) Ecological studies on the timberline of Mt. Fuji. II. Primary productivity of *Alnus maximowiczii* dwarf forest. Bot Mag Tokyo 100:349-363

Sakio H, Masuzawa T (1992) Ecological studies on the timberline of Mt. Fuji. III. Seasonal changes in nitrogen content in leaves of woody plants. Bot Mag Tokyo 105:47-52

Sasaki A, Nakatsubo T (2003) Biomass and production of the riparian shrub *Salix gracilistyla*. Ecol Civil Eng 6:35-44

Sasaki A, Nakatsubo T (2007) Nitrogen and phosphorus economy of the riparian shrub *Salix gracilistyla* in western Japan. Wetlands Ecol Manag 15:165-174

Sasaki A, Fujiyoshi M, Shidara S, Nakatsubo T (2001) Effects of nutrients and arbuscular mycorrhizal colonization on the growth of *Salix gracilistyla* seedlings in a nutrient-poor fluvial bar. Ecol Res 16:165-172

Schenk MK (1996) Regulation of nitrogen uptake on the whole plant level. Plant Soil 181. 131-137

Schmidt S, Stewart GR (1997) Waterlogging and fire impacts on nitrogen availability and utilization in a subtropical wet heathland (wallum). Plant Cell Environ 20:1231-1241

Schnitzler A (1997) River dynamics as a forest process: interaction between fluvial systems and alluvial forests in large European river plains. Bot Rev 63:40-64

Singh B (1998) Biomass production and nutrient dynamics in three clones of *Populus deltoids* planted on Indogangetic plains. Plant Soil 203:15-26

Small E (1972) Photosynthetic rates in relation to nitrogen recycling as an adaptation to nutrient deficiency in peat bog plants. Can J Bot 50:2227-2233

Smith SE, Read DJ (1997) Mycorrhizal symbiosis, 2nd edn. Academic Press, San Diego

Stachurski A, Zimka JR (1975) Methods of studying forest ecosystems: leaf area, leaf production and withdrawal of nutrients from leaves of trees. Ekol Pol 23:637-648

Stanford JA, Gaufin AR (1974) Hyporheic communities of two Montana rivers. Science 185:700-702

Tadaki Y, Hatiya K, Tochiaki K (1969) Studies on the production structure of forest. XV. Primary productivity of *Fagus crenata* in plantation. J Jpn For Soc 51:331-339 (in

Japanese with English summary)

Tadaki Y, Mori H, Mori S (1987) Studies on the production structure of forests. XX. Primary productivity of a young alder stand. J Jpn For Soc 69:207-214

Van der Heijden EW (2001) Differential benefits of arbuscular mycorrhizal and ectomycorrhizal infection of *Salix repens*. Mycorrhiza 10:185-193

Van der Heijden EW, Vosatka M (1999) Mycorrhizal associations of *Salix repens* L. communities in succession of dune ecosystems. II. Mycorrhizal dynamics and interactions of ectomycorrhizal and arbuscular mycorrhizal fungi. Can J Bot 77:1833-1841

Van Heerwaarden LM, Toet S, Aerts R (2003) Nitrogen and phosphorus resorption efficiency and proficiency in six sub-arctic bog species after 4 years of nitrogen fertilization. J Ecol 91:1060-1070

Viereck LA (1970) Forest succession and soil development adjacent to the Chena River in interior Alaska. Arc Alp Res 2:1-26

Von Fricks Y, Ericsson T, Sennerby-Forrse L (2001) Seasonal variation of macronutrients in leaves, stems and roots of *Salix dasyclados* Wimm. grown at two nutrient levels. Biomass Bioenerg 21:321-334

Walker LR, Chapin FS III (1986) Physiological controls over seedling growth in primary succession on an Alaskan floodplain. Ecology 67:1508-1523

White PS (1979) Pattern, process, and natural disturbance in vegetation. Bot Rev 45:229-299

Zavitkovski J, Stevens RD (1972) Primary productivity of red alder ecosystems. Ecology 53:235-242

14 The expansion of woody shrub vegetation (*Elaeagnus umbellata*) along a regulated river channel

Mari KOHRI

National Institute for Environmental Studies, 16-2 Onogawa, Tsukuba, Ibaraki 305-8506, Japan

14.1 Introduction

In the past few decades, propagation of woody vegetation in river channels has become conspicuous throughout the lowlands of the temperate zones of the Northern Hemisphere. When attempting to conserve or control populations of riparian woody plant species along regulated river environments, it is important to clarify their colonization patterns and processes and reproductive mechanisms. The expansion processes and mechanisms of many woody species, especially of Salicaceae, have been clarified (e.g., Niiyama 1990; Johnson 1994; Nakamura et al. 1997; Kamada & Okabe 1998; Dixon & Johnson 1999; Karrenberg et al. 2002; Shin & Nakamura 2005). Natural river channels are usually disturbance-prone areas where woody species merely survive to maturity and reproduce. However, anthropogenic factors such as dam construction and artificial river embankments have decreased the disturbance frequency and intensity. The stabilized water levels in springtime cause *Salix* seeds to germinate along water edges, rather than being washed downstream, and controlled flood discharges, in turn, enhance seedling survival and ultimately allow woodland expansion along river channels.

In the Yoshino River of southwestern Japan (Fig. 1), the riparian woodland area increased dramatically after the completion of dams in 1970s (Kamada et al. 1996). The dominant woody species in the lower stream reaches is *Salix chaenomeloides* Kimura, whereas in the middle and upper reaches the dominant woody species changes to *Elaeagnus umbellata* Thunb. (Kamada et al. 1999). Unlike Salicaceae species, the factors leading to domination and expansion of this species in riparian gravel bars has not been investigated in relation to fluvial processes, because *E. umbellata* is clearly ornithochorous and many researchers assume that its expansion is simply due to effective seed dispersal by birds.

Sakio, Tamura (eds) Ecology of Riparian Forests in Japan : Disturbance, Life History, and Regeneration
© Springer 2008

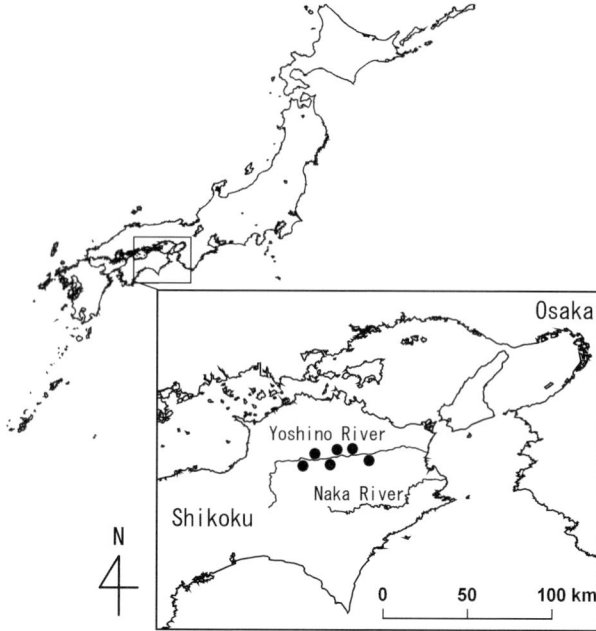

Fig. 1. Map of the study sites along the Yoshino River

Elaeagnus umbellata is a fleshy-fruited deciduous shrub endemic to eastern Asia that produces massive amounts of palatable fruits in late autumn. Each fleshy drupe contains only one seed, but a considerable number of seeds are consumed and dispersed (i.e., endozoochory) into riparian areas by frugivorous birds and mammals (Kohri et al. 2002). Based on observation along the Yoshino River, the fruits of *E. umbellata* are exploited more quickly than those of other fruiting trees (M. Kohri, unpublished data).

The primary habitats of *E. umbellata* are open pasture or rocky gorges of mountainous areas, and this shrub usually appears randomly in the pioneer vegetation. However, intensive domination of the riparian gravel bars in the Yoshino River by *E. umbellata* became apparent in the late 1980s (Kamada et al. 1999) and in the downstream reaches of the Naka River (Yuuki et al. 2000; Kohri et al. 2002), which flows parallel to the Yoshino River (Fig. 1). A considerable number of mature communities now exist in linear formations or in patches (Fig. 2a, b). A similar situation has been reported along the Joganji River (Ohta 1996) and the Kurobe River in Toyama Prefecture, north-central Japan. These four rivers all have steep gradients.

Studies on the expansion of fleshy-fruited shrubby vegetation in riparian habitats are limited. The cause(s) of the recent expansion of *E. umbellata* onto riparian gravel bars to such a considerable extent cannot be fully explained by effective frugivore seed dispersal or by a decrease of flood disturbances, as noted

Fig. 2. (a) An *Elaeagnus umbellata* community in the middle reaches of the Yoshino River (48 km from the river mouth, on the left bank); (b) a mature individual

for *Salix* species. Some kind of environmental change likely has occurred that allowed *E. umbellata* to become established and to dominate in the riparian gravel bars.

In this study, I first summarized the spatio-temporal distribution pattern of *E. umbellata* in riparian habitats by ascertaining the establishment years of trees in the population and measuring the relative elevation of establishment sites. I then analyzed the relationships between these parameters and autumnal flood-inundation patterns. Second, I investigated colonization and expansion mechanisms by testing seed germination and seedling survival characteristics, which suggested the possibility of hydrochory as the secondary means of seed dispersal. Finally, I identified regeneration safe-sites with regard to base material sizes and quantified the disturbance regime history that determined the distribution of *E. umbellata* communities on the riparian floodplains. On the basis of these findings, I discuss factors affecting the conservation of riparian biodiversity and the management of future riparian woodlands.

14.2 Study sites

The study was conducted at riparian gravel bars formed in the middle and upper reaches of the Yoshino River in Shikoku, southwestern Japan. The river stretches eastward for 194 km, and its drainage area is 3750 km². The studied section is 17.0 to 77.7 km from the river mouth, where most woody vegetation is distributed (Kamada et al. 1996). The Yoshino River has the largest flooding capacity in Japan, and its flood control has been a great issue for centuries. The stretch from the river mouth to Ikeda Dam (at 77.7 km) is under the management of the Ministry of Land, Infrastructure and Transport of Japan. Periodic river cross-section level surveys were conducted at 200-m intervals every year from late 1960s to 1992 and every 5 years since 1995. Hydraulic data are also recorded hourly at major gauging stations at 25.3 km from the river mouth in the lower reaches, 40.2 km in the middle reaches, and 74.8 km in the upper reaches, where the respective riverbed gradients are approximately 1/1100 m (0.09%), 1/800 m (0.125%), and 1/400 m (0.25%). The gradient of the Yoshino River is relatively steep, and its designed flood discharge at the 40.2-km point is 18,000 m³/sec.

The woody species that co-occur with *E. umbellata* along the Yoshino River are *Alnus serrulatoides* Callier, *Ulmus parvifolia* Jacquin, *Celtis sinensis* Persoon var. *japonica* (Planch.) Naka, *Albizia julibrissin* Durazz., *Mallotus japonicus* (Thunb. ex Murray) Mueller-Arg., *Melia azedarach* L. var. *subtripinnata* Miq., *Rosa multiflora* Thunb., *Celastrus orbiculatus* Thunb., and Salicaceae species such as *Salix gracilistyla* Miq. and *S. yoshino* Koidz. A precise flora for the Yoshino River was also provided by Ishikawa (1997).

14.3 Spatial and temporal distribution of the population

14.3.1 Vegetation mapping and positioning

To identify the population distribution pattern of *E. umbellata* along the water-course and to determine the species' habitat characteristics, a vegetation map was created by distinguishing every existing patch of *E. umbellata* and other major dominant riparian tree species and drawing these on a 1:2500 scale base map of the river. The precise location of each *E. umbellata* community was measured in the field by a portable distance meter (laser range finder, Impulse 200, Laser Tech Co., Centennial, USA), and the data were adjusted by referring to ground-control points set every 200 m on the levee by the government.

A total of 79 *E. umbellata* communities were identified within the study area. As noted by Kamada et al. (1996, 1999), woody species began to appear at 17 km from the river mouth, and they were composed mainly of *Salix chaenomeloides*, *S. subfragilis* Anders., and *S. yoshino*. In contrast, *E. umbellata* did not appear until 30 km from the river mouth, and most *E. umbellata* communities were established further upstream. *Elaeagnus umbellata* individuals were found far above 77 km from the Ikeda Dam, but at each gravel bar in the study sites, they tended to dominate in places with relatively elevated sites, compared to those sites inhabited by Salicaceae species.

14.3.2 Analysis of relative elevation from water

To clarify the topographic characteristics of *E. umbellata*'s habitat on the gravel bars, the relative elevation of 63 communities distributed adjacent to the cross-section survey lines was calculated. The river profile intersection measured in 1995 was divided into contour elevation ranges at 1-m intervals from the average low-water level, and the availability of land and actual occupation rate of *E. umbellata* for each 1-m elevation class for each cross-section were obtained. Habitat availability (%) and occupation rate (%) were determined for each relative elevation class by referring to the vegetation map and the level data (Fig. 3).

A variety of open niches exist along the river, and other woody species were distributed across a wide elevation range. However, the distribution of *E. umbellata* communities was restricted to an elevation range between 2 to 4 m above the average low-water level, with more than 70% of the patches concentrated in this range. *Elaeagnus umbellata* becomes established in sites at greater elevations than those of *Salix* species, but these establishment sites still get inundated during flooding, just less frequently than those of *Salix*. Based on the vertical and horizontal distribution patterns of *E. umbellata,* the mechanism for seeds to reach such elevated sites and seedlings to survive in such environments seemed to be strongly related to hydrogeomorphic events. The relative elevation of the establishment sites has increased owing to incision of the deepest part of the

(a)

(b)

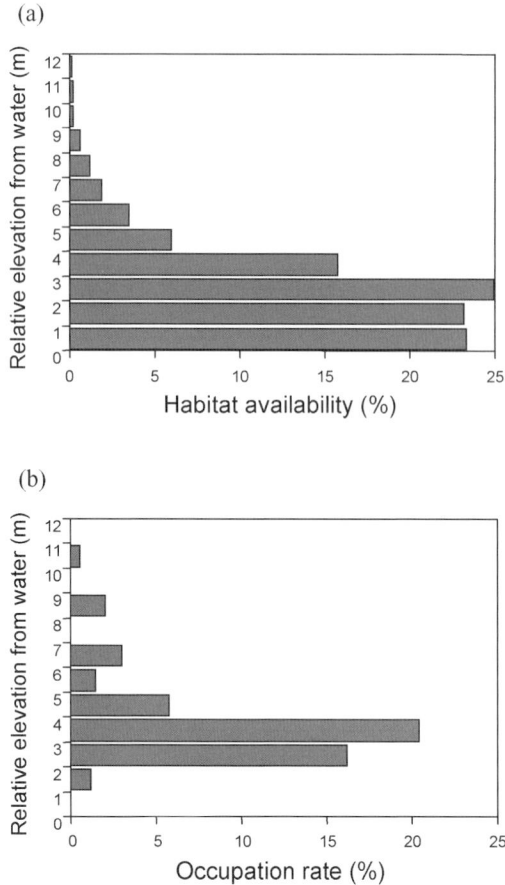

Fig. 3. (a) The percentage of each 1-m elevation class along the riverbed, as a measure of habitat availability; (b) actual occupation rate of *Elaeagnus umbellata* within each 1-m elevation class (after Kohri et al. 2000)

riverbed. As the channel deepens, the inundation frequency of establishment sites decreases, and this in turn protects established plants and enhances the survival of many woody species.

14.3.3 Size structure within the population and age distribution throughout the watercourse

To determine timing of the establishment events (history) of the *E. umbellata* population along the river, the age structure of *E. umbellata* communities along the Yoshino River was examined. The height of the tallest (oldest) individual from each dominant patch accessible at 44 cross-section lines was measured (Fig. 4), and a histogram of maximum vegetation heights along the watercourse was obtained (Fig. 5). Tree-ring sampling was conducted at six gravel bars dominated by *E. umbellata*, located at least 10 km apart from each other. At each site, some individuals were selected from each representative tree size class (marked A, B, C, and D in Fig. 5), and each tree was destructively sampled referring to the relationship between age and tree size obtained in Kohri et al. (2002). Discs for tree-ring sampling were obtained from the base; in some cases the base was buried in accumulated sediments and required excavation.

The tree-ring analysis revealed a similar age structure at all six sampling sites,

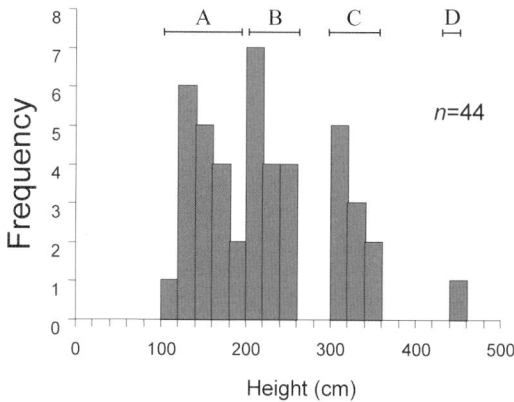

Fig. 4. Histogram of the tallest *Elaeagnus umbellata* individual measured in each patch at 44 cross-section lines along the Yoshino River (after Kohri et al. 2000)

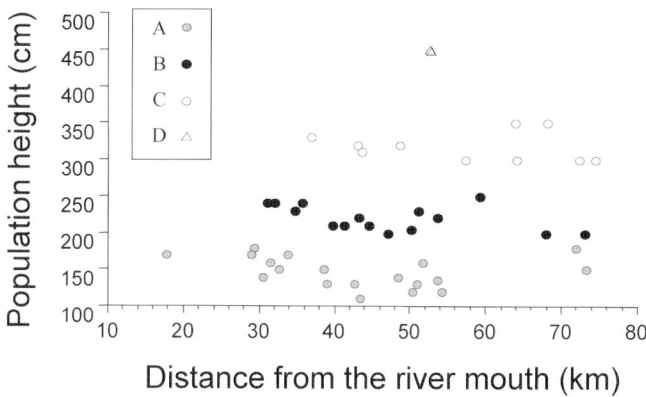

Fig. 5. Maximum height distribution of *Elaeagnus umbellata* along the Yoshino River. The older individuals are found rather upstream (after Kohri et al. 2000)

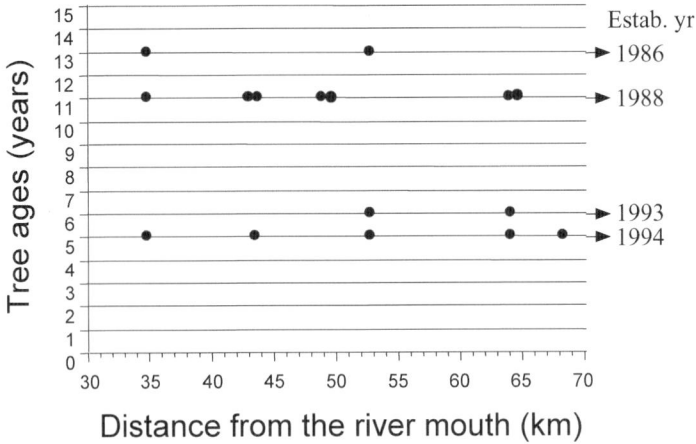

Fig. 6. The establishment years of trees in six *Elaeagnus umbellata* communities along the Yoshino River (after Kohri et al. 2000)

despite the fact that there is no clear masting in *E. umbellata*. The establishment years did not occur at constant intervals; instead, establishment was restricted to the years 1986, 1988, 1993, and 1994 (Fig. 6). Thus, establishment of individuals occurred synchronously at all six sampling sites. Younger individuals appeared to be more abundant in the middle reaches than in the upper reaches, indicating that suitable habitat for establishment was created synchronously by past hydro-geomorphic events throughout the watercourse from upstream to downstream.

14.3.4 Hydrological analysis of inundation pattern of the habitat

During the autumnal fruit-ripening season, rains often cause flooding on the Yoshino River. Some of these floods can reach establishment-site elevations. Thus, it is important to identify the disturbance magnitude necessary for *E. umbellata* seeds to reach and seedlings to become established at these sites. Toward this aim, the major autumnal floods that occurred between October 1 and December 31 from 1978 to 1995 were analyzed. The analysis focused on peak discharges exceeding 1000 m^3/sec because this discharge level is the threshold necessary for the river terraces to be inundated for this section of the Yoshino River (Okabe et al. 1996).

The occurrence of inundating autumnal floods in 1985, 1987, 1992, and 1993 (Fig. 7) was perfectly related to the tree ages (Fig. 6): the establishment events occurred simultaneously 1 year after the sites experienced a late autumnal inundation. These autumnal floods not only create open sites, the autumnal inundation pattern may be also an important feature of the seed dispersal of *E. umbellata*. The

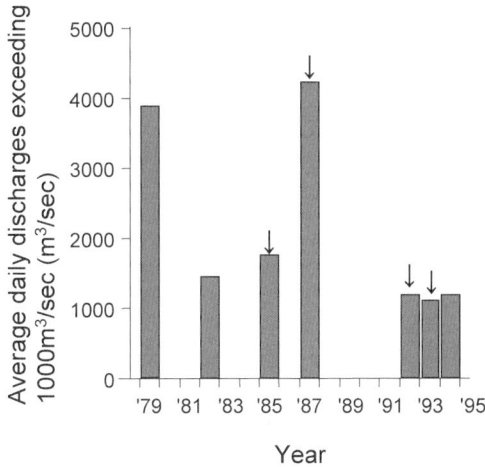

Fig. 7. The occurrence of autumnal floods between 1978 and 1995 (after Kohri et al. 2000). The peaks of average daily discharges exceeding 1000 m3/sec between October 1 and December 31 were analyzed. Arrows indicate years when *Elaeagnus umbellata* seedlings were established

autumnal floods recede quickly from the terraced gravel bars owing to the steep gradient of the Yoshino River, short duration of autumnal typhoon, and coarse bed material sizes. These floods also leave behind open sites facilitating deposited seeds to germinate and seedlings to establish at the elevated sites.

14.4 Seed germination and survival of the seedlings

14.4.1 Germination characteristics

To confirm the establishment years of the sampled trees, a seed germination experiment was conducted in the laboratory (Kohri et al. 2002). Seeds were gathered at the study sites in 1997 and exposed to 6 weeks of cold-moist treatment at 4 °C. More than 99% of the seeds from which the fruit pericarp (pulp) was removed and 61% of the seeds with pulp germinated under these conditions. Such germination rates are very high for woody species. Based on the germination regime of reported experiments (Washitani 1987), these findings suggest that once *E. umbellata* seeds were dispersed by autumnal floods, they would germinate in the following spring and not at other times of the year.

In addition, Nakagoshi (1985) studied the permanent seed bank in a temperate deciduous forest in Japan. No seeds of *E. umbellata* were found during the summer sampling, and no seeds were left after their first spring in the soil, indicating that *E. umbellata* does not create a permanent seed bank. Thus, it is valid to assume that *E. umbellata* seeds present in the soil were current-year seeds, and a current-year seedling resulted from a viable seed deposited at the site in the previous autumn. Whether the seeds were dispersed by endozoochory or by hydrochory, it appears that autumnal floods prime the habitat for the establishment of *E. umbellata* in river channels in spring.

14.4.2 Seedling survival

Observations were conducted on gravel bars at 43 and 49 km from the river mouth. Most seeds derived from bird feces recorded the previous autumn successfully germinated in spring. However, since a flood had not occurred in previous autumn when the seeds were deposited, only a few recruits survived to the second year, because most of the cohort could not survive subsequent summer floods, either being washed away or buried by sand sedimentation. The survivors in the cohort had emerged in areas not covered by other vegetation and in elevated sites with little disturbance and sufficient moisture. Such microhabitats are nearly always surrounded by some large boulders. This fact was also supported by the tree-ring analysis: when mature *E. umbellata* trees were excavated during sampling, there was always a large anchor stone within the main root system, indicating that establishment had occurred on strata made up of large boulders.

14.4.3 Riverbed material sizes at safe-sites

A high survival rate of current-year seedlings around boulders and adult trees carrying large boulders within their root systems were observed. Therefore, boulder sizes along the Yoshino River were measured to clarify the relationship between the availability of larger riverbed materials and the degree of *E. umbellata* population coverage. Size distributions of the bed materials were investigated along 76 cross-section lines, and about 20 of the largest boulders within each cross-section were randomly selected and measured.

The size of riverbed materials tended to get smaller toward downstream as usual. However, the cross-sections where *E. umbellata* was the dominant species were covered by relatively larger gravels or boulders compared to those sites dominated by other tree species. In particular, the boulder sizes were notably larger at *E. umbellata* dominated sites where the patches of shrubs exceeded 30 m (Fig. 8).

The Yoshino River has such unusually large boulders because chains of precipitous mountains run parallel to both the left and right banks, and the river's many steep tributaries bring relatively larger debris into the main stream. This coarsening of the riverbed materials, called the "armor-coating effect," became conspicuous during the 1980s (Kamada et al. 1997). The sabo dams constructed in the upstream reaches and tributary rivers entrapped sand and gravel such that their supply into the main stream decreased. Overconsumption of remaining gravels in

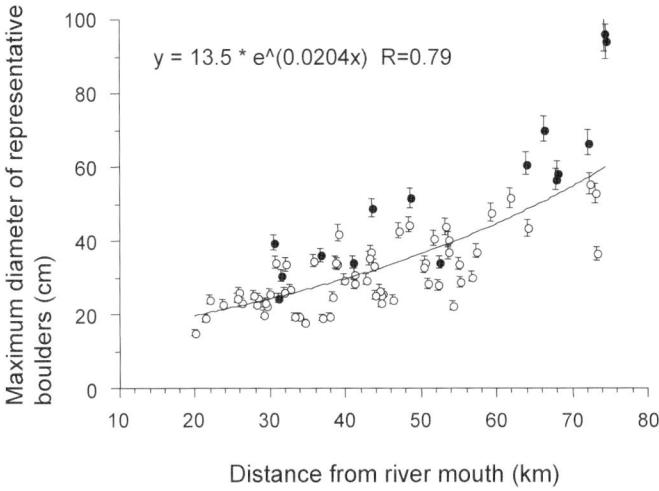

Fig. 8. The distribution of maximum boulder diameters (mean ± SD, $n \approx 20$ at each site) along 76 cross-section lines of the Yoshino River. The solid symbols represent sites dominated by *Elaeagnus umbellata* patches larger than 30 m in diameter (after Kohri et al. 2000)

the main channel occurred until the early 1970s, and fluvial processes caused large boulders to remain in this section of the river. Together, these phenomena provided new habitat and safe-sites throughout the watercourse, which helped *E. umbellata* to occupy the gravel bars. In addition, the tree size distribution indicates that younger individuals inhabit more downstream reaches (Figs. 4, 6), suggesting that the increased availability of safe-sites and coarsening of the river base materials have occurred from upstream toward downstream.

14.4.4 Hydrogeomorphic processes at seedling establishment sites

Hydrogeomorphic processes actively change the riverbed profile each year. To clarify the disturbance regime history of *E. umbellata* establishment sites, the positions of vegetation across the entire gravel bar at the 66-km post on the right bank were digitized. This map was then overlaid on the 20-year mean spatial contour distributions of cumulative annual bed movement, average elevation relative to the low-water level, and the disturbance magnitude of peak floods (expressed as the nondimensional tractive force). Finally, the spatio-temporal relationships were extracted.

The distribution of *E. umbellata* was concentrated on those sites where the absolute elevation and past elevation relative to the low-water level stayed relatively stable compared to other sites (Fig. 9), and the elevations have not changed for 20 years (from 1975 to 1995) (Fig. 10). This pattern may allow sufficient time for *E. umbellata* individuals to complete their life history. The stabilization of a riverbed is often caused by the establishment of woody vegetation, which causes sediments to accumulate and form mounds. In the case of *E. umbellata*, however, the stability of the substrate appears to be an initial habitat requirement, because the trees become established in large boulders exposed on the gravel bar.

The nondimensional tractive force of the habitat ranged between 0.00 and 0.05 (Fig. 11). At a value of 0.05, small particles like seeds would be washed away. This degree of force, however, does not provide enough shear velocity for larger bed materials to move during floods. Thus, the large boulders trap seeds and protect emergent seedlings from summer flood disturbances and droughts. These hydrogeomorphic characteristics seem to be important for *E. umbellata* seed dispersal and the establishment of seedlings.

14.5 Life-history strategies of *E. umbellata* in relation to the river's disturbance regime

The age structure of *E. umbellata* in the Yoshino River indicates that the major communities became established in the year following a late autumnal flood, and this same age structure is observed throughout the watershed basin. The distribution of the present population is restricted to 2 to 4 m above the low-water level across the length of the river. Thus, the peak floods that occur during the autumn fruiting season serve as a seed dispersal vector, carrying seeds dropped in the river either from endozoochory or by gravity from upstream to downstream. These autumnal floods also bring seeds to wider and elevated sites that may be less frequently inundated by subsequent summer floods, which allows for establishment to occur. In addition, the strong floods facilitate *E. umbellata* seedling establishment by clearing herbaceous vegetation already on the gravel bars.

Flood disturbances may also directly affect *E. umbellata* seed germination. In a seed germination experiment, without being ingested by birds the seeds still had a high germination rate when the pericarp was removed by hand. Thus, those fruits dropped into floodwaters by gravity or by birds during fruit handling would also have an enhanced germination rate due to the removal of pericarp in the turbulent water. Flood currents may also facilitate fruit removal from the twigs when there are no frugivore visitations.

A nondimensional tractive forces between 0.00 and 0.05 represent the unique disturbance regime where *E. umbellata* exists. At these sites, small particles such as *E. umbellata* seeds can be easily transported and entrapped by the flood current's shear velocity, but large boulders in the riverbed cannot be moved by such forces. Both seedling survival and the species' present distribution show high correlations with areas where geomorphic movement remained small for many

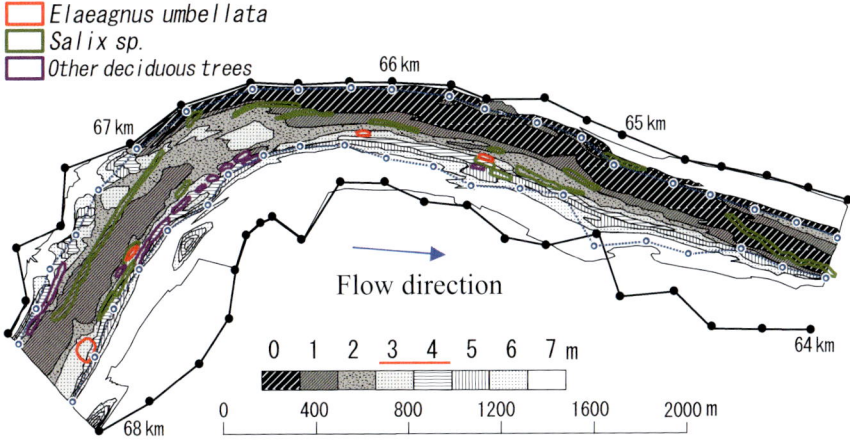

Fig. 9. Contour map of the 20-year mean elevation relative to the water level of a gravel bar formed 66 km from the river mouth. The red line above the elevation key indicates the elevation range of *Elaeagnus umbellata* sites

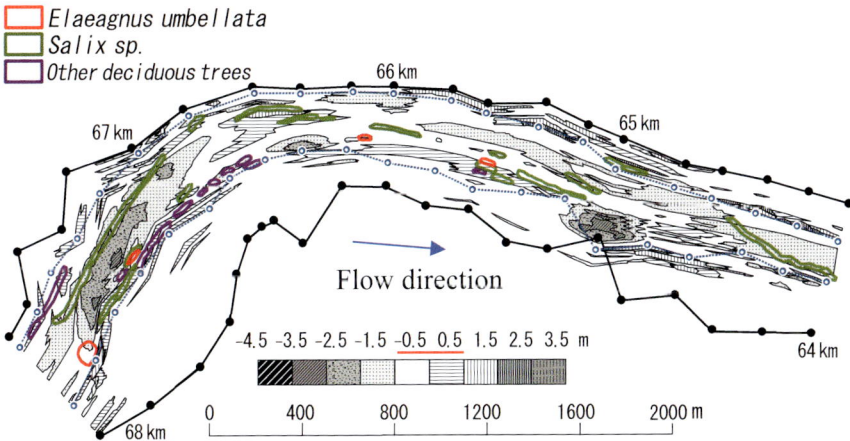

Fig. 10. The 20-year change of the cumulative bed movement of a gravel bar formed at 66 km from the river mouth. The red line above the key indicates the range of movement at *Elaeagnus umbellata* sites

years. Moreover, hydrochory becomes more effective where huge base materials exists, because these materials entrap more seeds and provide safe-sites for seed germination (Chambers et al. 1991), as well as allowing current-year seedlings to survive subsequent summer floods. The larger the boulder, the safer the site for

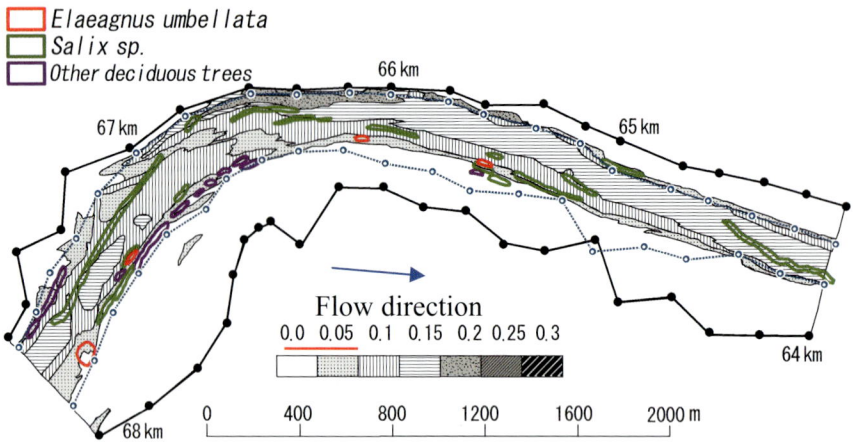

Fig. 11. The spatial distribution of the 20-year mean nondimensional tractive force experienced at a gravel bar formed at 66 km from the river mouth. The actual distribution of the woody vegetation on the gravel bars are shown in solid lines. The red line above the key indicates the tractive force range at *Elaeagnus umbellata* sites

the survival of emergent seedlings. Thus, *E. umbellata* colonizes places where the flood magnitude is strong enough for hydrochory yet moderate enough for seedling survival in the face of future disturbances.

Elaeagnus umbellata has multiple life-history strategies, including the ability to regenerate vegetatively and grow aggressively by sprouting and root sucker formation. Thus, in addition to reproducing via seed germination, broken twigs that are dispersed during floods may act as propagules. Such vegetative regeneration is likely made possible owing to the presence of nitrogen-fixing actinorhizal bacteria (*Frankia* sp.) on the plants' root nodules. The bacteria help *E. umbellata* recover quickly in nutrient-poor gravel bars after experiencing severe damage by floods and sedimentation, and the extra nitrogen helps the plants produce massive fruit crops. The production of massive fruit crops on open gravel bars is an effective reproductive strategy, because this enhances the conspicuousness of the display for wintering birds (Laska & Stiles 1994). Frugivorous birds carry seeds upstream and to places where floodwaters cannot reach. These multiple strategies enable *E. umbellata* to maintain high fitness in the riparian habitat.

14.6 Managing *E. umbellata* populations

Because *E. umbellata* can grow in a wide range of soil environments and tolerates disturbance, it was introduced to North America as greening material and for ornamental usage in 1830. The species spread widely, however, and has become a notorious invader of open pastures and disturbed sites in the United States. Many control measures have been conducted in the United States (e.g., Darlington 1994). *Elaeagnus angustifolia* L. (Russian olive) has produced a similar phenomenon along watercourses throughout the western United States (Knopf & Olson 1984; Knopf et al. 1988). The rapid spread of these species is unwelcome, not only because of its detrimental effects on the biodiversity of riparian areas but also from the point of view of flood control. Woody vegetation interrupts the flow of flood discharges and leads to the destruction of river embankments by skewing the channel flow (Okabe et al. 1996).

The environmental and biological parameters obtained in this study may provide parameters for estimating the future population distribution and expansion of *E. umbellata* along river channels. This study also revealed basic information necessary for the control of invasive plant species in ecosystems and for the conservation and management of riparian biodiversity.

14.7 Conclusions

Elaeagnus umbellata employs several highly adaptive life-history strategies. The species achieves effective seed dispersal to safe establishment sites using both endozoochory and hydrochory, and its fruiting season is synchronized with the occurrence of late autumnal floods. *Elaeagnus umbellata* has a high seed germination rate, and its ability to reproduce vegetatively allows it to withstand the damage caused by disturbance. These phenological traits and reproductive mechanisms enable *E. umbellata* to become established in wide and elevated areas along regulated river channels. Along the Yoshino River, the spatial distribution of *E. umbellata* is limited to places where there is little geomorphic movement of the riverbed and the elevation is just above the level reached by floods. The availability of elevated sites and the presence of large bed materials were especially important for seedling survival. Thus, woodland expansion and domination is strongly related to the flooding regime. The distribution and expansion of *E. umbellata* on gravel bars can be considered as a bio-indicator of physical changes, namely the geomorphic and compositional changes of the riverbed and disturbance regime of this regulated river.

Acknowledgments

I thank Dr. T. Okabe of the Department of Civil Engineering, Tokushima University, for performing hydrological calculations (especially of the nondimensional tractive force) and providing research funds. I also thank Dr. M. Kamada of the Tokushima University, for research advice and the field support of his laboratory students. Miss I. Kotera helped with the level survey, fieldwork, and programming for hydraulic calculations. Dr. N. Nakagoshi gave helpful advice on this study and served as my doctoral advisor at the Graduate School for International Development and Cooperation, Hiroshima University. I am also indebted to the Tokushima Construction Office of the Ministry of Land, Infrastructure and Transport of Japan for providing hydraulic data and precise maps. Partial funding was provided by the Foundation for River and Watershed Environment Management, Japan, and a Grant-in-Aid for Scientific Research from the Ministry of Education, Science, Sports and Culture (no. 11650531).

References

Chambers JC, MacMahon JA, Haefner JH (1991) Seed entrapment in alpine ecosystems: effects of soil particle size and diaspore morphology. Ecology 72:1668-1677

Darlington J (1994) Control of Autumn Olive, Multiflora Rose, and Tartarian Honeysuckle. USDA NRCS, Washington DC, USA

Dixon MD, Johnson WC (1999) Riparian vegetation along the middle Snake River, Idaho: zonation, geographical trends, and historical changes. Great Basin Naturalist 59:18-34

Ishikawa S (1997) Distribution behavior of riparian plants and species diversity of the vegetation on rocky river banks in the Yoshino River in Shikoku, Japan. Mem Fac Sci Kochi Univ Ser D (Biol) 18:1-7

Johnson WC (1994) Woodland expansion in the Platte River, Nebraska: patterns and causes. Ecol Monogr 64:45-84

Kamada M, Okabe T (1998) Vegetation mapping with the aid of low-altitude aerial photography. Appl Veg Sci 1:211-218

Kamada M, Ohta Y, Okabe T (1996) Interrelation between tree distribution in river and environmental change of basin due to human activity. Proceedings of the International Symposium-Interpraevent 1996 2: 245-252

Kamada M, Okabe T, Kotera I (1997) Influencing factors on distributional change in trees and land-use types in the Yoshino River, Shikoku, Japan. Environmental System Research 25:231-237 (in Japanese with English abstract)

Kamada M, Kohri M, Mihara S, Okabe T (1999) Distribution of *Salix* spp. and *Elaeagnus umbellata* communities in relation to their stand characteristics on bars in the Yoshino River, Shikoku, Japan. Environmental System Research 27:331-337 (in Japanese with English abstract)

Karrenberg S, Edwards PJ, Kollmann J (2002) The life history of Salicaceae living in the active zone of floodplains. Freshwater Biol 47:733-748

Knopf FL, Olson TE (1984) Naturalization of Russian-olive: implications to Rocky Mountain wildlife. Wildl Soc Bull 12:289-297

Knopf FL, Johnson RR, Rich T, Samson FB, Szaro RC (1988) Conservation of riparian ecosystems in the United States. Wilson Bull 100:272-284

Kohri M, Kamada M, Okabe T, Nakagoshi N (2000) Distribution pattern of *Elaeagnus umbellata* communities on the gravel bars in relation to hydrogeomorphic factors in the Yoshino River, Shikoku, Japan. Environmental System Research 28:353-358 (in Japanese with English abstract)

Kohri M, Kamada M, Yuuki T, Okabe T, Nakagoshi N (2002) Expansion of *Elaeagnus umbellata* on a gravel bar in the Naka River, Shikoku, Japan. Plant Species Biol 17:25-36

Laska MS, Stiles EW (1994) Effects of fruit crop size on intensity of fruit removal in *Viburnum prunifolium* (Caprifoliaceae). Oikos 69:199-202

Nakagoshi N (1985) Buried viable seeds in temperate forests. In: White J (ed) Handbook of vegetation science: the population structure of vegetation. Junk, Dordrecht, pp 551-570

Nakamura F, Yajima T, Kikuchi S (1997) Structure and composition of riparian forests with special reference to geomorphic site conditions along the Tokachi River, northern Japan. Plant Ecol 133:209-219

Niiyama K (1990) The role of seed dispersal and seedling traits in colonization and coexistence of *Salix* species in a seasonally flooded habitat. Ecol Res 5:317-331

Ohta M (1996) Community distribution and vegetative succession process of *Elaeagnus umbellata* in five major rivers in Toyama Prefecture. In: Toyama Science Museum (ed) Vegetation of Alluvial Rivers in Toyama Prefecture. Toyama Science Museum, Toyama, pp 37-47 (in Japanese)

Okabe T, Kamada M, Hayashi M (1996) Ecological and hydraulic study on floodplain vegetation developed on a bar. Proceedings of the International Symposium-Inter-praevent 1:235-244

Shin N, Nakamura F (2005) Effects of geomorphology on riparian tree species in Rekifune River, northern Japan. Plant Ecol 178:15-28

Washitani I (1987) A convenient screening test system and a model for thermal germination response of wild plant seeds: behavior of model and real seeds in the system. Plant Cell Environ 10:587-598

Yuuki T, Okabe T, Kamada M, Kohri M, Nishino K (2000) Growth of woody plants in the downstream of Nakagawa River and its hydraulic influence. Ann J Hydraul Eng JSCE 44:843-848 (in Japanese with English abstract)

Part 7

Riparian forests in wetland

15 Distribution pattern and regeneration of swamp forest species with respect to site conditions

Hiroko FUJITA and Yoshiyasu FUJIMURA

Botanic Garden, Field Science Center for Northern Biosphere, Hokkaido University, Kita 3 Nishi 8 Chuo-ku, Sapporo, Hokkaido 060-0003, Japan

15.1 Introduction

Japan, on the eastern margin of the Eurasian continent, has numerous and various types of wetlands owing to its temperate climate and abundant precipitation under the influence of the monsoon. Many of these wetlands have developed on floodplains. River dynamics have a great influence on the generation of topographic features such as mesic natural levees and poorly drained back swamps on floodplains. These floodplains support various wetland vegetation types such as reed swamps, sedge grasslands, swamp forests, and bog communities. In this chapter, we give an outline of swamp forests on floodplains in the cool temperate zone of Japan.

The predominant tree species on swamp forests in the cool temperate zone of Japan are alder (*Alnus japonica* (Thunb.) Steud.), ash (*Fraxinus mandshurica* Rupr. var. *japonica* Maxim.), and Japanese elm (*Ulmus davidiana* Planch. var. *japonica* (Rehder) Nakai). These genera are common to North American and European swamp forests; for example, *Alnus rugosa* (DuRoi) Spreng. (Parker & Schneider 1974; Turner et al. 2004), *Fraxinus nigra* Marsh. (Parker & Schneider 1974; Turner et al. 2004), *Fraxinus pennsylvanica* Marsh. (Turner et al. 2004), and *Ulmus americana* L. (Turner et al. 2004) in North America, and *Alnus glutinosa* (L.) Gaertn. (Lawesson 2000), *Fraxinus excelsior* L. (Lawesson 2000; Schnitzler 1994), *Ulmus minor* Mill. and *Ulmus laevis* Pall. (Schnitzler 1994), and *Ulmus glabra* Huds. (Lawesson 2000) in Europe. In Japan, the swamp species mentioned above grow along a soil moisture gradient: alder in wet sites, elm in mesic sites, and ash in sites intermediate between wet and mesic (Miyawaki 1977; Tatewaki et al. 1967; Tsuneya 1996). Alder forms pure stands in low peatlands and in extremely wet sites in back swamps. Ash grows near natural levees in back

Sakio, Tamura (eds) Ecology of Riparian Forests in Japan : Disturbance, Life History, and Regeneration
© Springer 2008

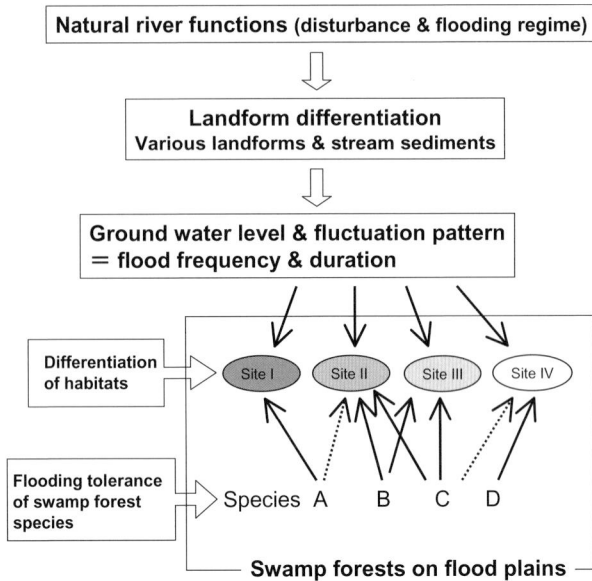

Fig. 1. Cause-and-effect relation between site conditions and the distribution of swamp forest tree species

swamps, often in association with alder. Elm usually grows on well-drained natural levees and fans. Elm is associated with ash or alder on wetter sites, but with *Acer mono* Maxim. var. *marmoratum* (Nichols.) Hara f. *dissectum* (Wesmael) Rehder, and *Cercidiphyllum japonicum* Sieb. et Zucc. on mesic sites (Kon & Okitsu 1995). The boundaries of the three swamp species overlap in their main habitat.

Their distribution on floodplains is restricted by the source of water, as determined by landform, flooding regime, soil physicochemical features, and groundwater fluctuation regime. Figure 1 shows the cause-and-effect relationship between site conditions and the distribution of swamp forest tree species. River behavior causes landform differentiation and forms alluvial fans, natural levees, and back-land. Each landform unit has a distinctive sediment composition and its own ground water level and fluctuation pattern. Tolerance to flooding differs among species, as reflected in each species' habitat. We deal with these factors—topographical features, ground water, soil, and features of swamp tree species—in turn.

15.2 Topographical features of swamp forests

Typical habitats where swamp forests are found are characterized by waterlogged soil, high groundwater levels, and flooding during some or most of the year. The primary factor generating these water-saturated conditions is the landform. An alluvial plain generally extends from an alluvial fan to natural levees accompanied by back swamps, and a delta. Alder and ash forests establish on waterlogged sites in back swamps, and elm forests on alluvial fans and natural levees, where overflowing rivers deposit coarse material. Alder forests on alluvial plains used to be found throughout Japan, but have diminished or disappeared as a result of reclamation of the land for rice paddies and urbanization, especially in the main islands and southward.

The floors of valley bottom plains in diluvial uplands and hills are also typical habitat of swamp forests. Valley floors are often filled with variable unconsolidated sediments such as gravel, sand, silt, clay, peat, and mixtures of these materials (Kadomura 1981). Elm and ash establish on coarse deposits, and alder on fine deposits such as clay, silt, muck, or peat.

In addition to these two typical landforms, swamp forests grow by lakesides, at springs, and in poorly drained depressions generated by landslides or mud flows.

15.3 Site conditions of swamp forests

15.3.1 Water fluctuation regime and its significance

Figure 2 shows a ground surface profile near the Tohoro River, a natural river in eastern Hokkaido, northeast Japan (Takada & Fujita, unpublished data). The river meanders slowly through the alluvial plain and generates natural levees and back swamps. Elm, ash, and alder are distributed across the floodplain from natural levees to back swamps in that order. Figure 3 shows the water table fluctuations measured in an elm stand, a mixed ash–alder stand, and a pure alder stand (Takada

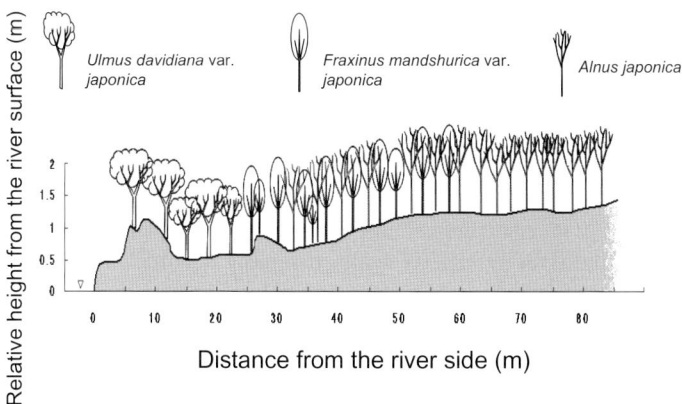

Fig. 2. Ground surface profile by the Tohoro River, eastern Hokkaido, and distribution of the three wetland tree species

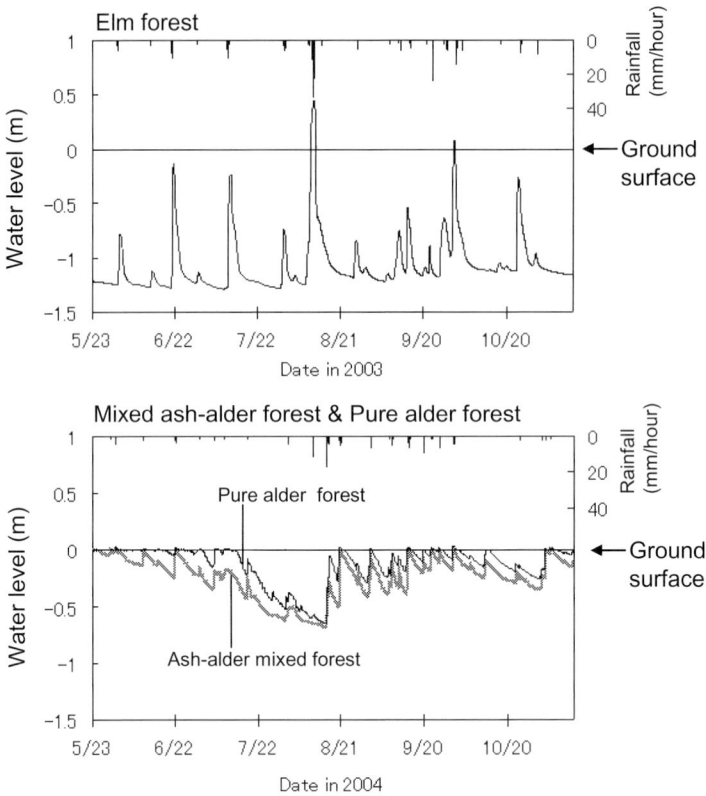

Fig. 3. Seasonal variations in water table and rainfall. The top figure shows water table fluctuation in an elm stand in 2003. The bottom figure shows the same in a mixed ash–alder stand and a pure alder stand in 2004. The water table was measured every half hour

& Fujita, unpublished data). The water table in the elm stand on a natural levee is usually lower than 1 m below the ground surface, because the ground surface is higher than the normal surface of the river and the coarse sediment deposits drain quickly (Figures 2, 3). It rises to 30–70 cm below the ground surface in response to rainfall, and immediately declines after the rain stops. When the river overflows, it rises up over the ground surface and flows into the ash and alder stands, then as the river level falls again, it rapidly declines. Thus, the saturated condition does not last long in elm stands on natural levees. In contrast, the water table in ash and alder stands on back swamps is constantly near the ground surface, because of poor drainage (Figure 3). However, it declines in the summer (from the middle of July to the middle of August) because of reduced rainfall and increased evapotranspiration.

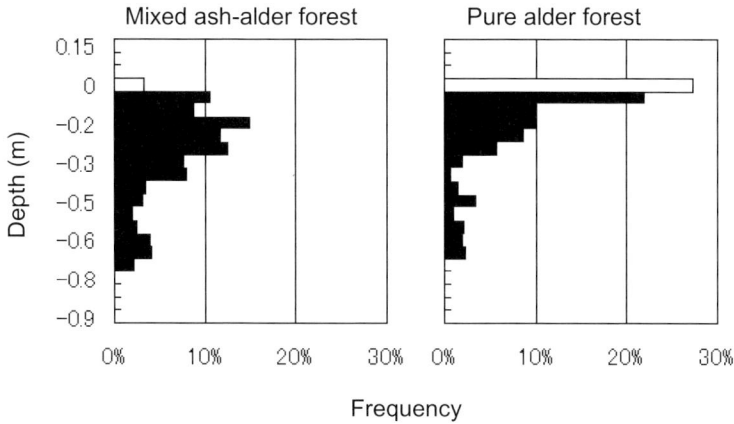

Fig. 4. Frequency distribution of water level. "0 m" means the ground surface. The open bars indicate that the water table was above the ground surface. The water table was measured every half hour from 23 May to 12 November 2004

Figure 4 shows the frequency distribution of the water table in a mixed ash–alder stand and a pure alder stand (Takada & Fujita, unpublished data). The frequency of the water table's occurrence above the ground surface clearly distinguishes between the stands: 3.7% in the ash–alder stand vs. 27.3% in the pure alder stand. The ecological optimum site for ash trees is wet soil without stagnant water, as revealed by studies of naturally regenerated and plantation ash stands (Nakae 1959; Nakae et al. 1960, 1961; Nakae & Manabe 1963; Nakae & Tatsumi 1964). Long periods of flooding can restrict ash establishment on extremely wet sites. Although ash has features adaptive to flooding (Nagasaka 2001), alder is more adaptable to stagnant water.

15.3.2 Soil types and corresponding species

Alder stands grow on various soil types, such as anaerobic wet peat soils, Grey soils, alluvial soils, and aerobic mesic or wet Brown Forest soils. Peat soils underlie swamp forests in Hokkaido and the cooler parts of Honshu (Fujita 2002). Around bogs or at lake shores where sphagnum moss grows well, sphagnum peat is sometimes found. Alder trees on these sites are dwarfed, measuring 0.5 to 2 m in height, and sprout many stems from the base.

Low-moor peat soil is widespread in back swamps of alluvial floodplains and sometimes on the floor of valley bottom plains. Alder trees on this soil type show moderate growth and often form pure stands. High water levels and anaerobic conditions characterize this soil type. Layers of fluvial deposits can sometimes be recognized in a low-moor peat soil profile. Field observations suggest that the soil horizon is formed only from peat, but a loss-on-ignition test can show silt and clay layers in the peat. This fact indicates that these sites are sometimes flooded, perhaps by snowmelt or typhoons.

On the other hand, in depressions generated by landslides or mud flow in mountainous or hilly regions, the soil under an alder forest often has a gleyed horizon and an anaerobic A-layer containing a large amount of humus or muck.

Alluvial soil is commonly found on or near natural levees and on the floor of valley bottom plains, and elms often grow there. The soil horizon shows complexity and has several coarse-textured fluvial deposit layers. Most deposited materials are transported by river flow. Occasionally, on valley bottom plains, the subsoil consists of coarse deposits brought to the site by mass movement (Makita et al. 1979). When the deposited sediments on the plain are either smooth or angular unconsolidated gravel, the water level is low and the soil drains well, resulting in the establishment of elm forest but not alder forest (Fujita & Kikuchi 1986).

Ash forests usually establish on natural levees or back swamps on alluvial plains and valley bottom plains. The soil types are usually alluvial soils, Grey soils, and sometimes peat soils. Ash forests are also found at mesic sites such as gentle slopes on diluvial uplands, at the lower end of hillsides, and along river terraces. The best sites for the growth and high productivity of ash receive enough water without stagnation. The growth of ash is greatly dependent on high soil moisture content, especially in early to mid-summer (Shiozaki et al. 1992), when the weather is dryer (Saito et al. 1990); drying of the soil reduces the growth of ash (Kikuchi et al. 1995). Growth is also reduced on poorly drained sites with a grey layer in the shallow part of the soil horizon (Abe 1989; Abe et al. 1989), because the roots of ash cannot grow through the grey layer (Abe 1989). The growth rate of ash seedlings in experiments improved as the oxygen concentration of the soils increased (Ohtomo & Nishimoto 1981). These facts suggest that a balance between high soil moisture and high soil oxygen is important for the growth of ash.

15.4 Features of swamp forest tree species in cool temperate zone in Japan

The differentiation of wetland forests according to site conditions is related to each swamp tree's life-history features, notably the requirements for germination, seedling growth, and regeneration. Physiological tolerance to flooding is also important (see chapter 16). The pertinent ecological characteristics of the three swamp species are shown in Table 1.

15.4.1 Alder (*Alnus japonica*)

Flowers of alder are catkins and bloom from November in warm regions and from April in cool regions. Seed production varies from large to small (Joint Research

Table 1. Main regeneration and photosynthesis characteristics of swamp forest trees

Species	Alnus japonica	Fraxinus mandshurica var. japonica	Ulmus davidiana var. japonica
Regeneration characteristics			
Sexuality	monoecious unisexual flower	dioecism	monoecious bisexual flower
Flowering and seed setting periodicity	masting exsistence, but no large mastirg (Mori 1991)	once in 3–4 years (Manabe 1982, Manabe & Ohkubo 1971)	irregular masting, medium crop coming once in 3.5 years (Sasaki 1985)
Fruit and seed	nucla (3-4 mm)	samara (2.5-3 cm length, 7-8 mm width)	samara (12-15mm) seed (5-6mm)
Seed dispersal	by wind, water	mainly bolochory and by partly wind; short distances	by wind
Seedbank	one year	few years	one year
Germination	next spring ; on moist soil	2 years later in spring; rarely 3 years later in spring	2 weeks after falling at the forest edge or gap (enough light), next spring (Apr.-May) at the forest understorey (before tree leaf expansion) (Seiwa 1997)
Flooding tolerance (2-year-old seedlings) (Nagasaka 2001)	very strong	strong	medium, moderate
Sprout	active production (birth)	only 10 % (Goto et al. 2002)	not birth
Photosynthetic performance (Koike 1991)			
* Light compensation point of photosynthesis (klux)			
Seedling	2.1-1.6	0.8-0.3	1.5-0.9
Adult	2.1-1.6	2.1-1.6	1.5-0.9
* Light saturation limits (klux)			
Seedling	40-21	20-10	20-10
Adult	60-41	40-21	40-21
Degree of leaf thinning under weak light	small change	large change	large change

* leaf temperature 20°C

Group of Public Forestry Research Institute 1983). The seeds are dispersed by wind and water.

Seeds can germinate when their temperature and moisture requirements are met, even if they lie on or beneath the water table (Murakami 1977; Fujita 2001). The length of exposure to light does not affect germination (Murakami 1977). However, few seedlings are found on alder forest floors because of inhibited germination and high mortality of seedlings. Possible explanations for the inhibited germination and high mortality are light deficiency, low temperature or desiccation just after germination, and allelopathy (Murakami 1977). Light deficiency is considered the most important factor, because light requirement of alder is high, like that of *Betula platyphylla* Sukatchev var. *japonica* (Miq.) Hara and *Alnus hirsuta* Turcz., which is a feature of pioneer species (Koike 1991). It is thus likely that seedlings under tall grass or sedge die from light deficiency. Further, the growth rate of alder seedlings is less than that of other deciduous tree species (Kubota 1979), suggesting a lower competitive ability for alders. Thus, the seedlings of alder are found at sites without competitive species, for example near lake shores or on exposed soil disturbed by flooding.

Alder forest regenerates mainly by sprouting of stems rather than by producing seedlings. The proportion of multi-stemmed alders is larger at extremely wet sites than at mesic sites (Nakamura et al. 2002; Iwanaga & Yamamoto 2007), suggesting that regeneration of alder depends on sprouting in wet conditions. This high ability to sprout contributes not only to regeneration, but also to adaptation to destructive environmental change such as a sudden rise in the water level. Sprouting to compensate for the destruction of stems was observed in mires where the water level rose drastically owing to river channelization: alders changed from tall trees to multi-stemmed shrubs (Fujimura, unpublished).

Alders, which have the highest tolerance of wet conditions, form pure forest stands at sites where other tree species cannot establish because of permanent high water levels. At these sites, alder forests reach a stable climax community, and regenerate by sprouting of stems in the absence of disturbances that promote seedling establishment of various tree species.

15.4.2 Ash (*Fraxinus mandshurica* var. *japonica*)

The fruits of ash are samaras, which are dispersed in autumn. Seeds germinate in the spring, usually two or sometimes three years later (Manabe 1982). The seedlings of ash are found on gloomy forest floors, in contrast to those of alder. The light compensation point of photosynthesis is lower in ash seedlings than in alder seedlings, showing the high shade tolerance of the ash seedlings (Table 1). Ash seedlings respond to shade by thinning of leaf tissue and spreading of leaf area (Koike 1991). They can survive on forest floors where the relative light intensity is around 7 % of full sunlight; they have a light–photosynthesis curve of the typical "shade leaf", whereas saplings have the "sun leaf" type (Koike et al. 1998). This change of photosynthetic rate with growth enables the ash to grow rapidly when the proper light environment is encountered. Ash can tolerate shade for more than 20 years while awaiting its chance to grow fast (Nakae et al. 1961).

Ash regenerates mainly by seedlings rather than by sprouting. Multi-stemmed trees account for as few as 10 % of trees in stands, and most of them have two trunks: Goto et al. (2002) suggested that when a terminal bud of a sapling is damaged, two lateral buds respond and grow.

15.4.3 Elm (*Ulmus davidiana* var. *japonica*)

Elm flowers in early spring and disperses its seeds in early summer. At the forest edge, in small gaps, and at bare sites formed after large disturbances, seeds germinate immediately after dispersion. When seeds land where light is insufficient, such as in the forest understory, they lie dormant and germinate in the following spring, when the canopy leaves have not emerged and enough light still reaches the forest floor (Seiwa 1997). Litter accumulation also inhibits seedling emergence (Seiwa 1997).

Elm seedlings have a light compensation point intermediate between those of alder and ash (Table 1). Elm seedlings grow well; for example, two-year-old seedlings reached 90 cm in height, whereas two-year-old alders reached 27 cm in a nursery (Kubota 1979). These features suggest that Japanese elm awaits gap formation or sediment accumulation caused by flooding or a landslide (Kon & Okitsu 1995). In unstable habitats with frequent disturbance, elm grows in the shrub layer (Fujita 2002; Kon & Okitsu 1995), but in stable habitats without disturbance, it grows exclusively in the canopy layer (Kon & Okitsu 1995; Haruki et al. 1992). The longevity of elm is as long as 180 years (Watanabe 1994), so elm trees in stable habitats can await their chance for regeneration until the next large disturbance that generates canopy gaps or causes sediment accumulation (Kon & Okitsu 1995).

15.5 Conclusions

River behavior causes landform differentiation and thus creates various sites. The sites' topography and soil deposits are the main determinants of the water level regime. Light conditions and disturbance are also important factors in determining the distribution of tree species. Tolerance to flooding differs among swamp species, as reflected in each species' habitat. Alder adapts to anaerobic wet conditions and grows at extremely wet sites. Ash shows intermediate tolerance to flooding, but high tolerance to low light, and grows at intermediate sites. Elm grows at mesic sites without stagnant water. It is adapted to large disturbances that generate canopy gaps and coarse soil deposits, which determine seed germination.

References

Abe N (1989) Classification of site productivity for ash (*Fraxinus mandshurica* var. *japonica*) plantation. Koshunai-Kiho, 74:3–7 (In Japanese)

Abe N, Goshu K, Usui G (1989) Height growth response of the ground water and of the vegetation of *Fraxinus mandshurica* var. *japonica* Maxim plantation. Trans Hokkaido Branch Jpn For Soc 37:128-130 (In Japanese)

Fujita H (2001) Germination test of *Alnus japonica* (Thunb.) Steud. on the assumption of the field condition. In: Association to Commemorate the Retirement of Prof. Dr. Shigetoshi Okuda, Yokohama (eds) Papers in Commemoration of Prof. Dr. Shigetoshi Okuda's retirement: Studies on the vegetation of alluvial plain, pp 33-36 (In Japanese with English summary)

Fujita H (2002) Swamp forest. In: Sakio H, Yamamoto F (eds) Ecology of Riparian Forests. Univ Tokyo Press, Tokyo, pp 95–137 (In Japanese)

Fujita H, Kikuchi T. (1986) Differences in soil condition of alder and neighboring elm stands in a small tributary basin. Jpn J Ecol 35:565–573.

Goto S, Takahashi Y, Kasahara H, Inukai M, Matsui M (2002) Sex expressions and ability of sprouting of a dioecious canopy tree species *Fraxinus mandshurica* var. *japonica*. For Tree Breed Hokkaido 45:18–21 (In Japanese)

Haruki M, Itagaki T, Namikawa K (1992) Forest construction in the Doran River watershed, Nakagawa Experimental Forest, Hokkaido University. Res Bull College Exp For, Hokkaido Univ 49:121–184 (In Japanese with English summary)

Iwanaga F, Yamamoto F (2007) Effects of flooding depth on growth, morphology and photosynthesis in *Alnus japonica* species. DOI 10.1007/s11056-007-9057-4 New For

Joint Research Group of Public Forestry Research Institute (1983) Technique for increasing useful deciduous tree yield, case studies. Forest Agency Japan, Tokyo (In Japanese)

Kadomura H (1981) Valley bottom plain. In: Machida T, Iguchi M, Kaizuka S, Satoh T, Kayane I, Ono Y (eds) Encyclopedia of Geomorphology. Ninomiya-Shoten, Tokyo

Kikuchi K, Shimizu H, Yamada K (1995) Effects of defoliation by insect and soil water content on growth of *Fraxinus mandshurica* var. *japonica*. Trans Hokkaido Branch Jpn For Soc 43:78–80 (In Japanese)

Koike T (1991) Photosynthetic characteristics of deciduous broad-leaved tree species. Tech Rep FFPRI Hokkaido 25:1–8 (In Japanese)

Koike T, Tabuchi R, Takahashi K, Mori S, Lei TT (1998) Characteristics of the light response in seedlings and saplings of two mid-successional species, ash and kalopanax, during the early stage of regeneration in a mature forest. J Sust For 6:73–84

Kon H, Okitsu S (1995) Structure and regeneration of *Ulmus davidiana* var. *japonica* forest at Mt. Asama and Mt. Togakushi, central Japan. Tech Bull Fac Hortic Chiba Univ 49: 99-110 (In Japanese with English summary)

Kubota Y (1979) Regeneration of deciduous hardwood trees by seedlings. Koshunai-Kiho 40:16-26 (In Japanese)

Lawesson JE (2000) Danish deciduous forest types. Plant Ecol 151:199-221.

Makita H, Miyagi T, Miura O, Kikuchi T. (1979) A study of an alder forest and an elm forest with special reference to their geomorphological conditions in a small tributary basin. Bull Yokohama Phytosociol Soc Jpn 16:237-244

Manabe I (1982) Growth of the ash at Shibecha region in the Kyoto University Forest in Hokkaido. Rep Kyoto Univ For 15:127-135 (In Japanese)

Manabe I, Ohkubo M (1971) Seed production and regeneration of ash forests. Trans Hokkaido Branch Jpn For Soc 20:73-75 (In Japanese)

Miyawaki A (1977) Vegetation of Japan. Gakken, Tokyo (In Japanese)

Mori T (1991) Seeds of deciduous broad leaved tree in northern Japan—treatments and afforestation characters. Northern Forestry, Japan, Sapporo (In Japanese)

Murakami K (1977) Ecological studies on the establishment of alder (*Alnus japonica*) forests. M Sci The Grad Sch Sci Tohoku Univ (In Japanese)

Nagasaka A (2001) Effects of flooding on growth and leaf dynamics of two-year-old deciduous tree seedlings under different flooding treatments. Bull Hokkaido For Res Inst 38:47-55 (In Japanese and English summary)

Nakae A (1959) Some knowledge from the field studies on natural ash forests. North For (Hoppo Ringyo) 11:120-123 (In Japanese)

Nakae A, Manabe I (1963) The silvicultural studies on Yachidamo in the Kyoto University Forest in Hokkaido, No. VII: Effects of water content of volcanic ash black soil on the growth of Yachidamo seedlings. Bull Kyoto Univ For 34:32-36 (In Japanese with English summary)

Nakae A, Tatsumi S (1964) The silvicultural studies on Yachidamo in the Kyoto University Forest in Hokkaido, No. VIII: Relation between the physical and chemical properties of soil and growth of Yachidamos in the artificial forest. Bull Kyoto Univ For 35: 157-176 (In Japanese with English summary)

Nakae A, Sakasegawa T, Tatsumi S (1960) The silvicultural studies on Yachidamo in the Kyoto University Forest in Hokkaido, No. I: The fundamental studies on the silviculture of Yachidamo (on the growth of Yachidamo in the natural old forest). Bull Kyoto Univ For 29:33-64 (In Japanese with English summary)

Nakae A, Tatsumi, S, Sakasegawa T (1961) On the silvicultural studies on Yachidamo in the Kyoto University Forest in Hokkaido, No. II: On the stand structure, the growing process and the land plant type in the middle aged forest of Yachidamo. Bull Kyoto Univ For 32:1-20 (In Japanese with English summary)

Nakamura F, Jitsu M, Kameyama S, Mizugaki S (2002) Changes in riparian forests in the Kushiro mire, Japan, associated with stream channelization. River Res Applic 18:65-79

Ohtomo R, Nishimoto T (1981) Relationship between oxygen concentration and growth of *Abies sachalinensis*, *Picea glehnii* and *Larix kaempferi* (1–0 age). Trans Hokkaido Branch Jpn For Soc 29:137-138 (In Japanese)

Parker GR, Schneider G (1974) Structure and edaphic factors of an alder swamp in northern Michigan. Can J For Res 4:499-508

Saito M, Goshu K, Terazawa K (1990) Optimum site for ash tree (*Fraxinus mandshurica* var. *japonica*) judging from height growth. Ann Rep Meet For Techn Hokkaido (1990): 118-119 (In Japanese)

Sasaki C (1985) Periodicity of seed production of broad-leaved trees and shrubs native to the central area of Hokkaido. Trans Hokkaido Branch Jpn For Soc 34:130-132 (In Japanese)

Schnitzler A (1994) European alluvial hardwood forests of large floodplains. J Biogeogr 21:605-623

Seiwa K (1997) Variable regeneration behaviour of *Ulmus davidiana* var. *japonica* in response to disturbance regime for risk spreading. Seed Sci Res 7:195-207

Shiozaki M, Sanada E, Ohta S (1992) Relationship between soil water condition and growth of an artificial stand of *Fraxinus mandshurica* var. *japonica*—A case of the Ashibetsu district forestry office. Trans Hokkaido Branch Jpn For Soc 40:47-49 (In Japanese)

Tatewaki M, Tohyama M, Igarashi T (1967) The forest vegetation on the lake-side of Abashiri, Prov. Kitami, Hokkaido, Japan. Mem Fac Agric Hokkaido Univ 6:283-334 (In Japanese with English summary)

Tsuneya F (1996) Distribution patterns and habitat conditions of the main tree species composing lowland forests in Oyafuru, Hokkaido, Japan. Jpn J Ecol 46:21-30 (In Japanese with English summary)

Turner MG, Gergel SE, Dixon MD, Miller JR (2004) Distribution and abundance of trees in floodplain forests of the Wisconsin River: environmental influences at different scales. J Veg Sci 15:729-738

Watanabe S (1994) Specia of Trees. Univ Tokyo Press, Tokyo (In Japanese)

16   Flooding adaptations of wetland trees

Fumiko IWANAGA, Fukuju YAMAMOTO

Department of Forest Science, Faculty of Agriculture, Tottori University, 4-101 Minami, Koyama, Tottori 680-8553, Japan

16.1 Introduction

A reduction in gas exchange between the air and the rhizosphere causes a major problem for terrestrial plants (Jackson & Drew 1984; Visser & Vosenek 2004). Soil flooding or submergence sets in motion a series of physical, chemical, and biological processes that profoundly influence the quality of soil as a medium for plant growth (Ponnamperuma 1984). In well-drained soils, the stability of the gas composition is maintained by rapid gas exchange between the soil and air, despite oxygen consumption, carbon dioxide production, and nitrogen fixation by soil organisms. In contrast, soil flooding or submergence causes oxygen depletion and carbon dioxide accumulation in the rhizosphere and plants (Jackson & Drew1984; Ponnamperuma 1984; Greenway et al. 2006). These events lead to an energy deficit in plants through the inhibition of aerobic respiration and disturbance of photosynthetic processes. The accumulation of phytotoxic compounds, including reduced forms of iron and manganese, ethanol, lactic acid, acetaldehyde, aliphatic acids, and cyanogenic compounds, is also a major problem for plants (Ponnamperuma 1984). The effects of such compounds on root metabolism cause the inhibition of root growth and development (Jackson & Drew 1984; Ponnamperuma 1984; Armstrong et al. 1996; Armstrong & Armstrong 1999; Pezeshki 2001; Greenway et al. 2006). Thus, vegetation in the peripheral zone of lakes and swamps typically consists of flood-tolerant species that have specific mechanisms to tolerate excessive water.

There is enormous variation in the flood tolerance of bottomland species (Kozlowski & Pallardy 2002). O_2 deficiency in the rhizosphere has negative effects on roots, leading to the injury of aboveground shoot systems, because it reduces the transportation of water, minerals, and plant hormones and the production of energy. Flood-tolerant plant species have anatomical and physiological

Sakio, Tamura (eds) Ecology of Riparian Forests in Japan : Disturbance, Life History, and Regeneration
© Springer 2008

features that allow them to tolerate anaerobic conditions. In this chapter, we discuss plant responses to soil flooding, with an emphasis on the relationship between flooding and the responses of woody species native to wetland forests in the cool temperate regions of Japan, particularly those in Kushiro Mire in Hokkaido.

16.2 Tree growth in flooded areas

Alnus japonica (Thunb.) Steud. and *Fraxinus mandshurica* Rupr. *var. japonica* Maxim. are dominant species in the wetland forests of Kushiro Mire, which is located in the northern part of Japan (Itoh 1987; Fujita & Kikuchi 1984, 1986). A field survey was used to determine the distribution and size of *A. japonica* and *F. mandshurica* trees in wetland forests of Kushiro Mire (Fig. 1B-D). The lowest areas of the study site have reducing soil conditions, as indicated by the low soil redox potential (Eh) and high water table (Fig. 1A). The elevation of the study site descends gradually toward Lake Takkobu, which is connected to the mire. Soil Eh decreased with decreasing ground height. The soil Eh decreased gradually and reached approximately 150 mV at 120 m from the shore edge. Under such deeply flooded conditions, the heights and stem diameters of these trees tended to decrease with the decrease soil Eh (Fig. 1B, C). The decreased development and distribution of *A. japonica* and *F. mandshurica* may be explained by the negative effects of low soil Eh on plant growth and physiology. For instance, soil Eh significantly affects energy conditions through inhibitory effects on various woody plants (DeLaune et al. 1998; Anderson & Pezeshki 2001; Pezeshki 2001). Flooded *Quercus nuttallii*, *Q. michauxii*, and *Taxodium distichum* seedlings show a decrease in photosynthetic rate and stomatal conductance under reducing conditions of approximately –200 mV soil Eh (Anderson & Pezeshki 2001). According to Pennington and Walters (2006), the inhibitory effects of low soil Eh are greater than those of hydrology on the growth and photosynthesis of several woody species. The total number of stems per individual tree of *A. japonica* increased with the decrease in soil Eh (Fig. 1D), whereas *F. mandshurica* completely disappeared from an area that had < 0 mV soil Eh. In controlled experiments using *A. japonica* seedlings, many epicormic shoots were observed on the stems under flooded conditions (Shinshoh 1985). These epicormic shoots were produced from dormant buds on the main stems or branches (Shinshoh 1985). Dormant buds are dominated by the terminal shoot and released by thinning or partial cutting of the shoot (Zimmermann et al. 1971; Kozlowski & Pallardy 1997a).

Suppressed growth in roots and shoots of flooded plants may affect the balance of plant hormones in relation to apical dominance. Under anaerobic conditions that cause declines in shoot and root growth, a regeneration system by coppicing is efficient for the maintenance, persistence, and development of individual trees. Shinshoh (1985) reported that most of the flooded *A. japonica* trees at Kushiro Mire have multiple stems that were presumably derived from epicormic shoots.

The rapid development of epicormic shoots in flooded environments would give a substantial advantage to *A. japonica* for survival.

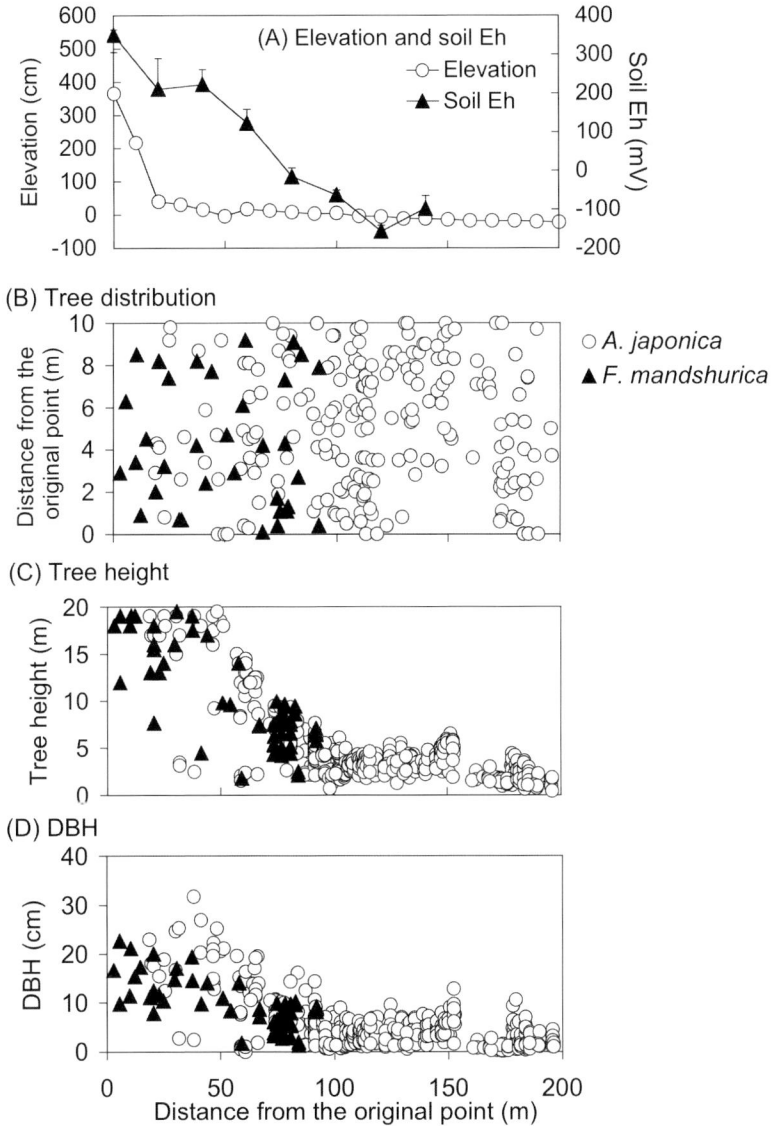

Fig. 1. Elevation, soil redox potential (Eh), (A), tree heights, (B), stem diameters (DBH), (C) and number of stems per individual, (D), in *A. japonica* and *F. mandshurica* trees growing in Kushiro mire. Error bars in figure (A) indicate standard errors (Iwanaga & Yamamoto 2007)

16.3 stem morphology and photosynthetic activity

Flooding stimulates many types of visible changes in the stem near the water level in various woody species (Penfound 1934; SenaGomes & Kozlowski 1980; Hook 1984; Tang & Kozlowski 1984; Harrington 1987; Yamamoto & Kozlowski 1987 a, b; Grosse et al. 1992; Yamamoto et al 1995b), including *A. japonica* (Yamamoto et al 1995a). In flooded *A. japonica* seedlings, stem lenticels on the submerged portions showed hypertrophic development a few days after the initiation of flooding (Figs. 2, 3). Within 2 weeks after the initiation of flooding, numerous adventitious roots had begun to develop. In seedlings flooded at 30 cm in depth, stem growth at the ground level decreased significantly, but increased near the water level (Iwanaga & Yamamoto 2007a). The growth of flood-tolerant species in oxygen-deficient environments requires the development of an extensive internal aeration system that allows the transport of atmospheric oxygen to the roots (Pezeshki 2002). Aerenchyma development is important because it facilitates the diffusion of oxygen to the roots, allowing some aerobic respiration and helping to detoxify the reduced rhizosphere (Pezeshki 2002). Hypertrophied

Fig. 2. Flooded *A. japonica* seedlings: A) flooded at 1cm above the ground level (GL), and B) at 30 cm above the GL, having developed adventitious roots and epicormic shoots. Arrow in each figure indicates water level (Iwanaga & Yamamoto 2007)

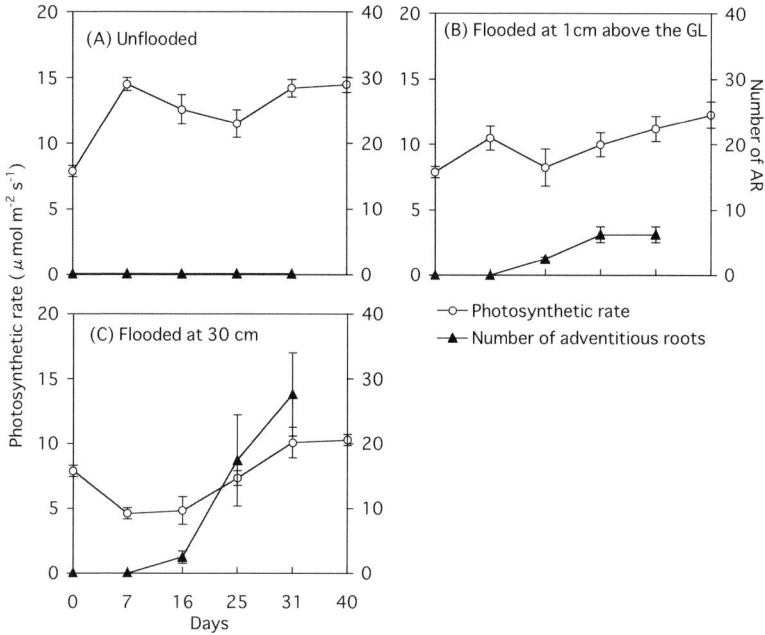

Fig. 3. Changes in photosynthetic rates and number of adventitious roots (AR) in flooded *A. japonica* seedlings; a), unflooded, b), flooded at 1 cm above the ground level, and c), flooded at 30cm. Error bars in each figure indicate S.E. Note the simultaneously occurred adventitious root formation and increase in photosynthetic rates in seedlings flooded at 30 cm depth (Iwanaga & Yamamoto 2007a)

lenticels, adventitious roots, and enlarged stems have abundant aerenchyma tissues. In controlled experiments using *A. japonica* seedlings, the inhibitory effects of flooding on photosynthesis varied with the depth and duration of flooding. Deep flooding greatly inhibited the photosynthetic activity of seedlings in comparison to shallow flooding (Fig. 3). Seedlings flooded to 30 cm in depth had lower photosynthetic rates than did unflooded seedlings within 25 days after the initiation of flooding. A previous study rates showed that there was no significant difference in photosynthetic rate among treatments by the 56th day after flooding began (Iwanaga & Yamamoto 2007a). Various studies have described decreased photosynthetic rates in relation to decreases in stomatal conductance (Jackson & Drew 1984; Kozlowski 1997; Jackson et al. 2003). Our studies of *A. japonica* also indicate a coincidental relationship between the formation of adventitious root and recovery from reduced photosynthetic rates, suggesting the importance of these morphological changes for the survival of *A. japonica* in wetlands.

  The importance of adventitious rooting on growth and photosynthesis in flooded woody plants varies with species (Gill 1975; Tang & Kozlowski 1984;

Terazawa et al. 1989). The removal of adventitious roots from submerged portions of stems reduces height and diameter growth in flooded *Platanus occidentalis* L. seedlings (Tsukahara & Kozlowski 1985). However, it had no significant effect on height growth in flooded *A. glutinosa* seedlings (Gill 1975).

16.4 Seasonal changes in morphology

As indicated above, various changes observed in *A. japonica* seedlings are important characteristics related to the flooding adaptability of this species. Flooded seedlings of *F. mandshurica*, another representative species of wetland forests in Kushiro Mire, exhibit morphological changes similar to those observed in flooded *A. japonica* seedlings (Yamamoto et al. 1995a, b) Figure 4 shows the anatomical characteristics of these seedlings subjected to flooding in three different seasons: April, July, and September. The acceleration of wood production and cell division by flooding in *F. mandshurica* seedlings was observed mainly in April. Further, adventitious root formation in *F. mandshurica* seedlings occurred in April and

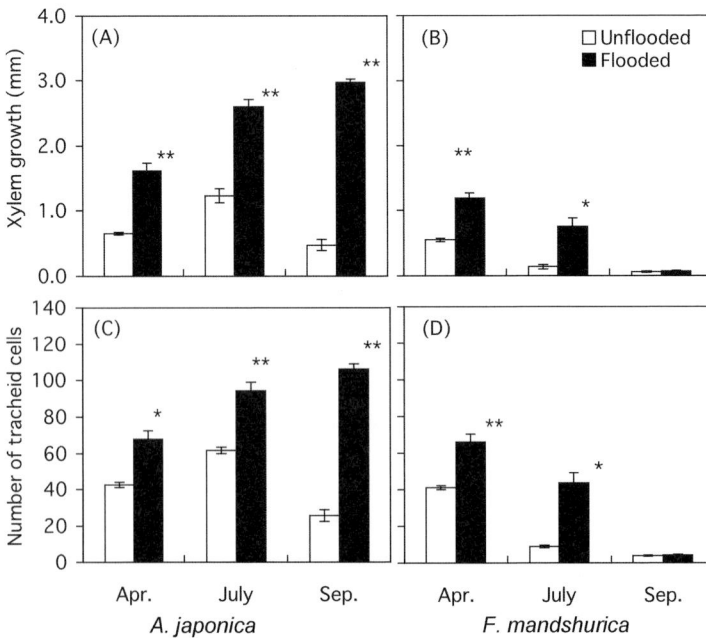

Fig. 4. Anatomical characteristics in *A. japonica* (A, C) and *F. mandshurica* (B, D) seedlings flooded in three different months; April, July, and September. Means with asterisks are significantly different from those of unflooded seedlings using t-test (n=7); *, at p<0.05 and **, p<0.01 (Iwanaga *et al.* unpublished data)

July, but not in September, whereas it was observed in every season in *A. japonica* seedlings. Similarly, there were differences between *A. japonica* and *F. mandshurica* in the seasonal responses of shoot growth to flooding (Figs. 4, 5). The shoot growth of *A. japonica* is indeterminate and occurs successively during the growing season, whereas that of *F. mandshurica* is determinate and limited to the early growing season (Fig. 5).

The different growth patterns in these two species may reflect changes in endogenous levels of auxin, a major growth regulator produced in growing shoots (Kozlowski & Pallardy 1997b), because expanding buds and leaves are regarded as important sources of endogenous auxin (Little & Pharis 1995; Kozlowski & Pallardy 1997a). Therefore, endogenous levels of cambial auxin related to xylem production are affected by changes in canopy size (Sundberg et al. 1993; Funada et al. 1987, 2001), debudding, defoliation and girdling (Little & Pharis 1995). Decreased auxin levels derived from artificial reductions in canopy size in *Pinus densiflora* Sieb. et Zucc. (Funada et al. 1987, 2001) suggest that the reduction of growing shoots as an auxin source may directly affect xylem production and cambial activity in stems. These investigations suggest that the decrease in activity in terms of adventitious root formation and hypertrophic stem growth in *F. mandshurica* seedlings in September may be explained by the cessation of shoot growth in relation to decreased levels of endogenous auxin in the late stage of the growing season.

16.5 Plant growth regulators in relation to stem morphology

Flooding affects internal factors of plants, including water relations, carbo-hydrates, mineral nutrients, and the production and translocation of endogenous plant hormones (Clements & Pearson 1977; Reid & Bradford 1984; Abeles et al 1992; Angelov et al. 1996; Kozlowski 1997; Vartapetian & Jackson 1997). Internal and external factors act as signals in plants to sense environmental changes and trigger sequences of changes in physiology, anatomy, and morpholo-gy in flooded plants (Visser & Vosenek 2004). Several changes in flooded plants have been attributed to the effects of flood-induced ethylene production, including reduced shoot growth, leaf epinasty, leaf senescence and abscission, hypertrophied lenticel development, and aerenchyma formation (Tang & Kozlowski 1984; Abeles et al. 1992; Kozlowski 1997; Kozlowski & Pallardy 2002).

Oxygen deficiency in the rhizosphere stimulates ethylene production as a consequence of the increase in production of an ethylene precursor (1-aminocyclo-propane-1-carboxylic acid, ACC) and activation enzymes involved in ACC synthesis (Abeles et al 1992). A number of investigations have shown flood-induced ethylene production in relation to morphological changes in various species (Tang & Kozlowski 1984; Abeles et al 1992; Kozlowski 1997; Kozlowski & Pallardy 2002; Visser & Vosenek 2004), including *A. japonica* and *F.*

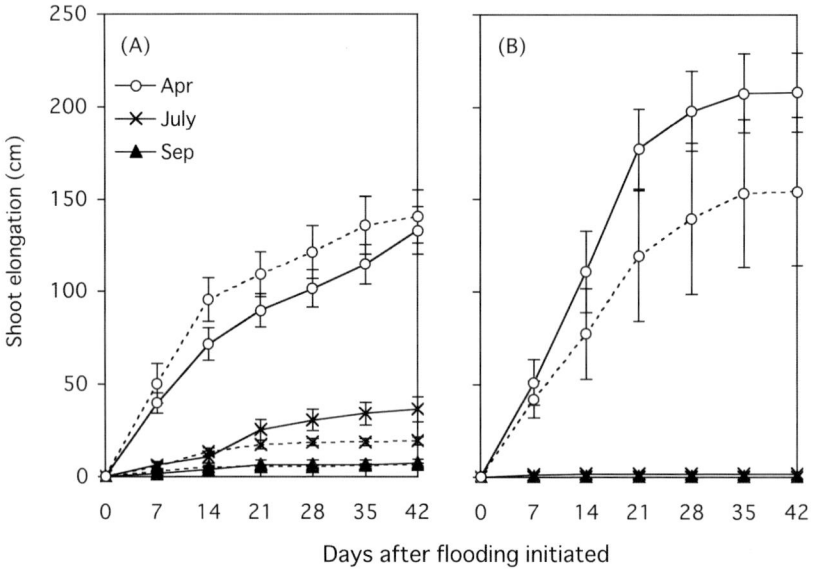

Fig. 5. Shoot growth of *A. japonica* (A) and *F. mandshurica* (B) flooded in 3 different months: April, ○; July,× ; September, ▲. Solid and dotted lines show unflooded and flooded seedlings, respectively. Error bars in each figure indicate SE. (Iwanaga *et al.* unpublished data)

mandshurica (Yamamoto et al. 1995a, b). However, flood-induced ethylene is not the sole factor regulating morphological changes in plants (Abeles et al 1992). Auxin (indole-3-actic acid, IAA) is also essential in adventitious root formation and xylem production in flooded woody plants (Yamamoto & Kozlowski 1987a; Bradford 1984; Vartapetian & Jackson 1997; Visser & Vosenek 2004). Yamamoto and Kozlowski (1987) reported that the application of an inhibitor of auxin translocation (*N*-1-naphtylphthalamic acid, NPA) reduced the number of adventitious roots and wood formation in flooded *Acer negundo* L. seedlings. The application of naphthaleneacetic acid (NAA), an artificially synthesized auxin, to the lower area where NPA had been applied resulted in the recovery of adventitious root formation (Yamamoto & Kozlowski 1987a). Further, some investigations indirectly demonstrated interactive effects of ethylene and auxin on physiological responses in woody plants, including wood formation. For instance, Eklund and Little (2000) emphasized that the application of ethrel, an ethylene-releasing compound, to stems caused IAA accumulation in the cambial region of the application site, indicating the importance of accumulated IAA in ethrel-induced wood formation in *Abies balsamea*. These investigations suggest that the accumulation of auxin caused by flood-induced ethylene would be highly related to morphological adaptations such as adventitious root formation and the acceleration of stem growth.

Seasonal differences between *A. japonica* and *F. mandshurica* seedlings in morphological adaptations may be caused by differences in the patterns of shoot growth between the two species. Morphological observations indicated that *A. japonica* maintained shoot development not only in April and July, but also in September. Whereas such modifications in flooded stem portions play important roles in supporting shoot growth during flooding, shoot development also assists submerged plant parts with morphological modifications where oxygen deficiency has occurred, with regard to metabolic and hormonal relations. Thus, the ability of *A. japonica* to develop shoots and adapt morphologically at any time during the growing season suggests that this species is capable of responding to soil flooding. This may support the growth of *A. japonica* in deeply flooded areas and be related to the wider distribution of this species in comparison to *F. mandshurica* in Kushiro Mire. Xylem production and adventitious root formation tended to decline in flooded *F. mandshurica* seedlings with the progression of seasons. The ability to modify the morphology of stems and roots in any season gives *A. japonica* an advantage in surviving soil flooding and may explain the distribution of this species in Kushiro Mire.

References

Abeles FB, Morgan PW, Saltveit ME (1992) Ethylene in plant biology. 2nd edn. Academic press, Inc, San Diego, California

Anderson PH, Pezeshki SR (2001) Effects of flooded pre-conditioning on responses of three bottomland tree species to soil waterlogging. J Plant Physiol 158:227-233

Angelov MN, Sung SS, Doon RL, Harms WR, Kormanik PP, Black CC Jr (1996) Long- and Short-term flooding effects on surviveal and sink-source relationships of swamp-adapted tree species. Tree Physiol 16:477-484

Armstrong J, Armstrong W (1999) *Phragmites* dieback: toxic effects of propionic, butyric and caproic acids in relation to pH. New Phytol 142:201-217

Armstrong J, AfreenZobayed F, Armstrong W (1996) *Phragmites* dieback: sulphide- and acetic acid-induced bud and root death, lignifications, and blockages within aeration and vascular systems. New Phytol 134:601-614

Bradford KJ (1984) Effects of soil flooding on leaf gas exchange of tomato plants. Plant Physiol 73:475-479

Clemens J, Pearson CJ (1977) The effects of waterlogging on the growth and ethylene content of *Eucalyptus robusta* Sm. (swamp mahogany). Oecologia 29:249-255

DeLaune RD, Pezeshki SR, Lindau CW (1998) Influence of soil redox potential on nitrogen uptake and growth of wetland oak seedlings. J Plant Nutri 21:757-768

Eklund L, Little CHA (2000) Transport of [1-14C]-indole-3-acetic acid in *Abies balsamea* shoots ringed with Ethrel. Trees 15:58-62

Fujita H, Kikuchi T (1984) Water table and neighboring elm stands in a small tributary basin. Jpn J Ecol 34:473-475

Fujita H, Kikuchi T (1986) Differences in soil condition of alder and neighboring elm stands in a small tributary basin. Jpn J Ecol 35:565-573

Funada R, Sugiyama T, Kubo T, Fushitani M (1987) Determination of indole-3-acetic acid levels in *Pinus densiflora* using the isotope dilution method. J Jpn Wood Res Soc 33: 83-87

Funada R, Kubo T, Tabuchi M, Sugiyama T, Fushitani M (2001) Seasonal variations in endogenous indole-3-acetic acid and abscisic acid in the cambial region of *Pinus densiflora* Sieb. et Zucc. stems in relation to earlywood-latewood transition and cessation of tracheid production. Holzforschung 55:128-134

Gill CJ (1975) The ecological significance of adventitious rooting as a response to flooding in woody species, with special reference to *Alnus glutinosa* (L.) Gaertn. Flora 164:85-97

Greenway H, Armstrong W, Colmer TD (2006) Conditions leading to high CO_2 (>5kPa) in waterlogged-flooded soils and possible effects on root growth and metabolism. Ann Bot 98:9-32

Grosse W, Frye J, Lattermann S (1992) Root aeration in wetland trees by pressurized gas transport. Tree Physiol 10:285-295

Harrington CA (1987) Responses of red alder and black cottonwood seedlings to flooding. Physiol Planta 69:35-48

Hook DD (1984) Adaptation to flooding with fresh water. In: Kozlowski TT (ed) Flooding and plant growth. Academic press, Inc, Orland, pp 265-294

Itoh K (1987) Vegetation of Hokkaido (in Japanese). Hokkaido University Press, Sapporo

Iwanaga F, Yamamoto F (2007a) Effect of flooding depth on growth, morphology and photosynthesis in *Alnus japonica* species. DOI 10.1007/s11056-007-9057-4 New Forests

Iwanaga F, Yamamoto F (2007b) Growth, morphology and photosynthetic activity in flooded *Alnus japonica* seedlings. J For Res 12:243-246

Jackson MB, Drew MC (1984) Effects of flooding on growth and metabolism of herbaceous plants. In: Kozlowski TT (ed) Flooding and plant growth. Academic Press, Inc, Orland, pp 47-128

Jackson MB, Saker LR, Crisp CM, Else MA, Jonowiak F (2003) Ionic and pH signaling fro root to shoots of flooded tomat plants in relations to stomatal closure. Plant Soil 253:103-113

Kozlowski TT (1997) Responses of woody plants to flooding and salinity. Tree physiol monograph No.1. Heron Pub, Victoria, Canada

Kozlowski TT, Pallardy SG (1997a) Growth control in woody plants. Academic press, Inc, San Diego, Tokyo

Kozlowski TT, Pallardy SG (1997b) Physiology of woody plants. 2nd edn. Academic press, Inc, San Diego

Kozlowski TT, Pallardy SG (2002) Responses of woody plants to environmental stresses. Bot Rev 68:270-334

Little CHA, Pharis RP (1995) Hormonal control of radial and longitudinal growth in the tree stem. In: Gartner BL (ed) Plant stems: physiology and functional morphology. Academic Press, Inc, San Diego, Tokyo, pp 281-319

Penfound W (1934) Comparative structure of the wood in the 'knees', swollen bases, and normal trunks of the tupelo gum (*Nyssa aquatica* L.). Am J Bot 21:623-631

Pennington MR, Walters MB (2006) The response of planted trees to vegetation zonation and soil redox potential in created wetlands. For Ecol Manage 233:1-10

Pezeshki SR (2001) Wetland plant responses to soil flooding. Env Exp Bot 46:299-312

Pezeshki SR (2002) Physiological ecology of trees in floodplain forests. In: Sakio H, Yamamoto F (eds) Ecology of riparian forests. Univ Tokyo Press, Tokyo, pp 169-189

Ponnamperuma FN (1984) Effects of flooding on soils. In: Kozlowski TT (ed) Flooding and plant growth. Academic press, Inc, Orland, pp 10-46

Reid DM, Bradford KJ (1984) Effects of flooding on hormone relations. In: Kozlowski TT (ed) Flooding and plant growth. Academic Press, Inc, Orland, pp 195-220

SenaGomes AR, Kozlowski TT (1980) Growth responses and adaptations of *Fraxinus pennsylvanica* seedlings to flooding. Plant Physiol 66:267-271

Shinshoh H (1985) Alder forests in Kushio Shitsugen marshes. Northern Forestry Japan 37:92-97 (in Japanese)

Sundberg B, Ericsson A, Little CHA, Nasholm T, Gref R (1993) The relationship between crown size and ring width in *Pinus sylvestris* L. stems: dependence on indole-3-acetic acid, carbohydrate and nitrogen in the cambial region. Tree Physiol 12: 347-362

Tang ZC, Kozlowski TT (1984) Water relations, ethylene production, and morphological adaptation of *Fraxinus pennsylvanica* seedlings to flooding. Plant Soil 77:183-192

Terazawa K, Seiwa K, Usui G, Kikuzawa K (1989) Response of some deciduous tree seedlings under water saturated soil condition (I) Growth, and morphological changes of stem and root (in Japanese). In: Abstracts of the 100th Transmeeting of Japanese Forest Society, University of Tokyo, Tokyo, pp 3-5 April 1989

Tsukahara H, Kozlowski TT (1985) Importance of adventitious root to growth of flooded *Platanus occidentalis* seedlings. Plant Soil 88:123-132

Vartapetian BB, Jackson MB (1997) Plant adaptation to anaerobic stress. Ann Bot 79 (Suppl A): pp 3-20

Visser EJW, Voesenek LACJ (2004) Acclimation to soil flooding; sensing and signal-transduction. Plant Soil 254:197-214

Yamamoto F, Kozlowski TT (1987a) Regulation by auxin and ethylene of responses of *Acer negundo* seedlings to flooding of soil. Env Exp Bot 27:329-340

Yamamoto F, Kozlowski TT (1987b) Effects of flooding on growth, stem anatomy and ethylene production of *Pinus halepensis* seedlings. Can J For Res 17:69-79

Yamamoto F, Sakata T, Terazawa K (1995a) Growth, morphology, stem anatomy and ethylene production in flooded *Alnus japonica* seedlings. IAWA J 16:47-59

Yamamoto F, Sakata T, Terazawa K (1995b) Physiological, morphological and anatomical responses of *Fraxinus mandshurica* seedlings to flooding. Tree Physiol 15:713-719

Zimmermann MH, Brown CL, Tyree MT (1971) Primary growth. In: Trees, structure and function; with a chapter on irreversible thermodynamics of transport phenomena. Springer-Verlag, New York Berlin, pp 1-60

Part 8

Species diversity of riparian forests

17 Diversity of tree species in mountain riparian forest in relation to disturbance-mediated microtopography

Takashi MASAKI[1], Katsuhiro OSUMI[2], Kazuhiko HOSHIZAKI[3], Daisuke HOSINO[4], Kazunori TAKAHASHI[1], Kenji MATSUNE[5], and Wajiro SUZUKI[1]

[1]Forestry and Forest Products Research Institute, 1 Matsunosato, Tsukuba, Ibaraki 305-8687, Japan

[2]Kansai Research Center, Forestry and Forest Products Research Institute, Kyoto 612-0855, Japan

[3]Akita Prefectural University, Akita 010-0195, Japan

[4]Tohoku Research Center, Forestry and Forest Products Research Institute, Morioka 020-0123, Japan

[5]Sumitomo Forestry Co., Ltd. Ibaraki 300-2646, Japan

17.1 Introduction

17.1.1 Riparian forests with high species richness

In riparian forests, concurrent disturbances of the canopy and ground is thought to be responsible for the high hardwood and herb species diversity, both in terms of species richness as well as equitability (Baker 1990; Nillson et al. 1989; Pabst & Spies 1999).

The riparian forests of northern Honshu Island in Japan are also characterized by a high diversity of tree species (Suzuki et al. 2002). Fig. 1 shows an example of a species-area curve for riparian forest in the Kanumazawa Riparian Research Forest (KRRF) (Hoshizaki et al. 1997, 1999; Masaki et al. 1999, 2005; Suzuki et al. 2002), compared to other temperate forests on Honshu Island for similar climatic conditions to those of KRRF (cf. Hara et al. 1995; Masaki 2002; Masaki et al. 1999; Nakashizuka & Numata 1982a, 1982b). The KRRF was observed to

Sakio, Tamura (eds) Ecology of Riparian Forests in Japan : Disturbance, Life History, and Regeneration
© Springer 2008

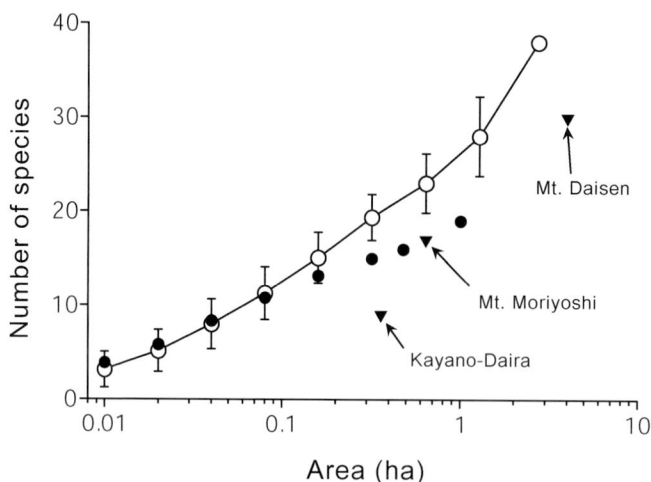

Fig. 1. The number of tree species (for stems ≥5 cm DBH) in a temperate riparian forest (KRRF: open circle) and other temperate non-riparian forests (closed symbols) as a function of stand area. Closed circles represent the number of species at the upper terrace KRRF. Closed triangles are based on the reports by Nakashizuka and Numata (1982a, 1982b) and Hata et al. (1995). Vertical bars show standard errors

have a higher associated species richness than non-riparian forests at spatial scales greater than 0.1 ha. The majority of the non-riparian forests examined had a beech (*Fagus crenata* Blume) and an oak (*Quercus crispula* Blume) as dominant species, and were also characterized as having a lower overall richness of tree species. Conversely, riparian forests were characterized as having a greater diversity of species, including *Cercidiphyllum japonicum* Sieb. et Zucc., *Aesculus turbinata* Blume, *Pterocarya rhoifolia* Sieb. et Zucc., in the KRRF (Suzuki et al. 2002) as well as elsewhere (e.g., Kaneko et al. 1999; and also see Chapter 6).

One possible explanation for the higher species richness in riparian forests is the greater heterogeneity of the riparian forest microenvironment, which provides riparian species with a variety of specific habitats to which they are better adapted. While this is a deterministic interpretation underlying the high species richness, a number of stochastic mechanisms are also possible. For example, multiple and complex disturbance regimes, such as canopy gaps and flooding may provide less adapted species the opportunity to recruit (cf. Chesson 1986; Masaki et al. 2007).

Among these two mechanisms, in this chapter, we examine the following deterministic hypothesis: heterogeneous microenvironments should facilitate the coexistence of numerous species in riparian forests, as well as be responsible for the higher diversity of tree species *in situ*. The data for the analyses was obtained during tree censuses in the KRRF.

17.1.2 Diversity at different scales

Given that species richness is usually dependent on spatial scale (e.g., Pollock et al. 1998), we examined diversity patterns at two spatial scales. One is at the local stand-level while the other is on a larger scale, such as at the landscape level. Diversity at the larger level refers to a change in the species assembly along the gradient from the riparian to the higher altitude area (i.e., gamma diversity). This will provide us with the information on how riparian ecosystems function as species pools in the landscape. Conversely, diversity at the local level means how tree species differently colonize microhabitats within riparian areas (i.e., alpha diversity). To examine species diversity at these different scales, the large-plot studies covering this wide range of habitats within the KRRF have provided useful information.

Previously, Suzuki et al. (2002) briefly analyzed species diversity within KRRF at these spatial scales. They showed that the high species richness in the KRRF was largely attributed to greater number of riparian species and the increased occurrence of locally rare species. In addition, many of the species that prefer xeric environments, such as the upper terraces, also occurred in the riparian area. However, these authors restricted their analysis of tree communities to only include trees of a certain minimum size (diameter at breast height (DBH) ≥5 cm). In forests, difference in species characteristics that will affect mature tree distribution will often occur during the seedling and sapling stages of tree life history. Masaki et al. (2005) reported that water stress experienced by seedlings on the upper terraces often inhibited growth and survival of species that prefer riparian habitats (e.g., *Acer mono* Maxim), and that this would have a lasting influence on the distribution of the mature trees of these species. It is therefore important to consider the younger stages in analyses of diversity.

In addition, in the study of Suzuki et al. (2002), classification of the microtopographic characteristics within the riparian region was too simplified. The authors only distinguished between two types of microtopographic sites based on relative elevation above water level. However, in practice, lower elevation sites can be further divided into several types depending on the occurrence of the last river disturbance (i.e., flooding, erosion, sedimentation, etc.), each of which reflect the stability of the forest floor and developmental stage of organic soil. Such heterogeneity in forest floor types can often affect the success of seedling establishment in riparian forests (Kubo et al. 2000, 2004; Sakio 1997; Sakio et al. 2002), and may consequently promote coexistence among species in riparian forests. It is thus necessary to classify the microtopography of riparian forests more comprehensively in analyses of the relationships between the diverse species characteristic of this environment. Consequently, in this chapter, we analyze community structure by also considering the earlier stages of life history, and by characterizing microtopographic sites in greater detail.

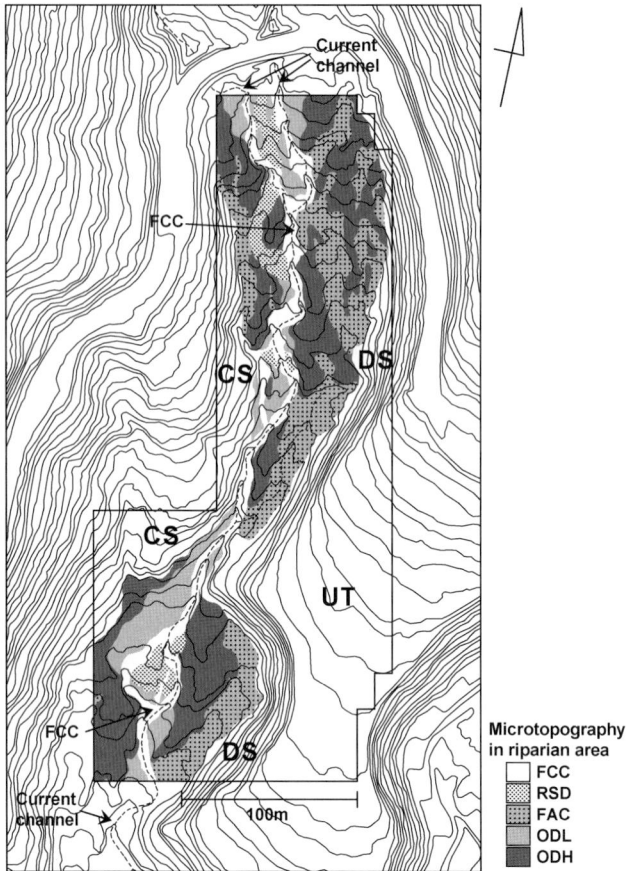

Fig. 2. The topography of KRRF. The 4.71-ha plot is shown by the polygon in solid line. The contour interval is 2 m. The dashed line represents the current channel. Abbreviations are as follows: FCC: floodplain around the current channel, RSD: recent sediment deposition, FAC: floodplain around the abandoned channel, ODL: older deposition at lower elevation, ODH: older deposition at higher elevation, CS: colluvial slope, DS: denudation slope, UT: upper terrace

17.2 Methods

17.2.1 Census methods

The study plot (4.71 ha; Fig. 2) was established in KRRF (39°N, 141°E, 400 to 460 m a.s.l.) in 1993. Mean annual temperature is 9.2 °C, while mean annual precipitation is 2060 mm and a maximum snow depth in winter of up to 2 m. This area has never been logged and appears to exhibit the mature conditions associated with old-growth forests. The plot includes a riparian area and an adjacent upper terrace with a relief distance of ca. 20 m. These two major topographic units are separated by a denudation slope. The other side of riparian area is bounded by a colluvial slope.

The dominant species within the riparian area of this forest is *Cercidiphyllum japonicum* (26% of total basal area at breast height), followed by *Aesculus turbinata* (19%), *Fagus crenata* (15%), *Quercus crispula* (13%), *Pterocarya rhoifolia* (9%), and *Acer mono* (8%) (Suzuki et al. 2002). All of these species are tall, deciduous trees. In terms of maximum size, *C. japonicum* reaches >150-cm DBH, followed by *A. turbinata*, *F. crenata* and *Q. crispula* (100 to 120-cm DBH) and *P. rhoifolia* and *A. mono* (ca. 80-cm DBH). Of these, *F. crenata* and *Q. crispula* were distributed in both the riparian and upper terrace areas, while other species were specific to the riparian area.

In this plot, censuses were conducted on several occasions for trees with DBH ≥5 cm (referred as 'mature trees' for convenience) during the eleven years from 1993 to 2003. The mean DBH was calculated for each of the standing trees for the period that the tree was alive, and transformed into basal area (BA) using the equation: $BA = DBH^2\pi/4$. For the census of saplings, 315 quadrats (2×2 m^2) were made in 10×10 m grids of the plot: 192 for the riparian area, 110 for the upper terrace, and 13 for the slopes. In this study, a sapling was defined as a tree with stem length ≥10 cm and DBH <5 cm. The number of trees was used to create an index of the specific dominance at the sapling stage because most of the saplings did not reach breast height.

17.2.2 Topography and microtopography

As mentioned above, four major topographic units were recognized: riparian area (RA), colluvial slope (CS), denudation slope (DS), and upper terrace (UT) (Fig. 2). The upper terrace is drier than the riparian area, with the summer soil matrix potentials at a 10-cm depth from the surface decreasing to -0.14 MPa in the riparian area and -0.25 MPa in the upper terrace (Masaki et al. 2005).

Within the riparian area, identification of microtopographic sites was also visually assessed and as a result, five microtopographic units were recognized: a floodplain around the current channel (FCC), a floodplain around an abandoned channel (FAC), recent sediment deposition (RSD), and older deposits. Older deposits were also classified into those at a lower elevation (ca. <2 m above water level (a.w.l.); ODL) and those at a higher elevation (ca. 2-4 m a.w.l.) (ODH), which were identified visually by apparent steps. To summarize, in order of ascending elevation above water level, microtopographic sites within the plot were as follows: FCC, RSD, FAC, ODL, ODH, CS, DS, and UT.

17.2.3 Analysis

To test for any distribution bias to any topographic or microtopographic sites for each species, parametric tests such as the χ^2-test are not useful because the carrying capacity of these sites differ with respect to each other (see Fig. 7 below). Consequently, the null hypothesis that the probability of a species occurring is proportional to the area of each (micro) topographic site is thus not valid.

We therefore permutated species randomly among tagged stems. This procedure was repeated 10,000 times and the confidence interval of expected dominance was estimated at each site. For representing specific dominance, basal area was used for mature tree populations and the number of stems was used for sapling populations.

17.3 Diversity on a larger scale

17.3.1 Topographic preference of mature trees

Based on the distribution patterns of mature trees, species were categorized into the four representative categories shown in Fig. 3. Certain species, such as *Cercidiphyllum japonicum* and *Pterocarya rhoifolia*, exhibited a biased distribution toward the riparian area. *Aesculus turbinate* and several other species preferred both the riparian area and the colluvial slope. As in other riparian forests

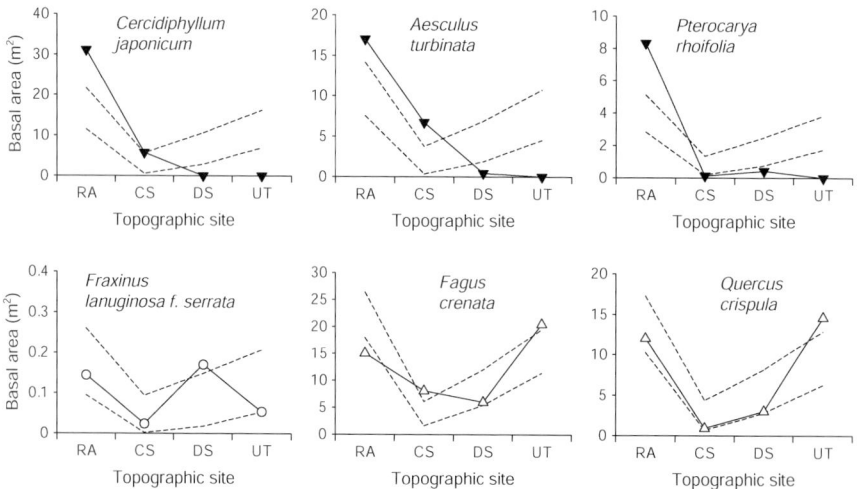

Fig. 3. Basal area of some representative species at each of the topographic sites within the plot. Data with solid lines are observed values and those with dashed lines represent 95% confidence envelopes. Abbreviations are as follows: RA: riparian area, CS: colluvial slope, DS: denudation slope, UT: upper terrace

Table 1. Species ordination based on preference of mature trees and saplings for topographic sites

Preference of mature trees	Preference of saplings				
	RA	CS	DS	UT	No significance
RA	Acer mono, Ulmus laciniata, Cercidiphyllum japonicum, Zelkova serrata, Pterocarya rhoifolia, Aesculus turzinata, Kalopanax pictus, Magnolia obovata, Morus bombycis, Acer palmatum var. amoenum, Swida controversa	-	-	-	-
CS	Euonymus oxyphyllus	-	-	-	Acer nipponicum, Tilia maximowicziana
DS	Fraxinus lanuginosa, Carpinus laxiflora	-	-	-	Sorbus alnifolia, Prunus sargentii, Alnus pendula
UT	Quercus crispula	Acanthopanax sciadophylloide	Acer distylum, Hamamelis japonica var. obtusata	Prunus grayana, Magnolia salicifolia, Acer japonicum, Rhus trichocarpa, Clethra barvinervis	Acer sieboldianum, Acer micranthum, Fagus crenata
No siginificance (rare species)	Clerodendrum trichotomum	Benthamidia japonica	-	-	Castanea crenata, Euonymus alatus f. ciliatodentatus, Ilex macropoda, Phellodendron amurense, Styrax obassia, Salix caprea, Alnus hirsuta var. sibirica, Betula maximowicziana, Acer rufinerve, Tilia japonica, Salix dolichostyla, Aralia elata, Sorbus commixta, Euonymus macropterus

in Japan (e.g., Kaneko 1999), these above species appeared to be specific to the riparian area and were collectively referred to as the RA-preferring species. Other species, such as *Fraxinus lanuginosa* Koidz. preferred the denudation slope (DS-

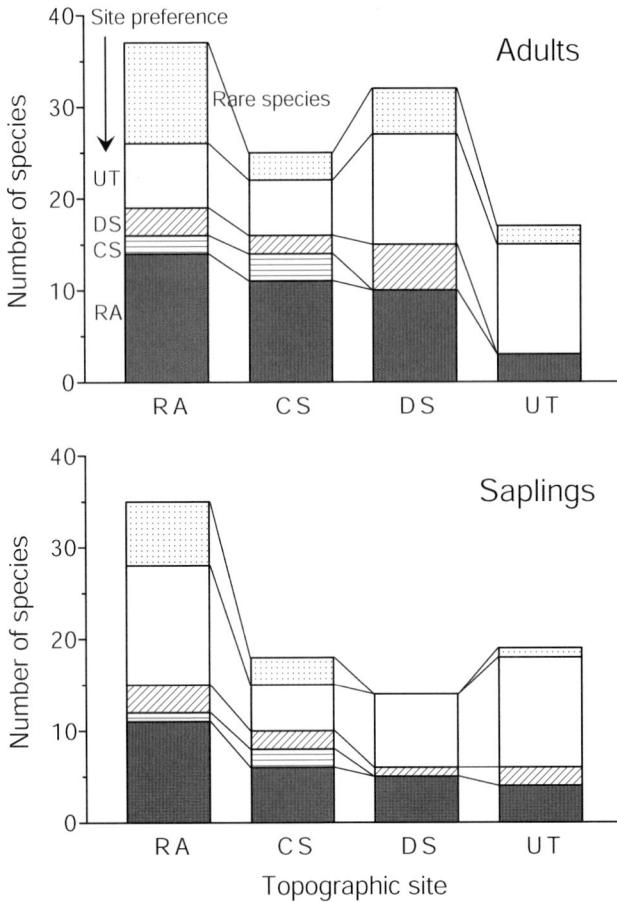

Fig. 4. The number of species at each of the topographic sites within the plot for mature trees (≥5-cm DBH; upper) and saplings (≥10-cm length; lower). The data are shown for each of the distribution types. Abbreviations for topographic sites are given in the caption for Fig. 3

preferring species), while others such as *Fagus crenata* and *Quercus crispula*, preferred the upland upper terraces (UT-preferring species).

Based on this analysis, specific preference could be summarized as shown in Table 1. The table shows that eleven of forty-seven species were RA-preferring. For other categories of distribution, there were three CS-preferring species, five DS-preferring species, and 12 UT-preferring species.

The other 16 species did not exhibit any significant bias ($p>0.05$). Since none of the species had basal areas exceeding 0.07 m^2 ha^{-1} (0.18 % of the total) in this plot, these sixteen species were considered to be too scarce to test for distribution patterns (i.e. rare species).

17.3.2 Occurrence of mature trees

Using this classification of species, the occurrence of species on each topographic site was examined. As shown in Fig. 4, the number of species at the mature tree stage was highest in the riparian area and lowest on the upper terraces. DS had a comparable number of species compared to RA, even though the sample area of DS was smaller than RA. However, as discussed below, DS was characterized by having the lowest basal area of all the sites due to repeated disturbance by denudation attributed to the gliding action of snow.

The riparian area was characterized by an abundance of rare species as well as by RA-preferring species. Seven of the UT-preferring species inhabited the RA. Conversely, the upper terrace was characterized mostly by UT-preferring species. Only three RA-preferring species inhabited the upper terrace. In addition, species preferring slopes were found to inhabit the riparian area but not the upper terrace. These patterns appear, in part, to be caused by the sporadic removal of evergreen shrubs and dwarf bamboos by river disturbances in the RA. These results mean that the riparian area functions as the species pool for this region.

17.3.3 Topographic preferences of saplings

We then tested whether the observed site preferences of mature trees could be extended to the distribution of saplings (Fig. 5, Table 1). It is clearly shown that saplings of RA-preferring species are more commonly found in the riparian area.

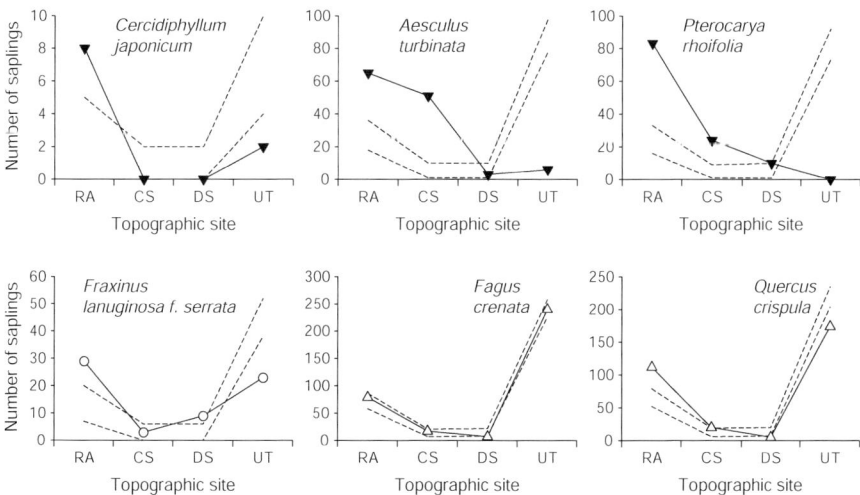

Fig. 5. Number of saplings for several representative species at each of the topographic sites within the plot. Data with solid lines are observed values and those with dashed lines represent 95% confidence envelopes. Abbreviations of topographic sites are the same as those in Fig. 3

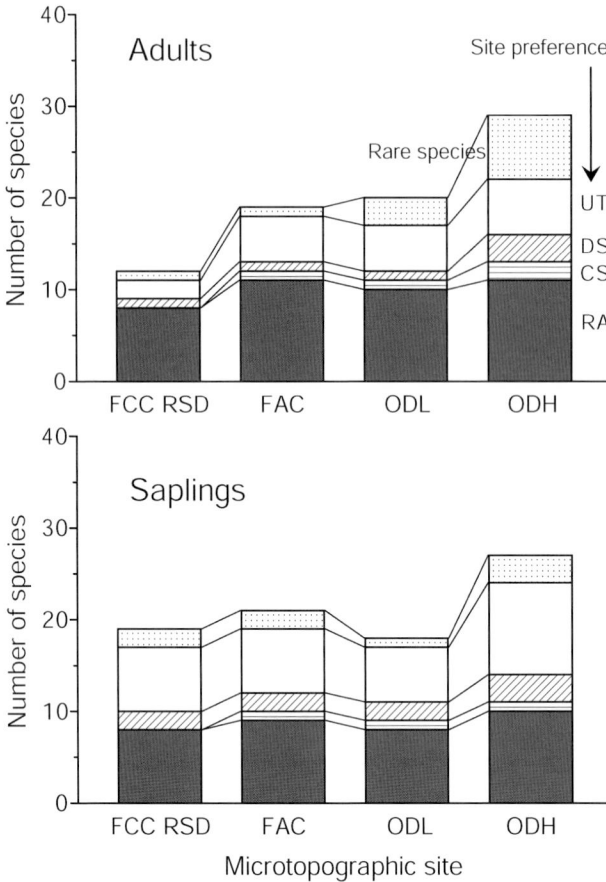

Fig. 6. The number of species at each of the microtopographic sites within the riparian area of the plot for mature trees and saplings. The data are shown for each of distribution types. Abbreviations for topographic sites are given in the caption for Fig. 2

Meanwhile, saplings of UT-preferring species varied in their preference at the sapling stage. For example, saplings of *Quercus crispula* were more commonly found in the riparian area, while distribution of *Fagus crenata* appeared to be a site-generalist. As a result, more than half of these species did not exhibit any significant bias to the upper terrace ($p > 0.05$).

This means that the saplings of many species prefer mesic sites to xeric sites. This is probably because, as saplings grow into mature trees, they will be selected by factors such as soil condition, resulting in a more apparent differentiation with respect to site preference in mature trees. The riparian area is likely to have favorable conditions for the saplings of most tree species. However, during the latter life-history stages of the trees, successful RA-preferring species may out-

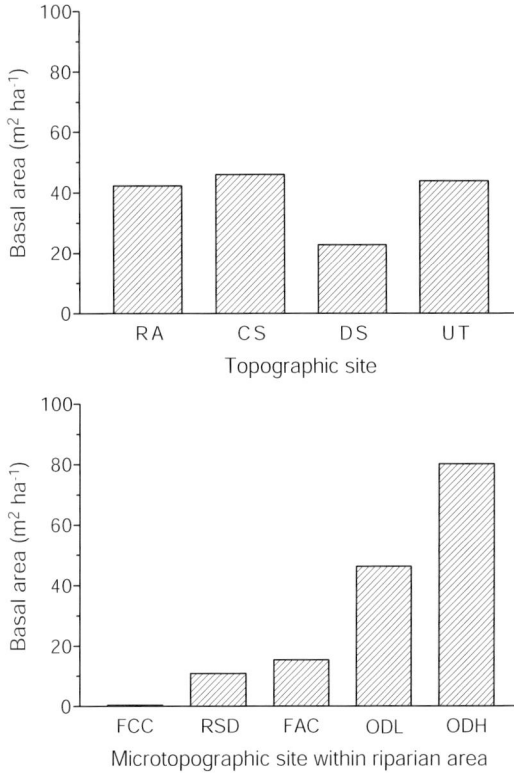

Fig. 7. Basal area of standing trees at each of topographic sites (upper) and of microtopographic sites (lower). See caption of Fig. 2 and 3 for the abbreviations of sites

compete, by superior growth, survival, resistance to external disturbance, etc., other species in the riparian area to a certain degree, resulting in niche differentiation in mature trees.

17.4 Diversity at the local stand level

17.4.1 Occurrence of hardwood species within the riparian area

The next theme is how tree species coexist within the riparian area. The occurrences of species based on the distribution categories are shown in Fig. 6, with similar patterns observed in mature trees and saplings. The number of species increased monotonically with the elevation above water level. The lowest sites (i.e., FCC and RSD) had the least number of species (12 species) while the highest site had a three-fold increase in the number of species.

The RA-preferring species had an almost even distribution at any of the micro-topographic sites surveyed within the riparian area. Conversely, UT- and DS-preferring species, as well as rare species tended to occur in greater numbers at more stable sites. Thus, riparian-specific species occur anywhere within the riparian area, but species specific to other topographic sites tended to inhabit relatively more stable sites.

17.4.2 Specific microtopographic preferences

Specific differences in microtopographic site-preference, and whether such differences, if they existed, could fully explain the patterns described above, were also investigated. Before assessing site preference however, we examined the stability and productivity of microtopographic sites using basal area (Fig. 7).

The BA for the entire riparian area (40 to 45 m^2 ha^{-1}) was comparable to other topographic sites, except for DS where BA was only half that observed for the others (due to repeated surface disturbance and soil denudation, caused primarily by the glide of heavy snow). Conversely, BA was very heterogeneous within the riparian area. BA increased monotonically with elevation sequence from FCC (<0.5 m^2 ha^{-1}) to ODH (nearly 80 m^2 ha^{-1}), implying that site stability at ODH was relatively high. At the same time, this also suggests greater productivity at ODH because for UT, which is also free from channel disturbance, had a BA of only ca. 40 m^2 ha^{-1}. Thus, the riparian area had a mosaic distribution of microtopographic sites, each of which differed with respect to stability and productivity (e.g., Pabst & Spies 1999).

Tables 2 and 3 show the distribution bias for each life-history stage of RA- and UT-preferring species, respectively. Other categoeis (CS- and DS-preferring species) did not exhibit any specific preferences to any of the microtopographic sites. Rare species, which were mostly found in the riparian area, exhibited no significant bias except for *Alnus hirsuta* Turcz. var. *sibirica* (Fischer) C. K. Schn. (more mature trees at FCC and RSD), and *Clerodendrum trichotomum* Thunb. and *Salix caprea* L., formerly *Salix bakko* Kimura (more saplings at ODL).

For mature trees of the RA-preferring species, some exhibited a biased distribution toward ODH, while others showed no significant bias. It appeared that the occurrence of RA-preferring species was not significantly biased toward unstable microtopographic sites such as FCC, FAC and RSD. However, the distribution of saplings exhibited markedly different patterns, with some preferring less stable sites and others preferred more stable sites. Niche differentiation appeared to exist at the sapling stages among the RA-preferring species. Potentially, this has long term effects on the coexistence of RA-species within the riparian area, but destructive disturbances at unstable microtopographic sites can disrupt this pattern and consequently, result in a greater number of mature trees colonizing more stable microtopographic sites. This shift in the distribution of RA-species was analogous to those exhibited by UT-preferring species in larger-scale analyses (Table 1).

Table 2. Ordination of RA-preferring species based on preference of mature trees and saplings for microtopographic sites

Preference of mature trees	Preference of saplings				
	FCC, RSD	FAC	ODL	ODH	No significance
FCC, RSD	-	-	-	-	-
FAC	-	-	-	-	-
ODL	-	-	-	-	-
ODH	*Acer mono*	-	*Cercidiphyllum japonicum*	-	*Kalopanax pictus*
No siginificance	*Zelkova serrata*	*Morus australis, Acer palmatum var. amoenum*	*Ulmus laciniata, Aesculus turbinata*	*Magnolia obovata*	*Pterocarya rhoifolia*

Table 3. Ordination of UT-preferring species based on preference of mature trees and saplings for microtopographic sites

Preference of mature trees	Preference of saplings				
	FCC, RSD	FAC	ODL	ODH	No significance
FCC, RSD	-	-	-	-	-
FAC	-	-	-	-	-
ODL	-	-	-	-	-
ODH	-	-	-	*Quercus crispula*	*Acer japonicum, Fagus crenata, Hamamelis japonica var. obtusata*
No siginificance	-	-	-	*Acer micranthum*	*Prunus grayana, Acanthopanax sciadophylloides, Acer sieboldianum, Magnolia salicifolia, Acer distylum, Clethra barvinervis, Rhus trichocarpa*

On the other hand, saplings of UT-species exhibited no significant preference ($p>0.05$) to any of the microtopographic sites, except for *Quercus crispula* and *Acer micranthum* Sieb. et Zucc. which had more saplings at the ODH. For mature trees, four species exhibited a significant bias toward ODH, while others did not. Thus, UT-species appeared to be generalists with respect to microtopography during the early stages of their life history, subsequently being more specific with respect to habitat preference as some of these species exhibit a bias toward ODH.

17.5 Conservation of diversity at different scales

In the riparian forest described here, riparian disturbances formed a micro-topographic mosaic, and provided various sites that were favorable for species within the riparian area. It has been reported that riparian forests experience both canopy disturbance by tree deaths and ground disturbance (flooding, erosion, deposition, of debris, etc.), create a diverse mosaic of vegetation, consequently resulting in a high diversity of plant species (Hughes & Cass 1997; Nillson et al. 1989). This study contributed positively to this hypothesis and corroborated previously published reports.

This study showed that microtopographic differentiation at the sapling stage occurs in riparian-specific species. In general, different soil condition preferences during the seedling and saplings stages often determine the structure of forests (e.g., Clark et al. 1999; Liang & Seagle 2002; Palmiotto et al. 2004). In other riparian forests in Japan, the seedlings of some species become established in abandoned channels and floodplains (Sakio 1997). Our results are consistent with these findings. Such differentiation would help promote coexistence of tree species within the riparian forests at a local scale.

Meanwhile, the microtopographic preferences of mature trees did not differ as clearly among species, including UT-species. In the riparian area, many of the mature trees exhibited a biased distribution at more stable sites such as sites of older deposits at higher elevation. As a result, species richness was greatest at stable microtopographic sites compared to other relatively unstable sites. Similar patterns have been reported elsewhere in Japanese riparian forests (Nakamura et al. 1997). As shown in Fig. 7, older deposits at higher elevation is most favorable for tree growth, probably reflecting soil moisture and more abundant organic content (Nakamura et al. 1997). Such stability at microtopographic sites may play a more critical role in the maintenance of locally higher species richness, than river-disturbed sites such as floodplains.

At a larger scale, this riparian forest appeared to act as a pool of tree species diversity in the landscape (e.g., Metzger 1997). Many species were found to be specific to the riparian area (i.e., RA-species), with rare species most commonly found within the riparian area. Consequently, the riparian forest was demonstrated to be important for regional species diversity. This is deterministic conclusion, also shown previously (see Suzuki et al. 2002).

In addition, the riparian forest can also act as a sink of diversity because, as opposed to RA-species which live in stable and fertile sites in the riparian area, UT-, CS-, DS- and rare species were less selective there. In the riparian forest, the populations of these species are thus expected to be less stable and closer to local extinction than when compared to the RA-species.

This implies that some dynamic processes within the riparian forest can greatly alter species diversity in the riparian forest (Sakio et al. 2002). For example, a rare chance event such as flooding may promote colonization by rare species (cf. Chesson 1986) while concomitantly causing the local extinction of some UT-species. Therefore, the diversity in the landscape cannot be maintained by the potential of the riparian forest alone as a species pool. From the dynamic perspective, both the riparian and the surrounding forests are considered to be important for the conservation of woody flora at the level of the landscape. Long-term monitoring of the community dynamics at KRRF will illustrate how valid these expectations actually are.

Acknowledgements

The authors thank the Iwate-Nambu District Forestry Office for the use of their facilities. Thanks are also due to Dr. H. Tanaka, Dr. T. Ota, Dr. T. Katsuki, and Dr. S. Abe for their helpful comments on the manuscript and to Dr. W. Murakami for his critical support in identifying microtopographic sites. This study was partly funded by the Ministry of Agriculture, Forestry and Fisheries, Japan (BioCosmos Project).

References

Baker WL (1990) Species richness of Colorado riparian vegetation. J Veg Sci 1:119-124

Clark DB, Palmer MW, Clark DA (1999) Edaphic factors and the landscape-scale distributions of tropical rain forest trees. Ecology 80:2662-2675

Chesson PL (1986) Environmental variation and the coexistence of species. In: Diamond J, Case TJ (eds) Community ecology. Harper & Row, New York, pp 240-256

Hara T, Nishimura N, Yamamoto S (1995) Tree competition and species coexistence in a cool-temperate old-growth forest in southwestern Japan. J Veg Sci 6:565-574

Hoshizaki K, Suzuki W, Sasaki S. (1997) Impacts of secondary seed dispersal and herbivory on Seedling Survival in *Aesculus turbinate*. J Veg Sci 8:735-742

Hoshizaki K, Suzuki W, Nakashizuka T (1999) Evaluation of secondary dispersal in a large-seeded tree *Aesculus turbinata*: a test of directed dispersal. Plant Ecol 144:167-176

Hughes JW, Cass WB (1997) Pattern and process of a floodplain forest, Vermont, USA: predicted responses of vegetation perturbation. J Appl Ecol 34:594-612

Kaneko Y, Takada T, Kawano S (1999) Population biology of *Aesculus turbinata* Blume: A demographic analysis using transition matrices on a natural population along a riparian environmental gradient. Plant Species Biol 14:47-68

Kubo M, Shimano K, Sakio H., Ohno K (2000) Germination sites and establishment conditions of *Cercidiphyllum japonicum* seedlings in the riparian forest. J Jpn For Soc 82:349-354 (in Japanese with English summary)

Kubo M, Sakio H, Shimano K, Ohno K (2004) Factors influencing seedling emergence and survival in *Cercidiphyllum japonicum*. Folia Geobot 39:225-234

Liang SY, Seagle SW (2002) Browsing and microhabitat effects on riparian forest woody seedling demography. Ecology 83:212-227

Masaki T. (2002) Structure and dynamics. In: Nakashizuka T, Matsumoto Y (Eds) Diversity and Interaction in a Temperate Forest Community, Ogawa Forest Reserve. Springer, Tokyo, pp 53-65

Masaki T, Tanaka H., Tanouchi, H, Sakai T, Nakashizuka T (1999) Structure, dynamics and disturbance regime of temperate broad-leaved forests in Japan. J Veg Sci 10:805-814

Masaki T, Osumi K, Takahashi K, Hoshizaki K. (2005) Seedling dynamics of *Acer mono* and *Fagus crenata*: an environmental filter limiting their adult distributions. Plant Ecol 177:189-199

Masaki T, Osumi K, Takahashi K, Hoshizaki K, Matsune K, Suzuki W (2007) Effects of microenvironmental heterogeneity on the seed-to-seedling process and tree coexistence in a riparian forest. Ecol Res 22:724-734

Metzger JP (1997) Relationships between landscape structure and tree species diversity in tropical forests of South-East Brazil. Landscape Urban Plann 37:29-35

Nakamura F, Yajima T, Kikuchi S (1997) Structure and composition of riparian forests with special reference to geomorphic site conditions along the Tokachi River, northern Japan. Plant Ecol 133:209-219

Nakashizuka T, Numata M (1982a) Regeneration process of climax beech forests. I. Structure of a beech forest with the undergrowth of Sasa. Jpn J Ecol 32:57-67

Nakashizuka T, Numata M (1982b) Regeneration process of climax beech forests. II. Structure of a forest under the influences of grazing. Jpn J Ecol 32:473-482

Nillson C, Grelsson G, Johansson M, Sperens U (1989) Patterns of plant species richness along riverbanks. Ecology 70:77-84

Palmiotto PA, Davies SJ, Vogt KA, Ashton MS, Vogt DJ, Ashton PA (2004) Soil-related habitat specialization in dipterocarp rain forest tree species in Borneo. J Ecol 92:609-623

Pollock MM, Naiman RJ, Hanley TA (1998) Plant species richness in riparian wetlands - a test of biodiversity theory. Ecology 79:94-105

Pabst RJ, Spies TA (1999) Structure and composition of unmanaged riparian forests in the coastal mountains of Oregon, U.S.A. Can J For Res 29:1557-1573

Sakio H (1997) Effects of natural disturbance on the regeneration of riparian forests in a Chichibu Mountains, central Japan. Plant Ecol 132:181-195

Sakio H, Kubo M, Shimano K, Ohno K (2002) Coexistence of three canopy tree species in a riparian forest in the Chichibu Mountains, central Japan. Folia Geobot 37:45-61

Suzuki W, Osumi K, Masaki T, Takahashi K, Daimaru H, Hoshizaki K (2002) Disturbance regimes and community structures of a riparian and an adjacent upper terrace stand in the Kanumazawa Riparian Research Forest, northern Japan. For Ecol Manage 157: 285-301

18   Diversity of forest floor vegetation with landform type

Motohiro KAWANISHI[1], Hitoshi SAKIO[2] and Keiichi OHNO[3]

[1]Open Research Center, Graduate school of Geo-Environmental Science, Rissho University, 1700 Magechi, Kumagaya, Saitama 360-0194, Japan

[2]Saitama Prefecture Agriculture & Forestry Research Center, 784 Sugahiro, Kumagaya, Saitama 360-0102, Japan (*Present address*: Sado Station, Field Center for Sustainable Agriculture and Forestry, Faculty of Agriculture, Niigata University, 94-2 Koda, Sado, Niigata 952-2206, Japan)

[3]Faculty of Environment and Information Science, Yokohama National University, 79-7 Tokiwadai, Hodogaya, Yokohama, Kanagawa 240-8501, Japan

18.1 Introduction

Two functional groups occur in the forest herbaceous layer: resident species and transient species (Gilliam & Roberts 2003). Resident species have life-history characteristics that confine them to aboveground strata of the herb layer. In contrast, transient species occur temporarily in the herb layer because they have the potential to grow taller, into higher strata. Thus, resident species include annuals, perennials, and small shrubs, whereas transient species include large shrubs and trees. Tree species' distributions are generally limited by various combinations of disturbances (Loehle 2000), but trees have the potential for rapid migration (Clark 1998). In contrast, the distributions of herbaceous plants are determined by the availability of suitable habitats, the likelihood of seed dispersal to these habitats, and successful germination of seeds and subsequent growth (Ehrlen & Eriksson 2000; Gilliam & Roberts 2003). In this chapter, we discuss the species diversity of herbaceous species as a community composed of resident species.

The herb layer plays a significant role in the species diversity of forest vegetation (Gilliam & Roberts 2003). For example, the species richness of herbaceous plants in eastern North America is greater than that of trees (Ricketts et al. 1999). Some studies have examined species diversity related to habitat heterogeneity and disturbance regimes in riparian areas (e.g., del Moral & Watson 1978; del Moral

Sakio, Tamura (eds) Ecology of Riparian Forests in Japan : Disturbance, Life History, and Regeneration
© Springer 2008

& Fleming 1979; Nilsson et al. 1989; Baker 1990). In Japan, various studies have explained the species diversity of trees on landslide scars (Sakai & Ohsawa 1993), flood plains, and terraces (Aruga et al. 1996; Suzuki et al. 2002). However, the diversity of herbs in these areas has received little attention.

Because disturbance regimes and ground conditions can differ greatly among landform types, species diversity is also expected to differ. However, we should not conclude that the pattern of species diversity of herbs is the same as that of trees, because the effects of particular disturbances differ for trees and herbaceous plants. Therefore, in this chapter we discuss the species diversity of forest floor vegetation in relation to the influence of landform. We specifically address the following two questions: What is the species richness of forest floor plants in a riparian area? How do plants coexist in a riparian area?

18.2 Landform type

We classified mountain slopes into several landform types according to inter-comparisons of ground forms, materials, and physical relationships (Table 1). The crest slope consists primarily of convex segments, including divides. The upper side slope is situated below the crest slope and above the erosion front. The lower side slope, which is steep and located below the erosion front, is composed of foot slopes, talus, and landslide slopes. Landslide slopes have evidence of old and new landslides of various sizes. The valley bottom is a flat site that is formed mainly

Table 1. Scale and system of landform type using in this chapter. Superscript letters indicate the study area composing the landform type; 1: Ooyamazawa, 2: Shikoku Mountains

Sub-small scale	Micro scale	Sub-micro scale
Crest slope[1]	unclassified	
Upper side slope[1]	unclassified	
Lower side slope[1,2]	Land slide slope[1,2]	Old landslide site[1]
		New landslide site[1]
	Foot slope[1,2]	
	Talus[1,2]	
Valley bottom[1,2]	Terrace of debris flow[1,2]	Higher surface[2]
		Lower surface[2]
		Terrace scarp[1]
	Alluvial fan[1]	
	Channel[1,2]	

by the deposition of debris flow. Deposition sites eroded by streamflow have terracing and terrace scarps. When a fan-like deposition site is situated at the confluence of small tributaries, it is termed an alluvial fan. We did not classify microscale and sub-microscale landforms on the crest slope and upper side slope.

18.3 Species richness of riparian forest floor plants

First, we discuss the number of species in the forest floor vegetation in riparian forest. Although several previous studies have noted that the species diversity of riparian forest in Japan is high (e.g., Maeda & Yoshioka 1952; Yamanaka 1962; Ohno 1986), tangible data about forest floor vegetation have not been presented. To recognize patterns of diversity, the species richness of riparian forest must be compared with that of other forest types. To solve this problem, Kawanishi et al. (2006) examined a series of neighboring communities in a headwater forest containing crest slope, upper side slope, lower side slope, and valley-bottom habitats. The study site was located in the Ooyamazawa River basin (35°57' N, 138°46' E; 1200–1670 m a.s.l.) along a small stream that flowed into the Nakatsugawa River, a branch of the Arakawa River in central Japan. The valley vegetation consisted of a riparian forest of *Fraxinus platypoda* Oliv. and *Pterocarya rhoifolia* Sieb. et Zucc.. *Fagus crenata* Blume–*Fagus japonica* Maxim. forests and *Tsuga sieboldii* Carrière forests were found on the upper side slopes and on the ridge, respectively (Maeda & Yoshioka 1952).

To examine patterns of species diversity, we used the hierarchical diversity model of Wagner et al. (2000), in which "within-quadrat diversity: d" is the sample quadrat diversity (Whittaker 1975), and "within-unit diversity: D" is the total diversity of a micro-landform·unit, which is comparable to the within-patch diversity of Wagner et al. (2000). A total of 230 species of vascular plants was found in the study plots. Species richness (d, D) differed between trees and forest floor plants (Table 2). The within-quadrat (d) and within-unit (D) richness of trees was greater on the upper side slope and crest slope than on the valley bottom and lower side slope. In contrast, the d of forest floor plants was very high on the valley bottom and lower side slope. Species diversity varied substantially among micro-landform types. Additionally, the effects of landform factors on species diversity differed between forest floor plants and trees.

We can explore the reason for the high diversity of trees using studies of forest ecology. For example, trees in beech forests generally regenerate in small canopy gaps (e.g., Nakashizuka 1982, 1983, 1984), and very few juvenile *Fagus crenata* are found in the beech forests on the Pacific Ocean side of Japan (included in this study site), although juveniles of many other species are present (Shimano & Okitsu 1993, 1994). The regeneration processes in *Tsuga sieboldii* forests show similar patterns (Suzuki 1980). As a result, many patches of small trees are located within a small area, producing a higher d for trees in the upper side slope and crest slope. These findings suggest that the long-term stability of these micro-landforms can support many climax tree species that require a long time to reach maturity.

In contrast, lower tree diversity on the valley bottom and lower side slope (Table 2) may occur because only tree species that are adapted to the natural soil disturbances (e.g., *Pterocarya rhoifolia*, *Fraxinus platypoda*) can grow on these micro-landforms. In addition, these trees are able to regenerate simultaneously in the wide openings created by large disturbances (Sakio et al. 2002). Therefore, selective pressures exerted by frequent soil disturbance probably result in low d for trees on the valley bottom and lower side slope. Aruga et al. (1996) obtained similar results in floodplain forest and reasoned that the diversity of trees in areas subject to frequent disturbance is low because specific trees regenerate simultaneously in sites created by large-scale disturbance. Suzuki et al. (2002) also showed that the number of species appearing on a terrace (a relatively stable site) at a spatial scale of less than 400 m^2 was larger than that in a riparian area. These findings indicate that the species diversity of trees is restricted by frequent disturbance at a scale of several hundred square meters.

In contrast to that of trees, the d of herbaceous plants on the valley bottom and lower side slope was very high. This indicates that frequent disturbances increase the diversity of forest floor plants. Thus, herbaceous plant diversity is high. Although various factors are important in the high diversity of herbs in a riparian area, the principal factors are likely as follows. First, the absence of strong competitors in a riparian area may allow many species to co-occur. Most herbaceous plants cannot grow in the upper slope area, where dwarf bamboo (*Sasamorpha borealis* (Hack.) Nakai) dominates overwhelmingly. Because light reaching the forest floor is very scarce in dwarf bamboo communities (Nakashizuka 1988), other herbs may not be able to grow. Consequently, many herbaceous plants become established on unstable sites where strong competitors

Table 2. Comparison of species richness for each life form among micro-landform types. Indices D and means of index d ($\underline{d}$) are shown with standard deviations. Different superscript letters indicate groups with significant differences ($P < 0.01$) between landform types, as derived with Sheffe's method. From Kawanishi et al. (2006)

		Landform types				
		Valley bottom	Lower side slope	Upper side slope	Crest slope	Total number of species
Number of plots		16	16	17	14	
Total quadrat area (m^2)		5725	5825	6825	3545	
Species richness						
All species	D	29.0	29.7	32.3	25.9	230
	$\underline{d}$	16.2±2.2 [a]	13.7±3.5 [ab]	10.6±4.5 [b]	10.9±3.4 [b]	
Forest floor plants	D	20.0	19.9	13.0	9.3	127
	$\underline{d}$	11.1±1.6 [a]	8.8±3.1 [a]	1.9±2.4 [b]	2.4±2.3 [b]	
Trees	D	9.0	9.8	19.3	16.6	103
	$\underline{d}$	5.0±1.2 [a]	4.9±1.2 [a]	8.8±2.4 [b]	8.6±1.7 [b]	

such as *Sasamorpha borealis* are absent. In addition, we expect that the land heterogeneity of such locations (Sakio 1997; Sakio et al. 2002) contributes to the high species diversity of forest floor plants. The various ground disturbances that occur in riparian areas are a source of heterogeneity, which operates through the heterogeneity of ground attributes and the sensitivity of herbs to microhabitat conditions. This issue will be discussed in the next section.

18.4 Effect of landform on forest floor vegetation

18.4.1 Correspondence of forest floor vegetation to micro-landform

Relatively active processes such as soil erosion, landslides, and slope failure occur frequently on lower slope segments (Onda et al. 1996). Debris flow caused by these processes forms micro-landforms by sedimentation of the ground material (Benda 1990; Gregory et al. 1991). Therefore, various landforms are formed in riparian areas. The relationships between forest floor vegetation and micro-

Table 3. Properties of matrix of deposits at a typical site of each micro-landform type in seven study plots in Shikoku Mountains. IB: Ibuki, NA: Nagoro, TS: Tsurugi, SA: Saorigahara, KU: Kuishi, IR: Irazu, JI: Jiyoshidani. From Kawanishi et al. (2005)

Micro-landforms	Study plot	Content by weight (%)						Mean grain size	Ignition loss
		-0.5φ	0.5φ	1.5φ	2.5φ	3.5φ	$4.5\varphi<$	(φ)	(%)
Channel	IB	38.4	23.7	15.4	7.8	4.7	10.0	1.0	6.8
	NA	46.0	24.5	15.2	6.1	2.4	5.9	0.6	5.0
	TS	33.5	14.8	11.9	9.5	7.2	23.2	1.6	15.0
	SA	45.6	23.1	12.0	6.0	2.9	10.3	0.8	8.1
	KU	11.7	9.9	12.5	10.9	4.9	50.1	2.9	78.2
	JI	16.4	18.8	23.2	12.5	5.1	24.0	1.9	7.2
Terrace of debris flow	NA	32.9	23.9	18.5	8.7	3.0	12.9	1.1	28.1
	TS	17.9	14.5	16.5	13.4	6.2	31.6	2.2	35.5
	SA (low)	48.4	19.7	9.6	5.3	3.2	13.8	0.9	9.0
	SA (high)	40.1	19.1	11.1	6.4	3.5	19.8	1.2	9.9
	KU (low)	27.9	15.6	10.9	10.2	7.1	28.4	1.9	16.2
	KU (high)	21.7	11.1	10.0	13.1	8.6	35.6	2.3	19.6
Foot slope	NA	41.1	19.7	9.4	5.9	3.7	20.2	1.2	10.0
	TS	73.1	16.6	3.7	1.6	1.1	4.0	0.0	6.6
	JI	22.0	20.1	19.1	12.7	6.5	19.7	1.7	11.9
Landslide slope	IB	16.9	10.4	9.3	10.1	11.4	41.9	2.6	19.9
	TS	23.9	11.6	11.7	9.9	10.9	32.0	2.2	18.1
	IR	14.1	14.0	17.6	14.4	10.0	30.0	2.3	30.5
Talus	IR (with humus)	-	-	-	-	-	-	-	71.7
	IR (with loam)	26.8	15.9	13.8	10.5	6.9	26.2	1.8	18.6
	JI (with humus)	-	-	-	-	-	-	-	82.5
	JI (with loam)	3.2	4.8	6.4	7.6	6.5	71.4	3.7	29.5

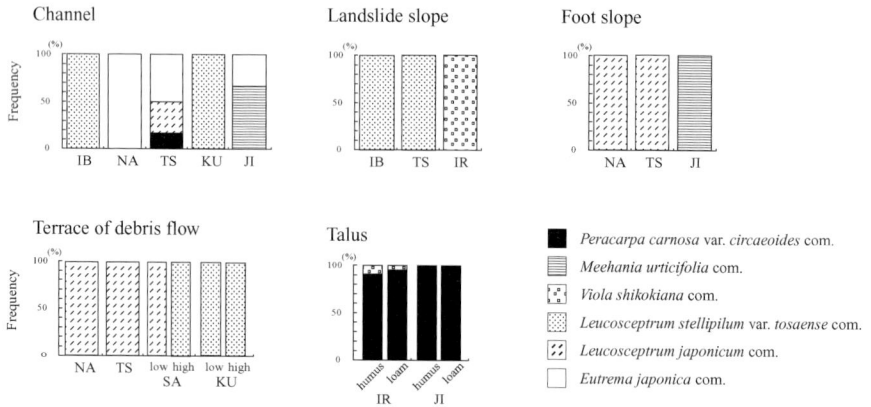

Fig. 1. Block graph showing the occurrence ratio of the community types on micro-landforms in each study plots. Abbreviations for each study plot are the same as in Table 3. From Kawanishi et al. (2005)

landforms in *Pterocarya rhoifolia* forests have been demonstrated in Shikoku, Japan (Kawanishi et al. 2005). Kawanishi et al. (2005) investigated the forest floor vegetation (<1.2 m in height) using phytosociological methods (1-m² quadrat size) and compartmentalized the vegetation in seven valleys.

There were five micro-landform types among the seven sites: channel, terrace of debris flow, foot slope, landslide slope, and talus. Terrace of debris flow was segmented into higher and lower surfaces. Talus was segmented into a humus type and a loam type according to the matrix. The sediments in each landform varied in grain size and organic matter content, which was measured as the ignition loss ratio (Table 3). Six types of forest floor vegetation were recognized. The percent occurrence of the forest floor vegetation in each micro-landform (Fig. 1) demonstrates that several types of forest floor vegetation became established on each micro-landform type. Only talus contained only one community: a *Peracarpa carnosa* (Wall.) Hook.fil. et Thomson var. *circaeoides* (Fr.Schm.) Makino community. Forest floor vegetation types were clearly related to landform factors if soil conditions were also considered. *Viola shikokiana* Makino communities occurred on landslide slopes that consisted of organic deposits. *Leucosceptrum stellipilum* (Miq.) Kitam. et Murat var. *tosaense* (Makino) Kitam. et Murata communities occurred on the higher terraces of debris flows and on landslide slopes unaffected by fluvial processes. *Meehania urticifolia* (Miq.) Makino communities occurred on foot slopes that had poor organic materials. *Leucosceptrum japonicum* (Miq.) Kitam. et Murata communities occurred on the foot slope and the lower terrace of a debris flow composed mainly of sand and gravel. *Eutrema japonica* (Miq.) Koidz. communities occurred only on channels in the riverhead (Table 3, Fig. 1).

From these results, it is clear that the establishment of forest floor vegetation was related to landform. The soil conditions, which vary with land-formation processes, affected the species composition of the forest floor vegetation. Several types of forest floor communities corresponding to micro-landforms were established within each study plot (Fig. 1). Thus, we recognize that the forest floor vegetation of the *Pterocarya rhoifolia* forest was a multi-unit of these herbaceous communities that was influenced not only by the micro-landform but also by soil conditions. In other words, the various forest floor plants have specific adaptations to ground conditions that vary with the landform. The heterogeneity of micro-habitat related to the landform, therefore, contributes to the high diversity of herbaceous plants in a riparian forest.

18.4.2 Characteristics of herbaceous plants

Forest floor vegetation is established in specific habitats related to landform, as stated above. Thus, the availability of suitable habitat determines the distributions of herbaceous plants. Given these findings, we sought characteristics of herbaceous plants that confer adaptation to various habitat conditions. The likelihood of seed dispersal to habitats and successful germination of seeds and subsequent growth determines the distributions of herbaceous plants (e.g., Ehrlen & Eriksson 2000; Gilliam & Roberts 2003). The habitat differentiation of herbaceous plants in riparian forests is caused by differences in breeding characteristics. In addition, it is clear that the number of species capable of vegetative reproduction in the forest floor vegetation in *Pterocarya rhoifolia* and *Fraxinus platypoda* forests is greater than that in *Fagus crenata* and *Quercus crispula* Blume forests (Oono 1996).

The vegetative reproduction of herbaceous plants plays important roles in maintaining the population (Silvertown 1982; Grime 2001) and recovery from damage caused by ground disturbance (Yano 1962). Therefore, in riparian forest, the role of vegetative reproduction among forest floor plants is particularly important for species coexistence.

Kawanishi et al. (2004) showed that the habitat differentiation of herbs in *Fraxinus playipoda* and *Pterocarya rhoifolia* forests is related to vegetative reproduction characteristics. Six landform types were distinguished along the Ooyamazawa valley: debris flow terrace, alluvial fan, terrace scarp, new landslide site, old landslide slope, and talus. Three major herbaceous species groups (clusters A, B, D) occurred at the site (Table 4). Cluster A was characterized by spring ephemerals, storage rhizomes, and anti-vegetative reproduction. This species group represented the forest floor vegetation inhabiting stable landforms such as debris flow terraces and alluvial fans. Muller (2003) showed that spring ephemerals appear to have significantly greater concentrations of nitrogen and iron than other herbs. The large concentration of these nutrients may relate to high anabolism. Storage organs have important roles in the effective distribution of carbohydrates or major nutrients in the plant body (e.g., Mooney & Billings 1961; Kimura 1970). We can infer from these observations that the distribution of this species group would be restricted by the efficiency of the storage organ because substances reserved in rhizomes sustain these species. The rich organic matter in soil on debris flow terraces and alluvial fans contributes to the maintenance of

Table 4. Mean coverage (%) of species in each landform type with maximum coverage of species in parenthesis. Superscript symbol '+' and '-' indicate desirable and undesirable site derived from $\chi2$ test (P < 0.001), respectively. GS: Growing season (sp: spring ephemeral, sm: summer period, f: three season), REP: reproduction types (th: annual species, bul: bulbil type, tub: tuber type, run: runner type, rhi: horizontal rhizome type and av: anti-vegetative reproduction type), and MR: morphology of rhizome (rs: storage type, rc: connector type and rr: replace type) are represented in column named 'life type' . TR: terrace of debris flow, AL: alluvial fan, SC: terrace scarp, LS: new landslide site, OS: old landslide slope and TL: talus. From Kawanishi et al. (2004)

Species name	GS	REP	MR	TR	AL	SC	LS	OS	TL
Cluster A									
Scopolia japonica	sm	rhi	rs	$8.1 (40)^+$	$7.4 (40)^+$	$2.1 (20)^-$	$1.6 (10)^-$	$3.2 (30)^-$	$9.5 (50)^+$
Corydalis lineariloba	sp	tub	rr	$7.0 (20)^+$	$9.0 (20)^+$	$4.3 (20)$	$1.7 (10)^-$	$3.6 (15)^-$	$3.0 (15)^-$
Spuriopimpinella nikoensis	f	av	rs	$1.5 (5)$	$5.5 (10)^+$	$2.2 (10)$	$0.5 (7)^-$	$0.6 (7)^-$	$1.4 (10)$
Aconitum sanyoense	f	tub	rr	$2.2 (15)^+$	$5.0 (30)^+$	$2.4 (20)^+$	$0.5 (10)^-$	$0.2 (3)^-$	$0.3 (7)^-$
Veratrum grandiflorum	sm	rhi	rs	$1.2 (10)$	$7.0 (20)^+$	$0.3 (5)^-$	$0.1 (2)^-$	$1.1 (20)$	$0.0 (0.1)^-$
Mitella pauciflora	f	rhi	rc	$4.5 (20)^+$	$2.4 (10)^+$	$0.4 (3)^-$	$0.7 (5)^-$	$0.4 (7)^-$	$0.4 (3)^-$
Dryopteris crassirhizoma	f	av	rs	$3.0 (10)^+$	$1.2 (10)$	$0.6 (5)^-$	$0.4 (5)^-$	$0.5 (10)^-$	$1.0 (10)$
Cornopteris crenulato-serrulata	f	rhi	rc	$2.4 (7)^+$	$1.9 (5)^+$	$0.3 (2)^-$	$0.5 (5)^-$	$0.3 (10)^-$	$0.4 (3)^-$
Adoxa moschatellina	sm	rhi	rr	$2.8 (10)^+$	$2.3 (15)^+$	$0.5 (2)^-$	$0.1 (2)^-$	$0.3 (5)^-$	$2.9 (15)^+$
Allium monanthum	sp	tub	rr	$3.1 (20)^+$	$0.5 (3)^-$	$0.7 (2)^-$	$0.7 (5)^-$	$0.7 (10)^-$	$1.3 (10)$
Asarum caulescens	f	run	rc	$2.4 (20)^+$	$0.9 (5)$	$0.9 (5)$	$0.3 (4)^-$	$1.3 (20)$	$0.2 (2)^-$
Cacalia yatabei	f	rhi	rr	$1.2 (15)^+$	$0.6 (2)$	$0.5 (7)$	$0.5 (10)$	$0.3 (3)^-$	$0.2 (3)^-$
Polystichum tripteron	f	av	rs	$1.4 (7)^+$	$0.1 (2)^-$	$0.7 (7)$	$0.4 (5)$	$0.0 (1)^-$	$0.8 (7)$
Chrysosplenium ramosum	f	run	rr	$1.0 (7)^+$	0^-	$0.4 (10)$	$0.4 (3)$	$0.3 (3)$	0^-
Polystichum ovato-paleaceum	f	av	rs	$0.4 (5)$	$0.1 (3)^-$	$0.6 (5)^+$	$0.8 (7)^+$	$0.2 (3)$	$0.1 (1)^-$
Diplazium squamigerum	f	rhi	rc	$1.6 (20)^+$	$0.2 (5)$	$0.2 (5)$	$0.0 (0.1)^-$	$0.3 (5)$	$0.2 (2)^-$
Cluster B									
Chrysosplenium macrostemon	f	run	rr	$0.6 (5)^-$	$0.1 (3)^-$	$1.0 (10)^-$	$4.5 (30)^+$	$3.8 (50)^+$	$0.1 (1)^-$
Elatostema umbellatum var. majus	f	bul/rhi	rc	$0.2 (3)^-$	$0.0 (0.1)^-$	$3.6 (20)^+$	$5.2 (20)^+$	$2.0 (25)$	$0.0 (1)^-$
Veronica miqueliana	f	rhi	rc	$0.0 (1)^-$	$0.1 (2)^-$	$1.0 (7)^+$	$2.0 (10)^+$	$0.8 (7)$	$0.0 (0.1)^-$
Persicaria debilis	f	th	rr	$0.0 (0.1)^-$	0^-	$0.4 (7)$	$2.3 (10)^+$	$0.9 (10)^+$	$0.0 (0.1)^-$
Cacalia farfaraefolia	f	rhi	rr	$0.0 (1)^-$	$0.1 (1)^-$	$1.6 (7)^+$	$0.7 (3)$	$0.4 (5)$	$0.2 (5)^-$
Laportea bulbifera	f	tub/bul	rr	$0.8 (7)^-$	$0.5 (3)^-$	$1.3 (5)$	$2.6 (10)^+$	$1.4 (7)$	$2.5 (7)^+$
Stellaria sessiliflora	f	run	rc	$0.1 (3)^-$	$1.1 (20)$	$0.5 (7)$	$1.0 (7)$	$0.8 (5)$	$1.5 (5)^+$
Impatiens noli-tangere	f	th	rr	$0.6 (7)$	$0.0 (0.1)^-$	$0.7 (3)$	$1.2 (5)^+$	$0.4 (5)^-$	$1.7 (7)^+$
Laportea macrostachya	f	rhi	rs	$1.0 (10)$	$0.3 (3)^-$	$0.1 (2)^-$	$0.8 (3)$	$1.1 (7)$	$3.9 (20)^+$
Deinanthe bifida	f	rhi	rs	$0.2 (3)^-$	0^-	$0.3 (3)^-$	$2.1 (10)^+$	$2.5 (20)^+$	$0.4 (5)^-$
Meehania urticifolia	f	run	rc	$1.0 (5)$	$0.8 (3)$	$0.4 (3)^-$	$0.5 (5)$	$0.6 (5)$	$1.5 (7)^+$
Dryopteris polylepis	f	av	rs	$1.1 (7)^+$	$0.0 (0.1)^-$	$0.2 (3)^-$	$0.6 (3)$	$1.0 (5)^+$	$1.3 (5)^+$
Cluster C									
Galium paradoxum	f	rhi	rr	$0.2 (5)$	0^-	$0.1 (1)$	$0.8 (7)^+$	$0.0 (0.1)^-$	$0.8 (7)^+$
Cluster D									
Hydrangea macrophylla var. acuminata	f	rhi	rc	$0.2 (3)^-$	0^-	$3.0 (40)^+$	$1.3 (7)^+$	$0.4 (5)^-$	$0.5 (5)$

Table 4. Continued

Species name	Life type			Mean coverage (%)					
	GS	REP	MR	TR	AL	SC	LS	OS	TL
Chrysosplenium album var. *stamineum*	f	run	rr	0⁻	0⁻	1.5 (7)⁺	1.0 (7)⁺	0.0 (2)⁻	1.3 (20)⁺
Astilbe thunbergii	f	rhi	rs	0⁻	0⁻	1.0 (10)⁺	0.8 (10)⁺	0.0 (0.1)⁻	0.0 (0.1)⁻
Cacalia delphiniifolia	f	rhi	rr	0.0 (1)⁻	0.3 (7)	0.6 (7)⁺	0.4 (5)	0.4 (3)	0.0 (0.1)⁻
Cluster E									
Chrysosplenium echinus	f	run	rr	0.6 (7)⁺	0.9 (10)⁺	0.2 (2)⁻	0.9 (15)⁺	0⁻	0.2 (3)
Cluster F									
Chrysosplenium pilosum var. *sphaerospermum*	f	run	rr	0.2 (7)⁻	0.1 (3)⁻	6.3 (30)⁺	5.5 (25)⁺	0.9 (15)⁻	0.7 (15)⁻
Cluster G									
Urtica laetevirens	f	rhi	rc	0.0 (1)⁻	0.0 (0.1)⁻	0.0 (0.1)⁻	1.3 (20)⁺	0.5 (15)	0.7 (10)⁺

these plants. In addition, the storage organ also allows vegetative reproduction (Jerling 1988; Westley 1993; Verburg et al. 1996; Grime 2001). We would expect these species groups to contain plants that propagate via vegetative reproduction; however, in this study this group was characterized by the anti-vegetative reproduction type (Table 4). Although this paradox has not been resolved, it appears that the attributes of organic matter storage are relevant in this habitat condition.

The other two species groups (B and D) characteristically comprised annual plants and those with replacement rhizomes. These were peculiar to terrace scarps, new landslide sites, old landslide slopes, and talus located at the foot of sideslopes. In particular, these characters were extreme on terrace scarps and new landslide sites with unstable soils. Annual plants such as *Impatiens noli-tangere* L. and *Persicaria debilis* (Meisn.) H. Gross are adapted to unstable sites that experience continual or annual disturbance (Silvertown 1982). Furthermore, the attributes of herbs that reproduce via bulbils (e.g., *Elatostema umbellatum* Blume var. *majus* Maxim., *Laportea bulbifera* (Sieb. et Zucc.) Wedd.) or separation from rhizomes of the mother plant (e.g., *Chrysosplenium macrostemon* Maxim., *Cacalia delphiniifolia* Sieb. et Zucc., *Cacalia farfaraefolia* Sieb. et Zucc.) resemble those of annual plants. These plants would be adapted to unstable habitat because their attributes are advantageous in maintaining populations subject to soil disturbance. In contrast, species lacking the abovementioned reproductive abilities (e.g., *Laportea macrostachya* (Maxim.) Ohwi, *Meehania urticifolia*, *Dryopteris polylepis* (Franch. et Savat.) C. Chr.) dominated relatively stable landforms such as talus and old landslide slopes.

We considered the influence of the landform on the life history of herbaceous plants. Our analysis indicates that the differentiation of herbaceous communities is related to the attributes of their storage organs and their vegetative reproduction. We therefore recognize that herbaceous plants are able to coexist in riparian areas because each species differs in the specific traits adapted to disturbance.

18.5 Conclusions

The species diversity of forest floor plants is greater in riparian forest than in forests on mountain slopes. An important factor contributing to the high diversity of forest floor plants is the variety of ground disturbance at various scales. The role of disturbance in the regeneration of riparian plants has been elucidated in *Pterocarya rhoifolia–Fraxinus platypoda* forest (e.g., Sakio 1997; Sakio et al. 2002), *Euptelea polyandra* Sieb. et Zucc. forest (Sakai & Ohsawa 1993, 1994), *Ulmus davidiana* Planch. var. *japonica* (Rehder) Nakai forest (Kon & Okitsu 1999; Sakai et al. 1999; Wada et al. 2004), and other floodplain vegetation (e.g., Johnson et al. 1976; Ishikawa 1991; Siebel & Bouwma 1998). These studies indicate that ground disturbances occur generally in riparian areas and contribute to the regeneration of trees. These disturbances in riparian areas create landforms that restrict specific resource conditions such as soil, water, and light availability. In addition, the types and cycles of disturbance differ among landform types because the effects of flood and debris flow are restricted by the relative height and spatial configuration of the ground surface. Therefore, at the microscale, habitats are diverse in both the amount of resources and the frequency of disturbance. Forest floor plants that have suitable traits, particularly vegetative propagation and material storage, form communities in specific microhabitats. Because forest floor plants respond to the small-scale heterogeneity of landforms and differentiate among habitats at the microscale, many species coexist in riparian areas through the establishment of plural communities.

References

Aruga M, Nakamura F, Kikuchi S, Yajima T (1996) Characteristics of floodplain forests and their site conditions in comparison to toeslope forests in the Tokachi River. J For Soc 78:354-362 (in Japanese with English summary)

Baker WL (1990) Species richness of Colorado riparian vegetation. J Veg Sci 1:119-124

Benda L (1990) The influence of debris flows on channels and valley floors in the Oregon coast range, U.S.A. Earth Surf Process Landf 15:457-466

Clark JS (1998) Why trees migrate so fast: confronting theory with dispersal biology and paleorecord. Am Nat 152:204-224

del Moral R, Watson AF (1978) Gradient structure in forest vegetation in the Central Washington Cascades. Vegetatio 38:129-48

del Moral R, Fleming RS (1979) Structure of coniferous forest communities in western Washington: Diversity and ecotope properties. Vegetatio 41(3):143-154

Ehrlen J, Eriksson O (2000) Dispersal limitation and patch occupancy in forest herbs. Ecology 81:1667-1674

Gilliam FS, Roberts MR (2003) Introduction, Conceptual framework for studies of the herbaceous layer. In: Gilliam FS, Roberts MR (eds) The herbaceous layer in forest of eastern North America. Oxford Univ Press, New York, pp 3-11

Gregory SV, Swanson FJ, McKee WA, Cummins KW (1991) An ecosystem perspective of riparian zones –focus on links between land and water–. BioScience 41(8):540-551

Grime JP (2001) Plant strategies, vegetation processes, and ecosystem properties (2nd ed.). John Wiley & Sons, Chichester

Ishikawa S (1991) Flood plain vegetation of the Ibi river in central Japan. II. Vegetation dynamics on the bars in the river course of the alluvial fan. Jpn J Ecol 41:31-43 (in Japanese with English summary)

Johnson WC, Burgess RL, Keammerer W R (1976) Forest overstory vegetation and environmental on the Missouri river flood plain in North Dakota. Ecol Monogr 46:59-84

Jerling L (1988) Population dynamics of *Glaux maritime* (L.) along a distributional cline. Vegetatio 74:161-170

Kawanishi M, Ishikawa S, Miyake N, Ohno K (2005) Forest floor vegetation of *Pterocarya rhoifolia* forests in Shikoku, with special reference to micro-landform. Veg Sci 22:87-102 (in Japanese with English summary)

Kawanishi M, Sakio H, Ohno K (2004) Forest floor vegetation of *Fraxinus platypoda-Pterocarya rhoifolia* forest along Ooyamazawa valley in Chichibu, Kanto District, Japan, with a special reference to ground disturbance. Veg Sci 21:15-26 (in Japanese with English summary)

Kawanishi M, Sakio H, Kubo M, Shimano K, Ohno K (2006) Effect of micro-landforms on forest vegetation differentiation and life-form diversity in the Chichibu Mountains, Kanto District, Japan. Veg Sci 23:13-24

Kimura M (1970) Analysis of production processes of an undergrowth of subalpine *Abies* forest, *Pteridophyllum racemosum* population 1 Growth, carbohydrate economy and net production. Bot Mag Tokyo 83:99-108

Kon H, Okitsu S (1999) Role of land-surface disturbance in regeneration of *Ulmus davidiana* var. *japonica* in a cool temperate deciduous forest on Mt. Asama, central Japan. J Jpn For Soc 81:29-35 (in Japanese with English summary)

Loehle C (2000) Strategy space and the disturbance spectrum: a life-history model for tree species coexistence. Am Nat 156:14-33

Maeda T, Yoshioka J (1952) Studies on the vegetation of Chichibu Mountain forest (2). The plant communities of the temperate mountain zone. Bull Tokyo Univ For 42:129-150 (in Japanese)

Mooney HA, Billings WD (1961) Comparative physiological ecology of arctic and alpine populations of Oxyria digyna. Ecol Monogr 31:1-29

Muller RN 2003. Nutricnt relation of the herbaceous layer in deciduous forest ecosystems. In: Gilliam FS, Roberts MR (eds) The herbaceous layer in forest of eastern North America. Oxford Univ Press, New York, pp 15-37

Nakashizuka T (1982) Regeneration process of climax beech forest II. Structure of a forest under the influences of grazing. Jpn J Ecol 32:473-482

Nakashizuka T (1983) Regeneration process of climax beech forest III. Structure and development process of sapling populations in different aged gaps. Jpn J Ecol 33:409-418

Nakashizuka T (1984) Regeneration process of climax beech forest IV. Gap formation. Jpn J Ecol 34:75-85

Nakashizuka T (1988) Regeneration of beech (*Fagus crenata*) after the simultaneous death of undergrowing dwarf bamboo (*Sasa kurilensis*). Ecol Res 3:21-35

Nilson C, Grelsson G, Johansson M, Sperens U (1989) Patterns of plant species richness along riverbanks. Ecology 70:77-84

Ohno K (1986) Riparian forest in mountain region. In: Miyawaki A (ed) Vegetation of Japan. Vol. 5. Kanto. Shibundo, Tokyo, pp 300-303 (in Japanese)

Onda Y, Okunishi K, Iida T, Tsujimura M (1996) Hydrogeomorphology: the interaction of hydrologic and geologic processes. Kokon-Syoin, Tokyo (in Japanese)

Oono K (1996) Life history of herb plants in summer green forest. In: Hara M (ed) Natural history of beech forest. Heibonsha, Tokyo, pp 113-156 (in Japanese)

Ricketts TH, Dinerstein E, Olson DM, Loucks C (1999) Who's where in North America? BioScience 49(5):369-381

Sakai A, Ohsawa M (1993) Vegetation pattern and microtopography on a landslide scar of Mt Kiyosumi, central Japan. Ecol Res 8:47-56

Sakai A, Ohsawa M (1994) Topographical pattern of the forest vegetation on a river basin in a warm-temperate hilly region, central Japan. Ecol Res 9:269-280

Sakai T, Tanaka H, Shibata M, Suzuki W, Nomiya H, Kanazashi T, Iida S, Nakashizuka T (1999) Riparian disturbance and community structure of a *Quercus-Ulmus* forest in central Japan. Plant Ecol 140:99-100

Sakio H (1997) Effects of natural disturbance on the regeneration of riparian forests in Chichibu Mountains, central Japan. Plant Ecol 132:181-195

Sakio H, Kubo M, Shimano K, Ohno K (2002) Coexistence of three canopy tree species in a riparian forest in the Chichibu Mountains, central Japan. Folia Geobot 37:45-61

Shimano K, Okitsu S (1993) Regeneration of mixed *Fagus crenata-Fagus japonica* forests in Mt. Mito, Okutama, west of Tokyo. Jpn J Ecol 43:13-19 (in Japanese with English summary)

Shimano K, Okitsu S (1994) Regeneration of natural *Fagus crenata* forests around the Kanto district. Jpn J Ecol 44:283-291 (in Japanese with English summary)

Siebel HN, Bouwma IM (1998) The occurrence of herbs and woody juveniles in a hardwood floodplain forest in relation to flooding and right. J Veg Sci 9:623-630

Silvertown JW (1982) Introduction to plant population ecology. (trans. Kawano S, Takada T, Oohara M 1992). Tokai Univ Press, Tokyo (in Japanese)

Suzuki E (1980) Regeneration of Tsuga sieboldii forest. II. Two cases of regenerations occurred about 260 and 50 years ago. Jpn J Ecol 30:333-346 (in Japanese with English summary)

Suzuki W, Osumi K, Masaki T, Takahashi K, Daimaru H, Hoshizaki K (2002) Disturbance regimes and community structures of a riparian and an adjacent terrace stand in the Kanumazawa Riparian Research Forest, northern Japan. For Ecol Manage 157:285-301

Verburg RW, Kwant R, Werger MJA (1996) The effect of plant size on vegetative reproduction in a pseudo-annual. Vegetatio 125:185-192

Wada M, Kikuchi T (2004) Emergence and establishment of seedlings of Japanese elm (*Ulmus davidiana* Planch. var. japonica (Rehder) Nakai) on a flood plain along the Azusa River in Kamikochi, central Japan. Veg Sci 21:27-38 (in Japanese with English summary)

Wagner HH, Wildi O, Ewald KC (2000) Additive partitioning of plant species diversity in an agricultural mosaic landscape. Landscape Ecology 15:219-227

Westley LC (1993) The effect of inflorescence bud removal on tuber production in *Helianthus tuberosus* L. (Asteraccae). Ecology 74:2136-2144

Whittaker RH (1975) Communities and ecosystems, 2nd ed. Macmillan, New York

Yano N (1962) The subterranean organ of sand dune plants in Japan. J Sci Hiroshima Univ Ser B Div 2 (Bot) 9:139-184 (in Japanese)

Yamanaka T (1962) Deciduous forests in the cool temperate zone of Shikoku. Res Rep Kochi Univ 11(2):9-14 (in Japanese)

Part 9

Endangered species and its conservation

19   Ecology and conservation of an endangered willow, *Salix hukaoana*

Wajiro SUZUKI and Satoshi KIKUCHI

Forestry and Forest Products Research Institute, 1 Matsuno-sato, Tsukuba, Ibaraki 305-8335, Japan

19.1 Introduction

Several riverine willow (*Salix*) species are known to occur in the floodplains of rivers in Japan. These willow species are characterized by their unique life-history features, as well as by the conditions of the habitats they occupy, e.g., soil texture and moisture, light conditions, and so on. These willow species often constitute typical riparian forests over open floodplains, representing important elements at the early stages of ecological succession (Ishikawa 1983, 1988; Niiyama 1987; Yoshikawa & Fukushima 1999). These *Salix* species have obviously adapted to changing riparian habitats, such as flooding caused by snow melting and typhoons with heavy rainfalls, since they can maintain their populations in such extremely changing environments (Niiyama 2002).

These willow species are characterized by similar ecological and physiological features with wide seed dispersal ranges, quick germination and fast growth after establishment. However, sometimes they coexist and sometimes they segregate from each other, spatially and temporally (Niiyama 1987, 2002). The coexistence and segregation mechanisms can be explained by the resource partitioning, and the spatial and temporal refuges provided by the differences in regeneration niches and by the storage effect (Silvertown & Lovett Doust 1993).

Salix hukaoana Kimura, endemic to Japan, is a rare willow species which occurs sporadically in several localities in central to northern Honshu, forming small populations in the upper reaches of some large river systems (Suzuki & Kikuchi 2006). Because of its limited geographical range, this species is regarded as a relic species which might have once been more widely distributed in the lowland floodplains of northern Japan during the last glacial periods (Kimura 1989), similar to *Salix rorida* Lackschewitz and *S. arbutifolia* Pall. (formerly *Chosenia arbutifolia* (Pallas) A. Skvortsov), and may have gradually reduced its range during the inter- or post-glacial period, approximately 10,000 years ago (Tsukada 1974). However, these small and fragmented *S. hukaoana* populations are now facing extinction due to extensive human disturbances and/or environmental changes in their habitats, directly or indirectly caused by river engineering projects such as the construction of check dams and artificial channels in the 60 years since the end of World War II (Sakio & Suzuki 1997). Today, *S. hukaoana* is referred to as a Class IB endangered species and is listed in the Red Data Book (Japanese Ministry of Environment 2000).

In the present paper, we first describe the taxonomy and morphology of *S. hukaoana*, and its geographical and ecological distribution. Then, we examine floristic compositions and stand structures in riparian habitats, including this *Salix* species, in an effort to understand how this rare willow species endemic to Japan can maintain its populations, through occasional disturbances in riparian habitats.

Based on the field data obtained from all of its known habitats in Japan, we will also consider how river engineering has changed the riparian disturbance regime, and is threatening populations of this willow species in such changing habitats.

19.2 Taxonomy and morphology

Salix hukaoana Kimura (Yubiso-yanagi, Japanese common name) was first discovered at the Yubiso River, a tributary of the Tone River Basin, Gunma Prefecture, central Japan in 1972 by Mr. Shigemitsu Hukao. This new willow species endemic to Japan was referred to as a new section of the genus *Salix* (Kimura 1973). Because of its morphological similarity, it was first considered as a sister species of *S. rorida*, which widely occurs in north-eastern Asia, including Hokkaido, as well as of *Salix kangensis* Nakai which occurs in northern regions of the Korean Peninsula to northeastern China. Then *S. hukaoana* was considered to be endemic to Japan, belonging to a new section (sect. *Hukaoanae* Kimura) of the genus *Salix*, because of its unique morphological feature of having two ovaries in each placenta (Kimura 1974). However, recent molecular data have indicated that phylogenetically, *S. hukaoana* is a species more closely related to *S. rorida* (Kikuchi, unpublished data).

Salix hukaoana is also morphologically similar to *S. udensis* Trautv. & Mey. (formerly *S. sachalinensis* Fr. Schmidt) which occurs in the same habitats as *S. hukaoana*. Therefore, this species has often been confused with *S. udensis*, which may explain why it had not been discovered until quite recently.

S. hukaoana has a characteristic leaf margin, while that of *S. udensis* is not clear. *S. hukaoana* also has a unique floral feature with two united stamens, while *S. rorida* has two separate and distinct stamens (Kimura 1989). However, similar to *S. rorida*, the inner bark of *S. hukaoana* is clear yellow, which has not been observed in any other willow species in Japan.

S. hukaoana is a tall tree species, with a maximum height of 25m and maximum dbh (diameter at breast height) of 73.2cm (Suzuki & Kikuchi 2006). Maximum age is estimated to be about 45 years old. Thus, compared to the larger-sized willows, such as *Salix cardiophylla* Trautv. & Mey. var. *urbaniana* (Seemen) Kudo (formerly *Toisusu urbaniana* (Seemen) Kimura) or *Salix dolichostyla* Seemen (formerly *S. jessoensis* Seemen), *S. hukaoana* could be recognized as a pioneer species with rapid growth and short life span (Suzuki, unpublished data).

S. hukaoana is dioecious, like many other willow species. Minimum diameter for flowering in *S. hukaoana*, however, is only 3-4cm at the stem base, and the age of such trees is estimated to be only three to four years old; thus, its pre-reproductive phase is exceedingly short for a woody species (Suzuki, unpublished data).

19.3 Distribution

19.3.1 Geographical distribution

Salix hukaoana was first believed to occur only in the floodplains along the Yubiso River, a tributary of the Tone River in Gunma Prefecture. Since then, however, this species was discovered at the upper reaches of the Naruse River in 1983, along the Ikusa River, a tributary of the Eai River in Miyagi Prefecture in 1985 (Takehara & Naito 1986), and at the upper reaches of the Waga River, Iwate Prefecture in 1993 (Takehara 1995). These findings show that *Salix hukaoana* is a rare woody species, having disjunct distributions in different river systems in central to northern Honshu (Fig.1).

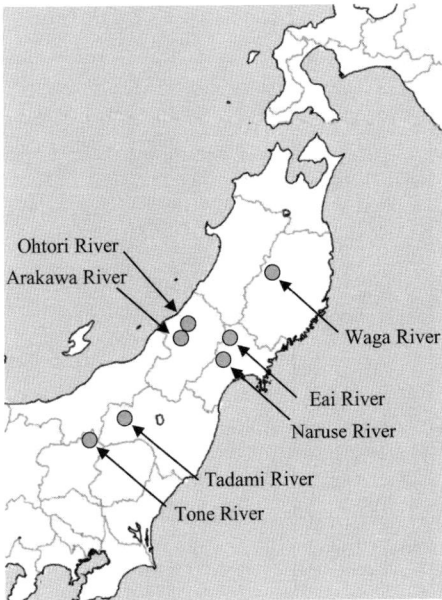

Fig. 1. Geographical distribution of *S. hukaoana*

Table 1. Site characteristics of *S. hukaoana* habitats

| Site | Altitude | | Warm index | Maximum snow depth | Basin area | | Width of valley bottom | | Slope degree | Rive system | |
	Upper limit (m)	Lower limit (m)		(cm)	Whole area (km²)	Till upper limit (km²)	Max. (m)	Min. (m)	(°)	Total length (km)	Till upper limit (km)
Waga River	460	260	69.8	174	10150	10.7	850	100	0.6	249	8.1
Eai River	360	320	72.5	160	10150	23.7	200	150	0.8	249	9.4
Naruse River	440	400	68.7	230	1130	10.1	150	50	2.0	89	6.7
Tadami River	630	360	78.8	231	7710	261.3	1050	150	0.4	210	24.2
Tone River	800	520	81.2	150	16840	17.7	200	50	2.1	322	8.5

The recent discovery of a new population of this species along the Ina River in Fukushima Prefecture, a tributary of the Tadami River between the Yubiso River and the Naruse River, provides additional evidence that supports our earlier assumption (Suzuki & Kikuchi 2006). First, it had been thought that this species occurred only on Japan's Pacific Ocean side. However, additional habitats were reported from the areas along the Ohtori River and at the upper reaches of the Arakawa River, Yamagata Prefecture, both located in the central mountain ranges on the Sea of Japan side of Honshu (Ohashi et al. 2007).

Table 1 shows environmental conditions of the five known habitats of *S. hukaoana* on the Pacific side of Honshu. The warmth index of the habitats is 70-80, and the maximum snow depth is 150~230cm, suggesting that its habitats extend over a cool temperate climate zone, but with very heavy snowfalls in winter. *S. hukaoana* habitats are located at the upper reaches of large-scale river systems in Japan, whose watersheds (except for the Naruse River) cover roughly more than 10000 km^2. In addition, the valley bottoms of the habitats are wider and the inclination of the riverbeds is very gentle, less than 2° (Table 1).

All *S. hukaoana* habitats can be characterized as wide floodplains at the upper reaches of large river systems at the foothills of central mountain ranges in northern to central Honshu. The geographical distribution of this willow species suggests that it belongs to the rare willow group that occurred widely in lowland floodplains during the last Ice Age, and then had been driven into the upper reaches which were once much cooler. Similar cases have been known in the isolated distribution patterns of *S. rorida* and *S. arbutifolia* along the Azusa River in Nagano Prefecture (Kimura 1989).

The largest known population of *S. hukaoana* is in the Yubiso River basin. This local population is composed of 2,820 individuals whose stem size is at least 5cm in dbh, including 143 trees larger than 30cm in dbh (Abe, unpublished data). On the other hand, the largest habitat is found in the Ina River basin, Fukushima Prefecture, 45km along the main stream and its tributaries (Suzuki & Kikuchi 2006).

Unfortunately, these habitats and riparian disturbance regimes have been directly or indirectly changed, or even destroyed by river engineering projects, such as the construction of check dams and artificial river banks. As a result, local populations have become increasingly fragmented, isolated, and dramatically reduced in size. *S. hukaoana* is now facing extinction, and thus is listed in the Red Data Book as a Class IB endangered species (Japanese Ministry of Environment 2000).

19.3.2 Ecological distribution

Floodplains that have developed on wide valley bottoms are a mosaic of various micro-topographies, each of which has vegetation patches with different stages of ecological succession (Shin et al. 1999).

A 300m wide floodplain at the upper reaches of the Yubiso River is one of the typical habitats of *S. hukaoana*. There are different types of forest vegetation,

Table 2. Floristic composition of stands developed on different micro-topographies from the valley bottom and the adjacent hill-slope at the Yubiso River, central Japan

Species name	Plot name					
	AC1	AC2	LF1	LF2	HF	HS
Salix hukaoana	(14800)*	(9200)	28.88	14.61		
Salix cardiophylla var. urbaniana			6.2	7.78		
Salix udensis	(400)	(3600)		0.71		
Salix jessoensis subsp. serissaefolia	(400)					
Pterocarya rhoifolia			8.68	15.41		
Ulmus davidiana var. japonica	(400)		6.25			
Betla platyphylla	(400)					
Morus australis			0.14	0.15		
Prunus grayana			0.18			
Aesculas turbinata			0.09	0.11	30.56	
Quercus crispula		(400)	0.35		9.8	10.02
Acer mono			0.12	0.15	0.39	
Phellodendron amurense				0.07		
Zelkova serrata				0.09		
Acer nipponicum				0.21		
Swida controversa				0.5		0.38
Juglans ailanthifolia					1.83	
Ulmus laciniata					0.52	
Fagus crenata					25.83	38.76
Acer palmatum					0.88	0.06
Acer japonica						0.09
Styrax obasstia						0.21
Acer distylum						0.81
Total basal area ($m^2 ha^{-1}$)			50.89	39.79	69.81	50.33
No. of species	6	3	9	11	7	7
Stand age(yrs)	2	4	20+α	20+α	100 >	100 >

*, No. of saplings larger than 30cm in tree height in parentheses.

reflecting the micro-topographies, from active channels to lower floodplains, higher floodplains and adjacent hill slopes (Table 2). At the active channel, 0.2~0.3m above the water level, there are 2-4 year-old seedling banks composed of *S. udensis*, *S. cardiophylla* and *S. hukaoana*. However, these banks will be disturbed frequently by seasonal changes and eventual flooding in the near future.

On the lower floodplain, distant from the main channel and 0.5m above the water level, a 20-year-old riparian stand exists. The lower floodplain was formed by infrequent riparian disturbances that have occurred once every several decades. It is more stable, and regenerated seedlings and saplings have been able to grow and form stands. The main components of these stands are the willow species mentioned above, and associated with *Salix hukaoana* are riparian species such as *Pterocarya rhoifolia* Sieb. et Zucc. and *Ulmus davidiana* Planch. var. *japonica* (Rehder) Nakai.

On the higher floodplain beside the lower floodplain, a more stable site formed by infrequent riparian disturbances occurring once in several hundred years, there occurs a riparian old-growth stand that is occasionally more than 100 years old. This riparian old-growth stand is mainly composed of late successional riparian components, such as *Aesculus turbinata* Blume and *Acer mono* Maxim., and occasionally main components of upland forests, such as *Fagus crenata* Blume and *Quercus crispula* Blume.

On the adjacent hill-slopes, the old-growth stands are mainly composed of *F. crenata*, more than one hundred years old. Thus, different types of forest communities develop there, each consisting of somewhat different micro-topographic elements at the valley bottom and/or on the adjacent hill-slope.

In general, *S. hukaoana* occurs in open floodplains that have developed over wide valley bottoms, and thus is a representative component of the riparian forests along with other willow and riparian elements. *Salix hukaoana* maintains its populations in unstable habitats with repeated disturbance and rebuilding regimes resulting from riparian disturbances, such as flooding and debris flows.

19.4 Community structures

To understand the stand structures and dynamics of riparian forests, including *S. hukaoana*, studies were conducted at 57 plots, 100-384 m^2 in area. Research plots were established at five different habitats (21 plots at the Tone River, 22 at the Tadami River, 4 at the Naruse River, 5 at the Eai River, and 5 at the Waga River).

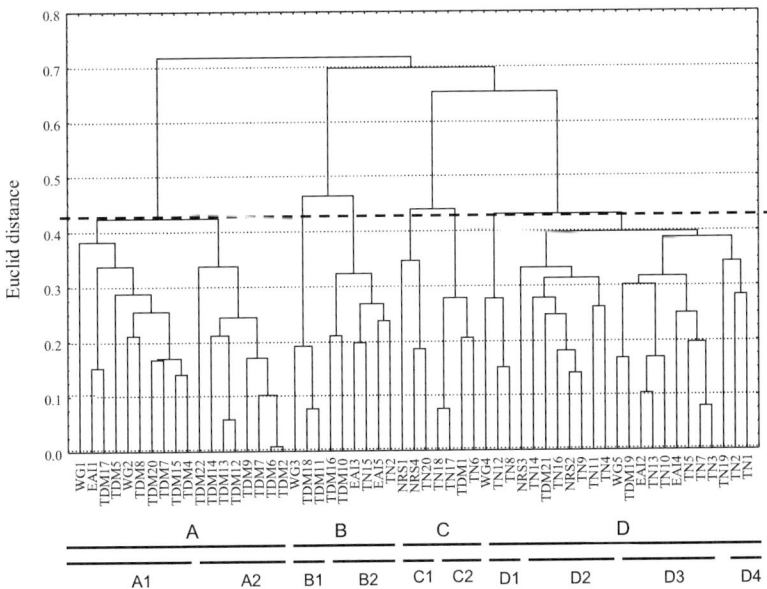

Fig. 2. Dendrogram obtained by the Group Average Method, using the data on tree community dissimilarity among the plots. TN, Tone River basin;, TDM, Tadami River basin;, NRS, Naruse River basin;, EAI, Eai River basin;, WG, Waga River basin

Then, investigations were conducted on floristic compositions, tree density, basal areas, stem size structures, stand age and habitat conditions, such as the distances from the active channel and elevations from water level and depth of the soil layer for each plot.

A total of 57 study plots including *S. hukaoana* were classified into four major community types (A, B, C and D) at the 0.5 level of Euclid distance by cluster analysis (Group average method). Each major type was further classified into several subtypes, i.e, A1 and A2, B1 and B2, C1 and C2, and D1, D2, D3 and D4, respectively, at the 0.4 level of distance (Fig. 2).

The floristic composition of each vegetation type is shown in Table 3. The four major types, A, B, C and D, were characterized by the degree of dominance of *S. dolichostyla*, *S. udensis*, *S. cardiophylla* and *S. hukaoana*, respectively. Type A was subdivided into Type A1, mainly composed of *S. dolichostyla* and *S. hukaoana*, and Type A2 dominated by *S. dolichostyla*. Type B was subdivided into Type B1, with a high relative dominance value of *S. udensis*, and Type B2

Table 3. Floristic composition of each stand type in the study plots

Type	A	A	B	B	C	C	D	D	D	D	Frequency
Sub-type	A1	A2	B1	B2	C1	C2	D1	D2	D3	D4	(%)
No. of plots	10	8	3	6	3	4	3	8	9	3	
Salix hukaoana	40.5	21.3	15.8	24.4	35.8	11.4	44.3	69.8	48.9	44.1	100.0
Salix udensis	6.8	6.0	84.2	44.2	1.3	10.0	2.7	2.3	22.6	0.0	61.4
Pterocarya rhoifolia	1.4	1.7		6.9	4.4	3.5	33.9	5.3	5.9	2.5	52.6
Salix dolichostyla	38.5	67.2		4.5				0.2	1.8		45.6
Salix cardiophylla var. urbaniana	1.0			4.2	40.9	68.3	14.4	7.4	10.1	10.3	43.9
Morus bombycis	1.1	0.3		1.2	0.4		0.3	0.1	0.0	0.3	28.1
Swida controversa	0.4			0.2	0.1	0.3	1.8	0.5	0.2	0.4	28.1
Alnus hirsuta	0.8			4.7	5.2			3.0	4.4		22.8
Betula grossa				0.6		2.2		1.3	2.0	9.4	22.8
Ulmus davidiana var. japonica	0.7			2.0		0.1		1.8	1.0	0.3	21.1
Acer mono				1.3	1.3		0.2	2.0	0.0	2.2	21.1
Aesculus turbinata	0.1				0.1		0.5	0.6	0.2	5.1	15.8
Juglans ailanthifolia	3.4	0.1		1.1		3.3				1.3	14.0
Robinia pseudoacacia	1.2				1.3			0.1	0.6	12.1	12.3
Quercus crispula		3.4					0.1	0.4	0.1		12.3
Acer palmatum	0.1			0.5		0.1		0.1		0.5	12.3
Prunus grayana	0.1					0.0		0.0		0.4	12.3
Zelkova serrata				0.2	0.3		0.1		0.1	4.8	10.5
Fagus crenata					1.5		0.1	0.7		0.4	8.8
Alnus pendula				0.9		0.2	0.5		0.5		7.0
Carpinus japonica					0.1				0.6	1.5	7.0
Populus maximowiczii	2.8				7.3			3.0			5.3
Salix integra	0.8										5.3
Cercidiphyllum japonicum	0.0			0.0						1.2	5.3
Ulmus laciniata				0.4				0.1	0.0		5.3
Parabenzoin praecox	0.1						0.3			0.5	5.3
Acer rufinerve						0.1		0.0	0.1		5.3
Acer sieboldianum									0.1	0.2	5.3
Salix jessoensis subsp. serissaefolia								1.2		1.7	3.5
Cryptomeria japonica									0.6	0.2	3.5
Carpinus laxiflora							0.7		0.0		3.5
Phellodendron amurense				0.3			0.1				3.5
Salix bakko								0.1		0.3	3.5
Elaeagnus multiflora				2.1							1.8
Alnus japonica	0.2										1.8
Betula ermanii						0.5					1.8
Castania serrata				0.3							1.8
Acer nipponicum							0.2				1.8
Malus baccata	0.1										1.8
Clethra barbinervis										0.1	1.8
Magnolia hypoleuca	0.0										1.8
Quercus serrata					0.1						1.8
Euonymus sieboldianus								0.1			1.8
Acer japonicum								0.0			1.8
Mean No. of species	5.9	3.1	2.0	7.3	7.7	5.8	9.7	7.0	6.9	11.3	

was composed of various tree species as well as *S. udensis*. Type C was sub-divided into Type C1, having a similar dominance value in both *S. cardiophylla* and *S. hukaoana,* and Type C2 having a high relative dominance value only in *S. cardiophylla*. Type D was further divided into four subtypes by the floristic composition: Type D1 dominated by *Pterocarya rhoifolia* and *S. hukaoana*, Type D2 dominated by *S. hukaoana*, Type D3 dominated by *S. hukaoana* and *S. udensis*, and Type D4 composed of many additional tree species, including *S. hukaoana*. These riparian community types containing *S. hukaoana* were similar to the "Salicetum jessoensis Ohba 1973" which is widely distributed in the flood-plains of north-eastern Honshu (Miyawaki 1987).

The largest number of study plots among the forest types distinguished in the field survey was 23 stands of Type D (40.4% of the total stands), followed by 18 stand of Type A (31.6% of the total). Among the subtypes, Type A1 characterized by the dominance of *S. dolichostyla* and *S. hukaoana* included the largest number of stands, 10 (17.5% of the total study stands), followed by Type D3 (9 stands, 15.8%), and types A2 and D2, with eight stands (14.0%) each. In regional terms, riparian forest communities containing *S. hukaoana* in the Tone River basin were dominated by type D, which accounted for 71.4% of the total stands, while those in the Tadami River basin were dominated by type A, which accounted for 68.2% (Fig.3).

Fig. 4 shows the relationship between stand age and BA. As stand age in-creases, BA increases. However, each forest type included a wide range of stand ages, and thus there was a significant difference in stand age among the four forest types A, B. C and D (Kruskal-Wallis test, p=0.028). The oldest mean stand age was Type D, followed by Type A, Type C, and the youngest, Type B. These results cannot be explained in terms of ecological succession, because *S. dolichostyla* and *S. cardiophylla* are much longer lived than the other two willow

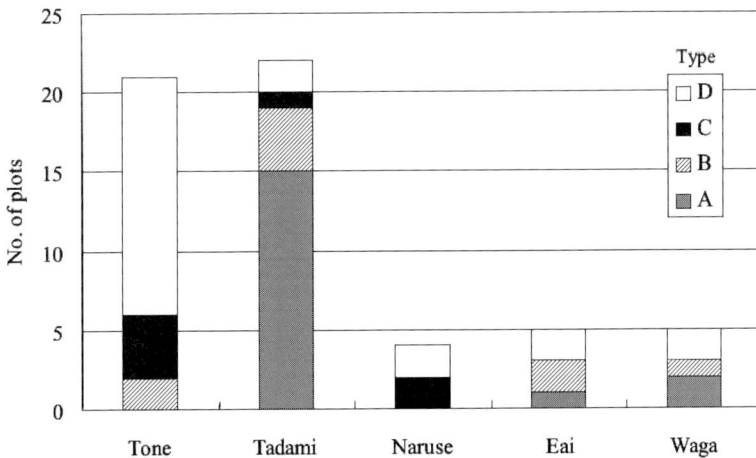

Fig 3. Forest type composition in a river basin

species (Niiyama 2002).

There was a significant positive correlation between the stand age and relative BA of *S. hukaoana* in the study plots (r_s=0.29, p=0.029), although the relative dominance values (basal area) of *S. hukaoana* varied widely from 5.4% to 90.7% (Fig.5). A significant difference in the relative BA among the main types and subtypes was also observed (Kruskal-Wallis test, p<0.001). These results suggest that forest types were basically determined by the relative dominance value of *S. hukaoana*. However, *S. hukaoana* will decline as the stand age exceeds 50 years old, and be replaced by long-lived willow species, such as *S. cardiophylla* var. *urbaniana*, *S. dolichostyla* and *S. jessoensis* subsp. *Serissaefolia* (Kimura) H. Ohashi (formerly *Salix serissaefolia* Kimura) (Niiyama 2002), and/or late successional riparian elements, such as *Cercidiphyllum japonicum*, *Aesculus*

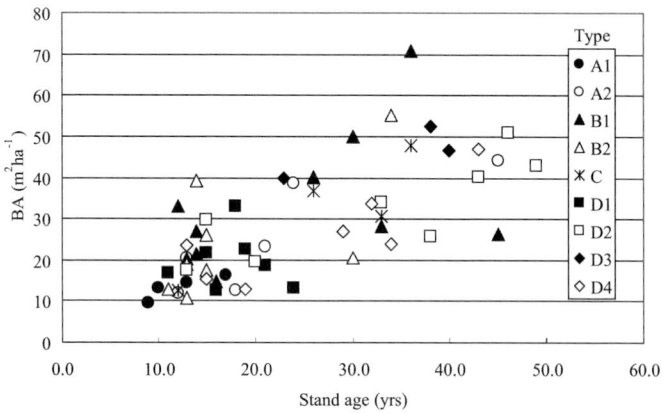

Fig 4. Relationship between stand age and total basal area (BA) in each of the riparian stands including *S. hukaoana*

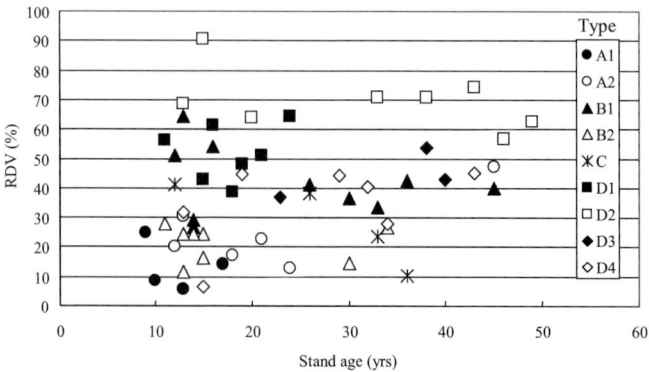

Fig 5. Relationship between stand age and relative dominance value (RDV) of *S. hukaoana* in each of the riparian stands

turbinata and *Ulmus davidiana* var. *japonica* (Shin et al. 1999).

19.5 Ecological succession and species diversity

Forty-four tree species (dbh >5cm) were observed in the study plots which were established in the habitats of *S. hukaoana*, and tree species showing the highest frequency were four willow species, *S. cardiophylla* var. *urbaniana, S. udensis, S. dolichostyla* and *S. hukaoana*. Riparian pioneers, such as *Pterocarya rhoifolia* and *Morus bombycis* Koidz., *Swida controversa* (Hemsl.) Soják, *Alnus hirsute* Turcz., and others were lower in frequency (Table 3).

Fisher's α index of species diversity varied widely from 0.5 to 10.2 (Fig.6). There was a positive correlation between stand age and Fisher's α (r_s =0.74, $p<0.001$). Significant differences in Fisher's α among the forest types (A, B, C and D), and major subtypes (A1, A2, B2, D2 and D3) were recognized (Kruskal-Wallis test, $p=0.003$).

The increment of species diversity indicated by Fisher's α with increasing stand age, was reflected in the ecological succession from young stands only being composed of willow species that had regenerated after riparian disturbance, to old-growth stands composed of late successional riparian species.

However, old stands including *S. hukaoana* with a high species diversity did not exist because of the short life span of *S. hukaoana* (less than 50 years) and/or due to frequent riparian disturbances.

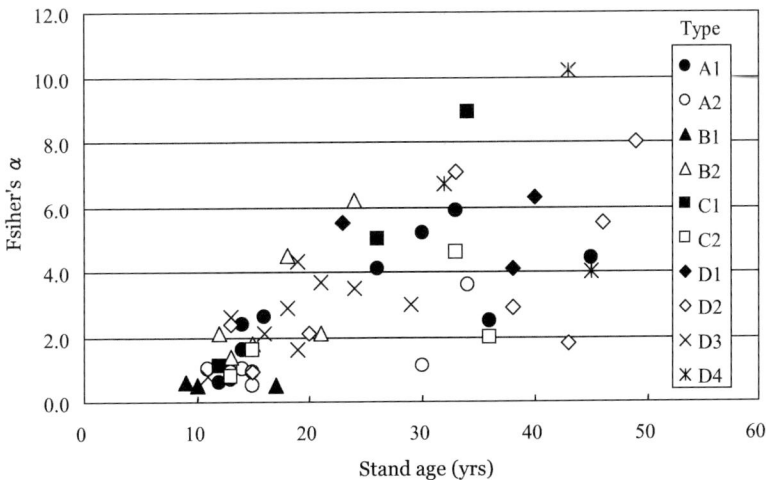

Fig. 6. Relationship between stand age and Fisher's α of each stand in all the study plots.

19.6 Life-history strategy of *S. hukaoana*

S. hukaoana flowers in middle to late April, one of the early blooming species of the Salicaceae, and is similar to *Salix gracilistyla* Miq. which occurs in the same habitats (Niiyama 2002). Seed dispersal of this species starts in late May after the snowmelt floods have ended, and the seeds with comospores are spread widely by wind. This species has a unique regeneration strategy, disseminating its seeds to newly disturbed sites created by snowmelt floods. Other willow species that coexist with *S. hukaoana* at the same sites, but disperse their seeds in different seasons, from mid-June to early September, have their own regeneration strategies that have been adapted to different riparian environments exposed to disturbances occurring in different seasons. This results in the spatial habitat segregation and different floristic composition of the stands (Niiyama 2002).

Dispersed seeds of *S. hukaoana* can germinate readily at a very high rate (almost 100% in a few days), regardless of whether the temperature regime is constant or fluctuating, and even without light if the soil is suitably wet (Ban & Ide 2004; Suzuki, unpublished data). However, seedlings grow faster under good light conditions. Hence, newly disturbed sites with sediment accumulation may be the most suitable habitats for this species. However, these sites are not stable due to repeated riparian disturbances such as occasional flooding, which tend to destroy the seedling and sapling populations of this species that have regenerated there.

Therefore, the habitats where this species can survive and grow are restricted to stable sites, somewhat distant from the active channel and/or much higher than the stream, such as higher floodplain deposits formed by infrequent large-scale riparian disturbances and/or around abandoned channels formed by channel migration.

These sites might also be favorable for other willow species, and thus mixed forests of some willow species occasionally develop there. It is believed that the main factors that determine the floristic composition are the differences in reproductive strategies, especially in seed dispersal and establishment, as well as the timing of riparian disturbances, and dispersal distances from the seed sources, but soil texture and/or soil moisture conditions seem to be minor factors.

Only *S. hukaoana* populations that have regenerated at stable sites in the riparian zones can grow faster and sometimes form mixed stands with other willow species. Through repeated successions, the short-lived (about 40-50 years) willow species, such as *S. udensis* and *S. hukaoana,* will disappear, while the long-lived willow species (more than 100 years) such as *S. dolichostyla* and *S. cardiophylla* var. *urbaniana,* can survive there. In contrast, the middle and late-successional riparian tree elements, such as *Pterocarya rhoifolia, Juglans ailanthifolia* Carr., *Ulmus davidiana* var. *japonica, Aesculus turbinata* and *Cercidiphyllium japonicum,* invade the sites, grow faster there, and increase their dominance at the next stage of the successional process. As a consequence, *S. hukaoana* disappears in response to the ecological succession in the riparian habitats of more than 50 years old.

Consequently, Japanese endemic willow species, including *S. hukaoana,* depend on riparian habitats subjected to frequent natural disturbances. Its strategy is to migrate into newly disturbed sites, *i.e.*, typical changing environments, and establish its populations. Therefore, its life-history characteristics, such as its fast growth rate, early maturation and wide seed dispersal ability for maintaining its populations, are unique but typical among species which have their niches in the riparian disturbance regimes.

19.7 Impacts of river engineering on species conservation

In this section, we will examine some examples of river engineering projects which threaten the maintenance of *S. hukaoana* populations, and the existence of this species at the local level. One of the greatest threats to *S. hukaoana* habitats is large-scale dams for hydroelectric power generation. Although it is not clear what the original habitats were before the first discovery of the species in 1972, there is a possibility that some of these habitats were inundated by the construction of Lake Okutone at the upper reaches of the Tone River in Gunma Prefecture, and Lake Tagokura in the upper reaches of the Tadami River in Fukushima Prefecture.

Another threat is the construction of check dams for erosion control. Direct

Fig. 7. River engineering works have drastically changed both riparian environments and the natural disturbance regime, and threaten many flora and fauna that depend on the riparian habitats at the upper reaches of the Waga River.

impacts to the habitats and/or populations of *S. hukaoana* include felling trees, and destroying habitats when dams and access roads are constructed (Fig.7). This is an ongoing problem for *S. hukaoana*. Indirect impacts include changes to the riparian disturbance regime that occur when check dams stabilize river-beds. This prevents the vegetation from regenerating. While gravel deposits that extend toward the backwater of the check dams may provide safe sites for regeneration and growth, the population dynamics of this species there are not well known.

However, the most significant problem associated with river engineering works is the construction of artificial river banks. Artificial river banks keep the active channel narrow and straight, and separate it from the floodplain. This type of alteration has been done widely along rivers, resulting in wide destruction of the populations and habitats of *S. hukaoana*. Therefore, the riparian forests including *S. hukaoana* are cut off from the riparian disturbances, and this willow species will gradually decline and be driven away from riparian forests as the ecological succession progresses.

Moreover, it may be impossible for this species to grow and form riparian stands in the gravel deposits within the artificial river banks, because of the narrowness of the gravel deposit within the banks and frequent floods. River engineering works such as check dams and artificial river banks usually cause great disturbance to habitats, but sometimes they provide safe sites for *S. hukaoana* to regenerate, and temporarily increase its population density. However,

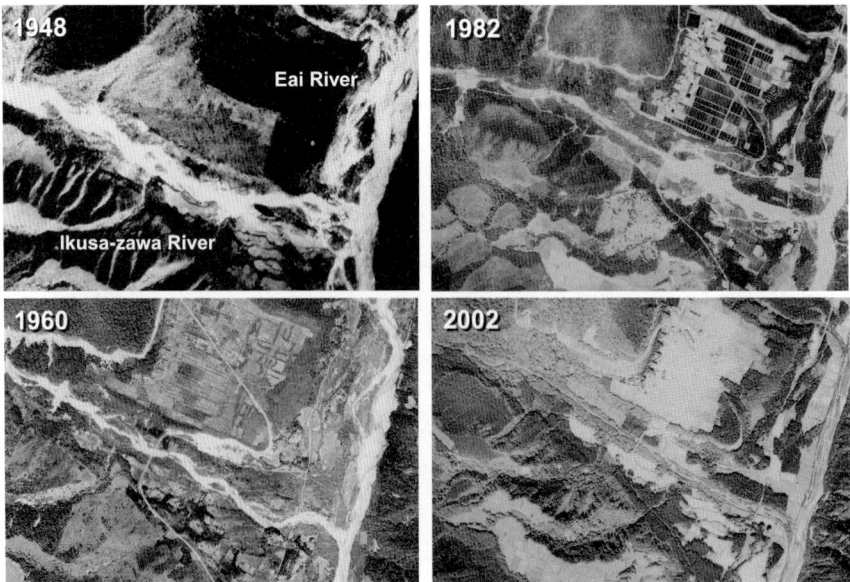

Fig. 8. Aerial photographs of the Eai River and the Ikusa-zawa River from 1948 to 2002. The photo in the upper left was taken by the US Air Force; the others were taken by the Forestry Agency of Japan

this population will not be able to maintain itself in such places over the long term, because of its relatively short life span. Therefore, a riparian disturbance regime which continuously creates safe sites for regeneration is needed for *S. hukaoana* to maintain its population.

Fig.8 shows aerial photographs of the Eai River and the Ikusazawa River, one of the *S. hukaoana* habitats from 1948 to 2002. These river systems have sometimes been disturbed by the construction of check dams and artificial river banks in the past 50 years. As a result, riparian forests including *S. hukaoana* have been repeatedly destroyed, and thus habitats of this species have been considerably changed since then.

19.8. Conservation strategy of *Salix hukuoana*

S. hukaoana is not a common willow species, but neither does it have any specific attractive floral and fruit characters, nor any particular commercial value.

However, this species is seldom encountered because of its restricted geographical and ecological distributions. Furthermore, it is easily confused with *Salix udensis,* which shares the same habitats and has similar morphological features. Nevertheless, *S. hukaoana* thus far has not been widely noticed.

The presence of *S. hukaoana* populations was discovered in 2001 along the Ina River in Fukushima Prefecture. It is possible that new habitats of this species will be discovered in the future. What we have to do is to conserve this endangered rare species, and gain a better understanding its geographical range, and distribution of local populations and their size. Further field surveys should be conducted in future at uninvestigated rivers with similar conditions for its distribution.

The next task is to learn about its life-history features, especially population dynamics in relation to ecological succession and riparian disturbances. Understanding the regeneration processes and maintenance mechanisms of this species is very important for preserving populations of *S. hukaoana* under strong human impacts.

We now must be aware of the current situations for most of the Japanese river systems facing serious artificial modifications. In order to protect the plants and animals, including *S. hukaoana,* etc., which are endemic to or indigenous to the flood plains of rivers, we have to take immediate measures to conserve their native habitats.

The main cause of these issues is a lack of understanding by those who engage in river management. This is one of the reasons why precise information about the distribution and ecology of organisms on the flood plains of rivers, including *S. hukaoana*, is essential now.

The next step involves measures to deal with river engineering works for erosion and flood control. Of course, unnecessary projects should not be undertaken, and actions which would severely affect *S. hukaoana* populations should be avoided. In the case of necessary projects, a concerted effort will be needed to conserve this species by preventing changes to and destruction of its habitats as well as unnecessary tree felling. Even after the completion of river projects, rehabilitation of the *S. hukaoana* habitats should be considered. For example, when riparian forests are separated from the riparian disturbances, work should be considered to improve artificial river banks to help restore the interaction between the riparian areas and the river.

To conserve *S. hukuoana* populations and their habitats, immediate action should be taken to designate the species as a "natural treasure" and/or set aside its habitats as "forest reserves".

Acknowledgments

We thank Dr. Shoichi Kawano, Emeritus Professor, University of Kyoto and Dr. Katsuhiro Ohsumi of the Kansai Research Center of Forestry and Forest Products Research Institute for his comments on an earlier draft of the paper. We are also grateful to Ayako Kanazashi of the Forestry and Forest Products Research Institute, Isamu Nikkuni of the Education Committee of Tadami Town, Fukushima Prefecture, and Naoko Ban of the University of Tokyo, for their help with the field study. The research was supported by a Japanese Ministry of Environment conservation project for rare tree species.

References

Ban N, Ide Y (2004) Life historical traits of *Salix hukaoana* along Yubiso River. Bull Tokyo Univ Forests 112:35-43 (in Japanese with English summary)
Ishikawa S (1983) Ecological studies on the floodplain vegetation in the Tohoku and Hokkaido Districts, Japan. Ecol Rev 20:73-114
Ishikawa S (1988) Floodplain vegetation of the Ibi River in central Japan. I. Distribution behavior and habitat conditions of the main species of river bed vegetation developing on the alluvial fan. Jpn J Ecol 38:73-84 (in Japanese with English summary)
Japanese Ministry of Environment, Wild life Management Section, (2000) Red-data plants in Japan. Japanese Wildlife Research Center, Tokyo 662p (in Japanese)
Kimura A (1973) Salicis nava species ex regione Okutonensi in Japonia. J Jpn Bot 48:321-326
Kimura A (1974) De *Salicis Hukaoanae* Kimura systematico positu. J Jap Bot 49:46
Kimura A (1989) Salicaceae. In: Satake Y, Hara H, Watari S, Tominari T (eds) Wild flowers of Japan. Heibon-sha, Tokyo, pp 31-51 (in Japanese)
Miyawaki A (1987) Vegetation of Japan, Tohoku. Shibundo, Tokyo, 605pp (in Japanese)
Niiyama K (1987) Distribution of Salicaceous species and soil texture of habitats along the Tokachi River. Jpn J Ecol 37:168-174 (in Japanese with English summary)
Niiyama K (2002) Floodplain Forests, In: Sakio H, Yamamoto F (eds) Ecology of Riparian Forests. Univ Tokyo Press, Tokyo, pp 61-93 (in Japanese)
Ohashi H, Kikuchi S, Sashimura N, Fujiwara R (2007) Distribution of *Salix hukaoana* Kimura (Salicaceae). J Jpn Bot 82:242-244
Sakio H, Suzuki W (1997) Overview of riparian vegetation: Structure, ecological function

and effect of erosion control works. J Jpn Soc Erosion Control Eng 49:40-48 (in Japanese)

Shin N, Ishikawa S, Iwata S (1999) The mosaic structure of riparian forests and its formation pattern along the Azusa River, Kamikouchi, central Japan. Jpn J Ecol 49:71-81 (in Japanese with English summary)

Silvertown JW, Lovett Doust J (1993) Introduction to plant population biology. Blackwell Sci Pub, Oxford, 210p

Suzuki W, Kikuchi S (2006) Floristic composition and stand structure of riparian forests in the Tadami River basin, and the ecological distribution of an endangered tree, *Salix hukaoana*. Jpn J Conserv Ecol 11:85-93 (in Japanese with English summary)

Takehara A (1995) Willow communities and the distribution of *S. hukaoana* in the upper reaches of the Waga River, Northern Honshu, Japan. Ann Rep Natural History 1:11-21 (in Japanese)

Takehara A. and Naito T. (1986) *Salix hukaoana* Kimura, newly found in Miyagi Pref. in northeastern Honshu. J Jpn Bot 61:127-128

Tsukada M (1974) Paleoichnology II. Kyoritsu Shuppan, Tokyo, 231p (in Japanese)

Yoshikawa M, Fukushima T (1999) Distribution and developmental patterns of floodplain Salicaceae family communities along the Kinu River, central Japan. Veg Sci 16:25-37 (in Japanese with English summary)

20 Strategy for the reallocation of plantations to semi-natural forest for the conservation of endangered riparian tree species

Satoshi Ito[1], Yasushi Mitsuda[2], G. Peter Buckley[3] and Masahiro Takagi[4]

[1]Division of Forest Science, Faculty of Agriculture, University of Miyazaki, Miyazaki 889-2192, Japan

[2]Forestry and Forest Product Research Institute, Tsukuba, Ibaraki 305-8687, Japan

[3]Faculty of Life Sciences, Imperial College, Wye, Ashford, TN25 5AH, UK

[4]Filed Science Center, Faculty of Agriculture, University of Miyazaki, Miyazaki 889-2192, Japan

20.1 Needs and problems for the conservation of rare riparian trees

Riparian forests provide habitat for rare or infrequent plant species (Sakio 1997; Suzuki et al. 2002; Sakio & Yamamoto 2002) including several endangered trees (Ito et al. 2003; 2004; Ito & Nogami 2005), and therefore have high conservation value. However, mountainous riparian forests in Japan, particularly in the warm-temperate region, have been heavily exploited or converted to plantations of evergreen conifers such as sugi (*Cryptomeria japonica* D. Don) or hinoki (*Chamaecyparis obtusa* Sieb. et Zucc.) because of their high site productivity (Ito et al. 2003). This has resulted in a severe decline in habitat for rare riparian trees (Ito et al. 2004; Ito & Nogami 2005). Most of the rare tree species dependent on riparian habitat cannot complete their life history beneath the dense canopy of planted conifers because their reproduction (flowering and fruiting) usually requires bright crown conditions. Thus, in addition to maintaining remnant populations in natural forest patches, an important strategy for conserving these rare riparian trees is to restore habitat by re-converting suitable portions of the conifer

Sakio, Tamura (eds) Ecology of Riparian Forests in Japan : Disturbance, Life History, and Regeneration
© Springer 2008

plantations to semi-natural forest (Ito et al. 2004).

In timber production areas, a perennial problem for forest managers is how to combine timber production and biodiversity conservation, including the maintenance of rare species. To manage these competing demands, a reallocation strategy is needed to convert parts of the plantation area to new semi-natural riparian forest covering the potential habitats of populations of target species with a view towards conservation efficiency. As riparian species often depend on specific soil conditions and/or disturbance regimes (Nakamura & Inahara 2007), the expected potential tree density based on prevailing soil and disturbance regimes is a criterion that can be used to select candidate sites for reallocation to semi-natural patches.

Proximity to remnant patches of natural riparian forests would also be an important criterion in the selection of sites for potential habitat. Generally, the connectivity of habitats is a principle for biodiversity conservation in managed landscapes (Kirby et al. 1999). The ecological corridor function of riparian forests has often been emphasized as an important conservation measure for managed landscapes. Moreover, remnant patches containing mother trees of rare species have an important role as seed sources for the restoration of populations utilizing natural regeneration processes. These considerations imply that the spatial arrangement of remnant patches of natural forest is a key factor in reallocating plantations to semi-natural forest in plantation-dominated landscapes.

In this chapter, we describe a systematic strategy for reallocating parts of a managed landscape to semi-natural forest through two case studies carried out for the estimation of potential habitat and examination of conservation efficiency for two rare riparian tree species in the warm-temperate region in southern Japan.

20.2 Case study 1: Estimation of potential habitat for an endangered riparian species

Riparian trees are often strongly dependent on specific topographic conditions. In the estimation of potential habitat for such species at a coarse scale, GIS-based analysis of tree distribution using digital elevation models is quite useful (Næsset 1997; Quine et al. 2002). In these analyses, the target points (pixels of a raster analysis) are usually treated as dispersed points, and their topographic attributes such as slope inclination are generally used without any relationship to other points (Franklin 1995, 1998; Guisan & Zimmermann 2000; Wimberly & Spies 2001a, 2001b). However, for riparian trees, which are dependent on fluvial and geomorphic processes, the attributes of river characteristics that affect the target point are also important in explaining the habitat condition for the species. Similarly, spatial relationships of the target point to the river, such as the horizontal and vertical distance from the channel, would also provide useful information.

In a recent study, the authors compared the usefulness of these different levels of topographic information in identifying potential habitat for *Lagerstroemia subcostata* Koehne var. *fauriei* (Koehne) Hatus. ex Yahara (LSF) (Ito et al. 2006). LSF is a rare riparian species native to mountainous riparian areas in Yakushima Island in southern Japan (Environmental Agency Japan, 2000). This species can be found on debris flow deposits along stream channels or on talus deposits in small catchments less than ca. 1000 ha. We measured the actual tree distribution of LSF including saplings and seedlings in a sample area (ca. 15 ha) within catchments A and B (60 ha in total) located in the western part of Yakushima Island (Fig 1). Based on this data, we modeled the distribution of tree density for 11,280 pixels with dimensions of 12.5 x 12.5 m covering the whole area of the two catchments by using the Poisson loglinear model and GIS-derived topographic factors. The modeling was performed by adding the four different levels of topographic factors (Table 1) in order to examine the validity of the factor levels: Level-1) topographic attributes of the target cell itself such as elevation or slope inclination; Level-2) cell position relative to the adjacent channel; Level-3) attributes of the adjacent channel; and Level-4) catchment differences.

We found that AIC (Akaike's Information Criterion) values of the obtained models decreased with the addition of factors from Level-1 to Level-4 (Fig. 1, Table 1). This indicated that the topographic attributes of the target cell itself (Level-1) cannot sufficiently explain the potential habitat for LSF (i.e., AIC remains high). Adding factors related to spatial relationship to the channel (Level-

Table 1. Summary of Poisson loglinear models. (After Ito et al. 2006)

Model	Inputted factors	Adopted variables*	AIC
Model-1	Level-1	(Level-1) log*DEM*, *Slope*, log*D2S*, log*FA*	14683
Model-2	Level-1 to Level-2	(Level-1) log*DEM*, *Slope*, log*D2S* (Level-2) *D2NNR*, *S2NNR*	13409
Model-3	Level-1 to Level-3	(Level-1) log*DEM*, *Slope*, log*D2S* (Level-2) *D2NNR*, *S2NNR* (Level-3) *NNRI*, *NNRD2C*	12730
Model-4**	Level-1 to Level-4	(Level-1) log*DEM*, *Slope*, log*D2S* (Level-2) *D2NNR*, *S2NNR* (Level-3) *NNRI*, *NNRD2C* (Level-4) *Catchment* (intercept)	8338

DEM, elevation; Slope; slope inclination; *D2S*, distance to the coastline; *FA*, flow accumulation; *D2NNR*, distance to the nearest channel; *S2NNR*, slope inclination to the nearest channel; *NNRI*, riverbed inclination of the nearest channel; *NNRD2C*, distance to the nearest junction of channels.

**Model-4 was constructed to have different intercept and partial regression coefficients for the two catchments.

a) Measured tree position   b) Model 1   c) Model 2

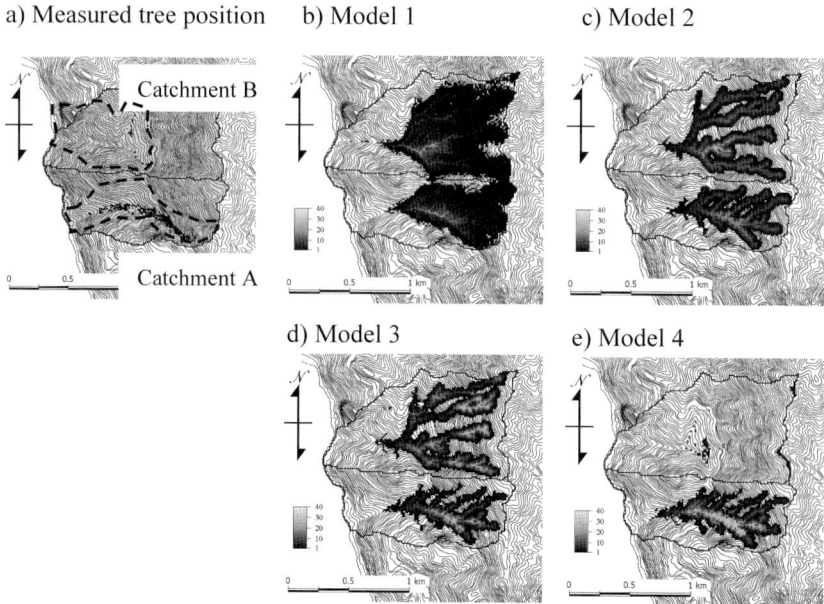

d) Model 3   e) Model 4

Fig. 1. Distribution of measured and estimated tree density of *Lagerstroemia subcostata* var. *fauriei* calculated by Poison loglinear models (modified from Ito et al. 2006). a) Field survey area (surrounded by broken lines) and measured tree position (dots), b-e) Tree density estimated by models. (After Ito et al. 2006)

2) and/or the channel characteristics (Level-3) clearly improved the accuracy of prediction (i.e., lower AIC) of habitat for LSF; this is presumably because these factors explain the occurrence of sedimentation along the channels. Moreover, the models adopting Level-4 factors showed the best performance (Fig. 1, Table 1) as they improved on the apparent over-estimation in catchment B of the other models. This might reflect differences between the catchments in the distribution of sediment sources and seed sources. Sediment sources such as small landslides upstream are closely related to the formation of downstream debris flow terraces that form a suitable habitat for LSF (Miyawaki 1980; Ito & Nogami 2005). An effective seed source is also critical for the establishment of a target species on physically suitable habitat (Ito et al. 2004). In the field survey, we observed more sediment sources and mature canopy (seed) trees in catchment A than catchment B.

We can assume that the geomorphic factors of Level-1 to Level-3 are less variable (or it is more difficult to detect differences from map information) within the time scale of LSF population dynamics. In contrast, the catchment characteristics (Level-4), such as sediment and seed sources, could vary drastically due to natural and human disturbances or artificial modification of the forest landscape. Therefore, the model-4 obtained for catchment A, where the sediment and seed sources are assumed to be sufficient, is the most appropriate for the prediction of "potential" habitat for LSF at a coarse resolution over a broad area. However, once we come to establish a conservation strategy or evaluate the effectiveness of possible countermeasures for conservation for a particular catchment, we have to put the catchment-specific characteristics, i.e., actual distribution of sediment and seed sources, into the prediction model.

20.3 Case study 2: Reallocation of plantation to semi-natural forests based on expected tree density

Once the potential habitat for the rare riparian species has been estimated based on identification of a suitable physical environment (i.e., suitable micro-topography), the success of tree regeneration in delineated areas after reallocating them to semi-natural forest would strongly depend on biological factors such as seed dispersal. The authors simulated the consequences of reallocating parts of a conifer plantation dominated landscape to semi-natural forest for the conservation of another riparian rare species, *Quercus hondae* Makino, taking into account the quality of the potential habitat and seed sources in the remnant natural forest patches (Ito et al. 2004). In this section, we provide a summary of our simulation results and discuss the importance of a strategic approach to the reallocation of plantations to semi-natural forest.

Quercus hondae is an evergreen oak found only in southeastern Kyushu and southern Shikoku (Mashiba 1973; Ito et al. 2004; Ito et al. 2007). The natural distribution of *Q. hondae* is limited to lower slopes along mountainous streams (Miyawaki 1981; Ito et al. 2000). This species has been designated as an endangered species (Environmental Agency of Japan, 2000) owing to its limited distribution and decreasing population due to the loss of natural habitats by the establishment of conifer plantations.

We used the following equation to estimate the expected tree density of *Q. hondae* on our study site (6784 pixels of 12.5 m x 12.5 m each) when the conifer plantation is converted to semi-natural forest:

$$De = Dmax \cdot f_1 \cdot f_2 \quad (1),$$

where *De* and *Dmax* are the expected tree density and maximum density, respectively, and f_1 and f_2 are constraint functions (varying from 0 to 1) of physical habitat quality (represented by micro-topography) and seed dispersal (distance from the natural forest patch), respectively. Based on the field survey of a natural forest (5.3 ha in total), we determined *Dmax* to be 45.1 per ha, which was the density of reproductively mature trees observed on the lower slope. The function f_1 was determined as the discrete variant for each of the three micro-topographies according to the observed ratio of tree density in natural forest to that of the lower slope: 1.000 for the lower slope, 0.358 for the upper slope and 0.040 for the slope crest, respectively (Table 2). The function f_2 consists of two factors, which are the

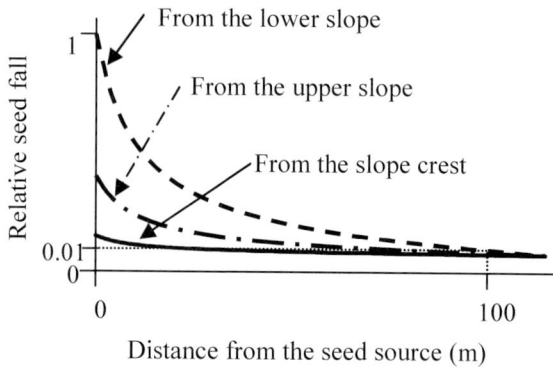

Fig. 2. Schematic drawing of the seed dispersal constraint function ($f2$) used to estimate tree establishment that would occur after reallocation of sites to semi-natural forest. (After Ito et al. 2004)

Table 2. Assumptions in predicting *De*. (After Ito et al. 2004)

Case	Habitat constraint function (f_1)	Seed-dispersal constraint as a function of distance from seed-source (f_2)
Case-1	1.000: 0.358: 0.040 (lower slope : upper slope : slope crest, based on measured relative density of target species on these elements in natural forest)	Exponential decrease from different values according to the micro-topography of the seed-source patch*
Case-2	Constant for all micro-topography	Linear decrease from a constant value

* Shown in Fig. 2. See the text for more details.

constraint functions of the relative amount of seed fall and quality of the seed source patch. The constraint function of relative amount of seed fall was basically defined as reducing from 1.0 at the edge of the seed source patch (distance = 0 m) to 0.01 at 100 m distant from the seed source assuming an exponential decrease. We also weighted this constraint score by the quality of the seed source patch according to micro-topography as described for f_1 (Fig. 2, Table 2).

A simulation was performed to reallocate parts of the site to semi-natural forest patches based on expected tree density. Different assumptions were adopted for the simulation, i.e., expected density was predicted with and without considering functions f_1 and f_2 (Table 2). In Case-1, f_1 and f_2 were fully adopted in predicting *De*, while in Case-2 we did not take into account site quality (f_1) and assumed that seed dispersal was simply proportional to the distance from the seed source and independent of the topographic position of the seed source patch. Sites for

a)   b)

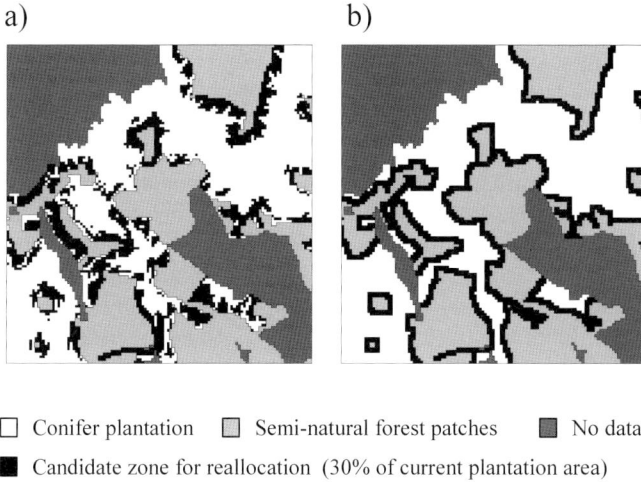

☐ Conifer plantation   ☐ Semi-natural forest patches   ■ No data

■ Candidate zone for reallocation (30% of current plantation area)

Fig. 3. Candidate zones for reallocation to semi-natural forest based on the estimated tree densities (*De*) of *Q. hondae* with different assumptions. a) Case-1: *De* based on distance, seed source quality and habitat quality for regeneration, b) Case-2: *De* based on distance from the seed source. (After Ito et al. 2004)

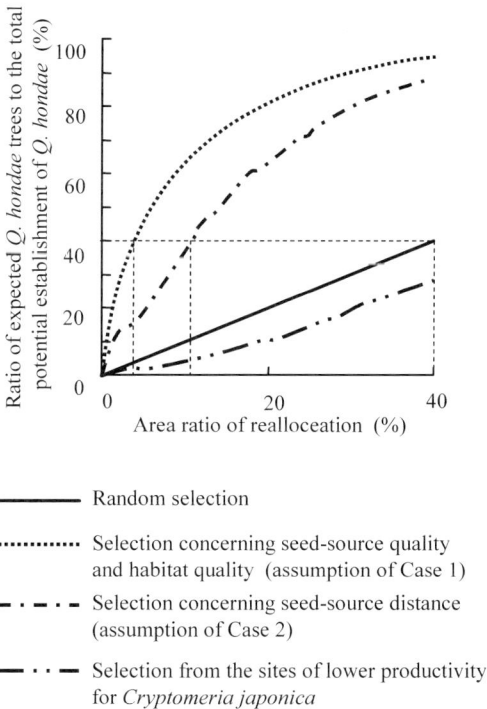

Random selection

Selection concerning seed-source quality and habitat quality (assumption of Case 1)

Selection concerning seed-source distance (assumption of Case 2)

Selection from the sites of lower productivity for *Cryptomeria japonica*

Fig. 4. Comparison of conservation efficiencies for different reallocation scenarios. Efficiency was examined by the relationship between the ratio of the area required for reallocation to the whole plantation area (horizontal axis) and the ratio of the expected *Q. hondae* trees to the total potential number of trees of *Q. hondae* (vertical axis). (After Ito et al. 2004)

reallocation were identified by selecting pixels in decreasing order of De until 30% of the total plantation area was selected.

The results of selection of sites for reallocation (Fig. 3) demonstrated that for Case-1, the candidate sites were distributed not only buffering the remnants of natural forest patches but also connecting them (Fig. 3a). For Case-2, where De was calculated based only on the distance from seed source, sites were selected in a uniform width of buffers surrounding the remnant natural forest patches (Fig. 3b). The result of Case-1 reflected the spatial distribution of the lower slope, indicating the indirect effect of site selection based on the micro-topography on the improvement of habitat connectivity. The continuity of semi-natural forest patches is significant not only for plant species conservation but for conservation of all aspects of biodiversity (Kirby et al. 1999). Thus, the establishment of new semi-natural forest patches taking into account the habitat quality based on micro-topography is expected to provide additional benefits for organisms other than the target tree species in this mountainous region.

Figure 4 compares these results with other scenarios from the viewpoint of conservation efficiency. When the reallocation sites were selected randomly, the ratio of the expected tree density to the total potential number of trees that would be achieved by conversion of the whole area of plantations to semi-natural forest (the vertical axis of Fig. 4) would increase proportionally to the increase in percent of the area required for reallocation (the horizontal axis of Fig. 4). The strategic selection of sites for reallocation based on seed source and habitat quality (Case 1 and 2) drastically improved the conservation efficiency, that is, a small percent of the total area was required to obtain the same number of trees. In particular, Case

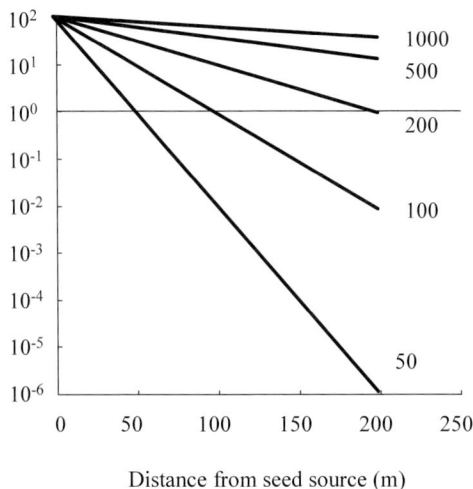

Distance from seed source (m)

Fig. 5. Effects of different hypothetical seed dispersal traits on the seed dispersal constraint. The figure beside each line denotes the distance (m) where the relative seed fall decreases to 1% of that at the edge of the seed source

Table 3. Assumed site preference characteristics for simulations for several hypothetical species

Degree of constraints	Relative tree density		
	Lower slope	Upper slope	Slope crest
Light	1.00	0.75	0.50
Medium-light	1.00	0.50	0.25
Medium-heavy	1.00	0.36	0.04
Heavy	1.00	0.20	0.01

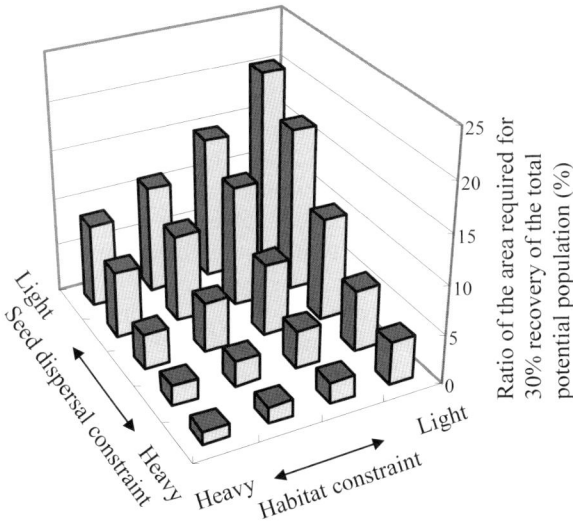

Fig. 6. Comparison of the effect of different hypothetical constraints of seed dispersal and habitat quality on the efficiency of strategic reallocation of plantation to semi-natural forest. The vertical axis indicates the ratio of the area required for reallocation to establish a number of trees equivalent to 30% of the total potential population. Lower values indicate that less area is required for restoration of 30% of the total potential population

1, concerning both the seed source and habitat quality, was more efficient than simple buffering (Case 2). Furthermore, reallocation of sites selected only for having low plantation productivity (c.f. Mitsuda et al. 2007) showed lower efficiency, requiring 10 times the area of Case-1, to achieve the same level of tree establishment.

These simulations indicated the effectiveness of strategic reallocation of plantation to semi-natural forest patches based on ecological factors for conservation planning of the target species. The more random and less strategic is the conservation planning, the lower the conservation efficiency, particularly when other management objectives for the forest landscape compete against the conservation of the target species. In addition, in the case of mountainous riparian species that inhabit sites of high plantation productivity, such as *Q. hondae*, simple site selection for reallocation aiming at minimizing the loss of productive area would not result in successful conservation.

Seed dispersal would be a relatively strong constraint for *Q. hondae*, while habitat constraint might not be so severe because it can regenerate on either slopes or river sediment provided they are located in lower slope positions. Other riparian species might have a less rigid seed-dispersal constraint, i.e., species with wind-dispersed seeds, but stronger habitat constraints limiting them to more specific micro-landforms such as the lower terraces along active channels. Thus, the approach introduced in this case study should be examined for a wider range of species characteristics. We applied the same approach as for *Q. hondae* to several hypothetical species with different levels of seed dispersal capacity (Fig. 5) and habitat specializations (Table 3). We found (Fig. 6) that for species with stronger constraints of habitat quality and/or seed dispersal, the same establishment ratio (e.g., 30% of the potential establishment in the case of Fig. 6) can be achieved by smaller areas of reallocation compared to species with fewer constraints. This suggests that strong constraints of seed sources or habitat quality would result in a low total expected tree density in the total landscape and raise the importance of an optimal strategy for reallocation of plantation to semi-natural forests. The rare riparian species would have strong constraints of habitat quality. For species having higher seed dispersal capacity (e.g., wind-dispersed seeds), we can examine the connectivity of their habitat in reallocating plantation to semi-natural forests without regarding the distribution of seed sources. In contrast, for species with low seed dispersal capacity (e.g., gravity-dispersal seeds), precise estimation of the distribution of seed-source trees will be critical for successful restoration.

It would not be particularly difficult to obtain from short field surveys the information required for estimating habitat constraints of target species. However, if not already well understood, the reproductive features of the target species require long-term and intensive study. Thus, further studies of the reproductive features, including observation of masting traits or determination of dispersal capacity using molecular techniques, of the rare riparian species are strongly recommended to provide a basis for conservation.

20.4 Conclusion

We illustrated a model explaining the spatial distribution of habitat suitability for endangered riparian species and its application for planning a semi-natural forest restoration strategy. In estimating the potential habitats of riparian species that have established under the strong influence of fluvial and geomorphic distur-bances, the spatial relationships between the target point and the river and the characteristics of the nearest channel can provide more precise predictions than those based only on the geomorphic features of each target site as isolated points. This habitat estimation, together with seed source proximity in the landscape, would help to determine efficient reallocation of plantation to semi-natural forest for the purposes of restoring populations of endangered riparian species.

References

Environment Agency of Japan (2000) Threatened wildlife of Japan.- Red Data Book 2nd ed.-. Vol.8 Vascular Plants. Japan Wildlife Research Center, Tokyo (in Japanese)

Franklin J (1995) Predictive vegetation mapping: geographic modeling of biospatial patterns in relation to environmental gradients. Prog Physic Geog 19:447-449

Franklin J (1998) Predicting the distribution of shrub species in southern California from climate and terrain-derived variables. J Veg Sci 9:733-748

Guisan A, Zimmermann NE (2000) Predictive habitat distribution models in ecology. Ecol Model 135:147-186

Ito S, Matsuda A, Nogami K (2000) Population structure and dynamics of *Quercus hondae* Makino 1. Size structure and spatial distribution in a small catchment area. Proceedings of the 43rd Symposium of the International Association for Vegetation Science, Nagano, Japan, pp 130

Ito S, Mitsuda Y, Buckley GP (2004) Reallocation strategy of semi-natural forest patches for conservation of *Quercus hondae* Makino, an endangered species (in Japanese with English summary). Landscape Ecol Manage 9(1):18-25

Ito S, Mitsuda Y, Gi N, Takagi M and Nogami K (2006) Modeling potential habitats for *Lagerstroemia subcostata* var. *fauriei*, a rare riparian species on Yakushima Island, by analysis of GIS-derived topographic factors. Veg Sci 23:153-161

Ito S, Nakagawa M, Buckley GP, Nogami K (2003). Species richness in sugi (*Cryptomeria japonica* D. Don) plantations in southeastern Kyushu, Japan. The effects of stand type and age on understory trees and shrubs. J For Res 8:49-57

Ito S, Nogami K (2005) Species composition and environments of riparian forests consisting of *Lagerstroemia subcostata* var. *fauriei*, an endangered species on Yakushima Island. Veg Sci 22:15-23 (in Japanese with English summary)

Ito S, Ohtsuka K, Yamashita T (2007) Ecological distribution of seven evergreen *Quercus* species in southern and eastern Kyushu, Japan. Veg Sci 24:53-63

Kirby KJ, Buckley GP, Good JEG. (1999) Maximising the value of new farm woodland biodiversity at a landscape scale. In: Burgess PJ, Brierley DR, Morris J, Evans J (eds) Farm woods for the future. Bios Scientific Pub, Oxford, pp 45-55

Mashiba S (1973) The *Cyclobalanopsis hondae* forests in Kyushu. J Geobot 21:36-41 (In Japanese)

Mitsuda Y, Ito S, Sakamoto S. (2007) Predicting the site index of sugi plantations from GIS derived environmental factors in Miyazaki Prefecture. J For Res 12:177-186

Miyawaki A (1980) Vegetation of Japan -Yakushima. Shibundo, Tokyo (in Japanese, with English abstract)

Miyawaki A (1981) Vegetation of Japan -Kyushu. Shibundo, Tokyo (in Japanese, with English abstract)

Nakamura F, Inahara S (2007) Fluvial geomorphic disturbances and life history traits of riparian tree species. In: Johnson EA, Miyanishi K (eds) Plant disturbance ecology: the process and the response. Academic Press, New York, pp 283-310

Næsset E (1997) Geographical information systems in long-term forest management and planning with special reference to preservation of biological diversity: a review. For Ecol Manage 93:121-136

Quine CP, Humphrey JW, Purdy K, Ray D (2002) An approach to predicting the potential forest composition and disturbance regime for a highly modified landscape: a pilot study of Strathdon in the Scottish Highlands. Silva Fenn, 36:233-247

Sakio H (1997) Effects of natural disturbance on the regeneration of riparian forests in a Chichibu Mountains, central Japan. Plant Ecol 132:181-195

Sakio H, Yamamoto F eds. (2002) Ecology of Riparian Forests. Univ Tokyo Press, Tokyo, Japan. 206p (in Japanese)

Suzuki W, Osumi K, Masaki T, Takahashi K, Daimaru H, Hoshizaki K (2002) Disturbance regimes and community structures of a riparian and an adjacent stand in the Kanumazawa Riparian Research Forest, northern Japan. For Ecol Manage 157:285-301

Wimberly MC, Spies TA (2001a) Influences of environment and disturbance on forest patterns in coastal Oregon watersheds. Ecology 82(5):1443-1459

Wimberly MC, Spies TA (2001b) Predicting spatial patterns of understory conifer regeneration in a Pacific Northwest forest landscape. Appl Veg Sci 4:277-286

Part 10

Conclusion

21   General conclusions concerning riparian forest ecology and conservation

Hitoshi SAKIO

Saitama Prefecture Agriculture & Forestry Research Center, 784 Sugahiro, Kumagaya, Saitama 360-0102, Japan (*Present address*: Sado Station, Field Center for Sustainable Agriculture and Forestry, Faculty of Agriculture, Niigata University, 94-2 Koda, Sado, Niigata 952-2206, Japan)

21.1 Riparian forest research in Japan

The aim of our research was to understand the dynamics and coexistence mechanisms of riparian forests with respect to natural disturbance and the life-history strategies of plant species and to contribute to the conservation and restoration of riparian ecosystems (see Chapter 1). There are many types of riparian vegetation in Japan (see Chapter 4). We have studied the vegetation of more than 20 riparian forests across Japan, from Hokkaido to Kyushu, and from headwater streams to lowland rivers. The studies were relatively broad, and included geography, vegetation science, population ecology, landscape ecology, and ecophysiology. In this chapter, I describe (1) the natural disturbance regime in riparian areas, (2) the life-history strategies of riparian trees, (3) the coexistence mechanisms of riparian trees, and (4) the conservation and management of riparian forests.

21.2 Disturbance regime, life history, and dynamics

Natural disturbance plays an important role in the regeneration of riparian forests (Gregory et al. 1991; Malanson 1993; Nakamura 1990). The disturbance regime in riparian zones varies in type, frequency, size, and magnitude (Ito & Nakamura 1994; Sakio 1997). In particular, there is variation attributable to the location of

Sakio, Tamura (eds) Ecology of Riparian Forests in Japan : Disturbance, Life History, and Regeneration
© Springer 2008

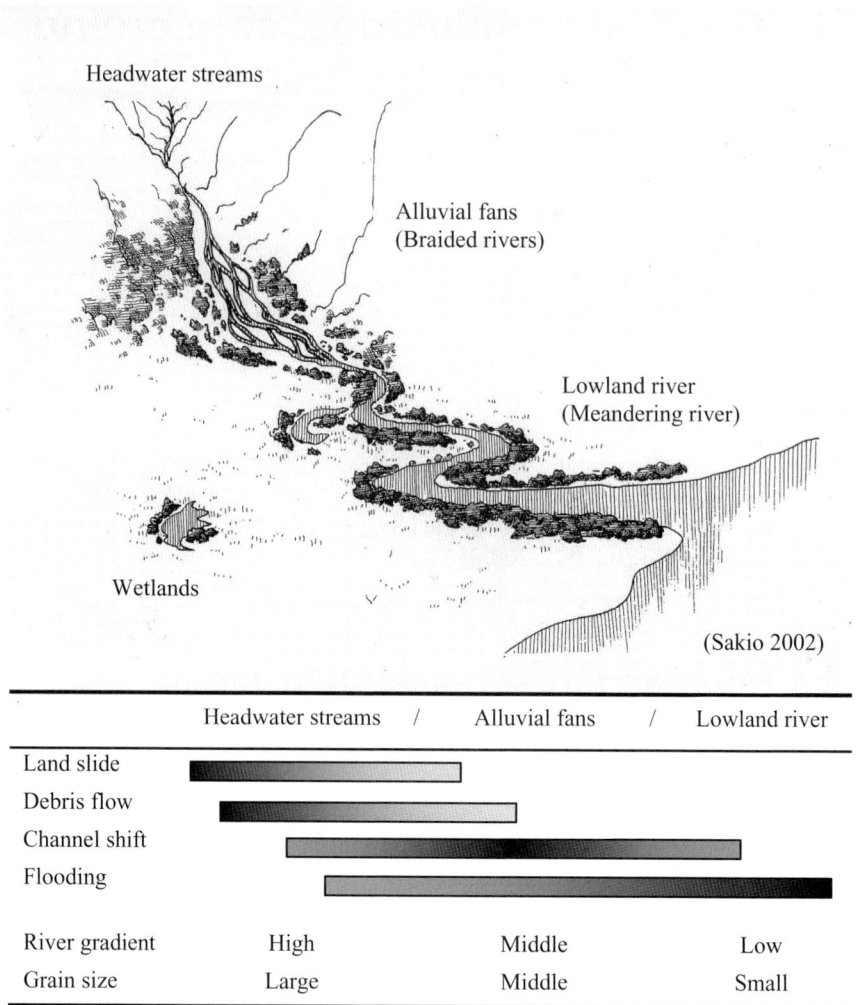

Fig. 1. Disturbance frequency and river characteristics according to the location within the river basin. Black to white shading indicates high to low disturbance frequency.

the river basin. For example, the disturbance regime differs between a headwater stream and a low-gradient meandering river (Fig. 1).

The coexistence of tree species is evident in the riparian zone. The life-history strategies of trees are also diverse and are closely related to the disturbance regime of the riparian zone (White 1979). Riparian tree species are well adapted to the disturbance regime throughout their life history.

Fig. 2. Deep landslide with destruction of the canopy layer and the emergence of large gaps. This landslide occurred during heavy rain from a typhoon in September 2007 in the Chichibu Mountains, central Japan

21.2.1 Headwater streams

21.2.1.1 Disturbance regime

Disturbance of the riparian zone in headwater streams is divided into hillslope and fluvial processes (see Chapter 2). Landslides are representative hillslope processes and are the dominant disturbance in headwater streams. Deep landslides are unpredictable and occur at low frequencies. Large-scale landslides (Fig. 2) may occur on the order of every 100–1000 years (Sakio 1997; Sakio et al. 2002). These landslides are accompanied by the destruction of the forest canopy layer and the emergence of large gaps (Fig. 2). Their major causes are large typhoons and earthquakes. Shallow landslides occur more frequently and periodically on the order of every 1–10 years and are caused by heavy rain in the rainy season or by typhoons. Such disturbances are small in size and do not destroy the canopy layer or cause the emergence of large gaps. Soil movement of the A_0 and A layers is a dominant disturbance in headwater streams.

Debris flow is a typical fluvial process and is closely related to hillslope processes. Debris flow in a small valley is considered to follow the occurrence of shallow landslides on adjacent hillslopes (see Chapter 2). Massive debris is accumulated in a channel by repeated shallow landslides; this debris flows down the channel with flooding from heavy rain during large typhoons. Debris flow may occur on the order of every 10–100 years. The size and magnitude of debris flow vary; for example, during large flows, canopy trees are destroyed and riparian landforms are altered considerably. The movement of gravel and sand at small scales occurs with high frequency.

The high diversity of disturbance regimes is a cause of the variety of micro-topography in riparian zones (Gregory et al. 1991; see Chapter 2). The active channel, including the channel-way and bottomland, is the direct product of the erosion and deposition of sediments, and fluvial terraces are bordered by lower side slopes. The characteristics of microtopography in riparian zones are mosaic in a narrow area and regeneration. The existence of streams and slopes produces the mosaic of microtopography, and high frequency of disturbance results in the repetitive destruction and regeneration of microtopography and substrates (White 1979).

21.2.1.2 Life-history strategies of riparian trees

In headwater streams, the main riparian canopy tree species are *Pterocarya rhoifolia* Sieb. et Zucc. (see Chapters 5, 6, and 17), *Aesculus turbinata* Blume (see Chapters 6, 7, and 17), *Cercidiphyllum japonicum* Sieb. et Zucc. (see Chapters 5 and 17), and *Fraxinus platypoda* Oliv. (see Chapter 5) in the cool-temperate zone of Japan. These species have different life histories and ecological characteristics.

The sizes of fruits differ greatly among the tree species. The fruit of *Aesculus turbinata* is the largest among these species (approximately 10–20 g fresh weight). In contrast, *Cercidiphyllum japonicum* fruit is the smallest of these species (approximately 0.001 g fresh weight).

The fruits of *Pterocarya rhoifolia*, *Cercidiphyllum japonicum*, and *Fraxinus platypoda* have wings and are dispersed by wind (see Chapter 5). The fruits of *Pterocarya rhoifolia* and *Fraxinus platypoda* are also dispersed by water. In contrast, the large-sized fruits of *Aesculus turbinata* are dispersed by certain animals (see Chapter 7). The distance of dispersal of *Cercidiphyllum japonicum* is the greatest among these species (> 600 m; Sato et al. 2006).

The survival ratio of seedlings differs among these species. The survival ratio of current-year seedlings for the next year of emergence is approximately 60% in *Aesculus turbinata*. Hoshizaki et al. (1997) reported a 51% survival ratio during the first growing season. Shoot clipping by rodents has the greatest negative effect on the survival ratio of current-year seedlings. In contrast, the main cause of death of other species may be from shading by leaves and the herb layer or desiccation stress from the litter layer. The survival ratio of *Pterocarya rhoifolia* is 27% in the Sea of Japan climate (see Chapter 6), but is < 10% in the Pacific Ocean climate (Sakio et al. 2002). The survival ratio of *Fraxinus platypoda* is < 20%, and that of *Cercidiphyllum japonicum* is negligible. In the case of *Cercidiphyllum japonicum*, seedlings are washed away by rain or typhoons (Kubo et al. 2000).

The waterlogging tolerance also differs among these species. *Aesculus turbinata* has the highest flooding tolerance, whereas *Pterocarya rhoifolia* and

Cercidiphyllum japonicum incur damage by flooding (Sakio 2005). The effects of flooding on seedling growth differed among tree species within the same area of the riparian zone.

The lifespans of *Aesculus turbinata* and *Pterocarya rhoifolia* are approximately 250 and 150 years, respectively (see Chapter 6). *Fraxinus platypoda* has a lifespan of approximately 300 years. Although the lifespan of a single stem of *Cercidiphyllum japonicum* is approximately 300 years, individuals seem to have a lifespan > 500 years. After the main stem of *Cercidiphyllum japonicum* dies, sprouts around the main stem begin to grow rapidly and replace the main stem. As a result, there is a life-history trade-off between sprouting and seedling regeneration in this species (Bellingham & Sparrow 2000).

In warm-temperate parts of western Japan, little information has been reported on species distributions, micro-landforms, and disturbance regimes. In Chapter 8, Ito and Ito described the habitat segregation and diversity patterns of tree species along a sedimentation-dominated mountainous stream. However, the ecological characteristics and life histories of riparian trees in the warm-temperate zone are mostly unknown for major species such as *Quercus gilva* Blume and *Machilus thunbergii* Sieb. et Zucc. The disappearance of natural riparian forest by afforestation may exacerbate this problem. Hereafter, the elucidation of ecological characteristics of riparian trees will be emphasized for riparian forest restoration in the warm-temperate zone.

21.2.1.3 Dynamics and coexistence mechanisms

One of the causes of the coexistence of canopy trees in riparian zones is the heterogeneous habitat resulting from various natural disturbances (Baker 1990; Denslow 1980). These disturbances, which vary in type, frequency, and magnitude in riparian zones, offer establishment sites for riparian species. The ecological characteristics of riparian trees differ by life-history stage among species such that the species have contrasting habitat requirements.

The habitat of a tree species changes depending on its relationships with other species. In the Sea of Japan climate (see Chapter 6), the area closest to the water is dominated by *Pterocarya rhoifolia*. In contrast, this area is dominated by *Fraxinus platypoda* in the Pacific Ocean climate, whereas *Pterocarya rhoifolia* is distributed in sites created by large landslides (Sakio et al. 2002).

When disturbances are small in scale, for example, small gaps and landslides without canopy destruction, the coexistence of canopy trees is maintained through some niche partitioning. There are tradeoffs among the ecological characteristics of riparian species. For instance, tree species with large seeds have mast years, whereas those with small seeds have yearly seed production.

If disturbances consist of unpredictable large-scale events within heterogeneous habitats, competition among tree species occurs on the disturbed site. These disturbances supply ample light to invading trees. In a large disturbance site, pioneer tree species such as *Euptelea polyandra* Sieb. et Zucc. can establish along with the canopy trees of riparian species. The seeds of this species are supplied by wind dispersal and by soil seed banks (Kawanishi et al. 2007; Kubo et al. 2008).

Euptelea polyandra grows on unstable ground such as very steep slopes with thin soil, and its sprouts may be a means of vegetative regeneration (Sakai et al. 1995). After the pioneer trees die, their seeds remain in the soil and canopy tree species replace the pioneer tree species. Kalliola and Puhakka (1988) showed that floodplain vegetation heterogeneity includes the spatial and temporal character-istics of riparian vegetation dynamics along a meandering river in peatlands in northern Finland. Therefore, the heterogeneity of unpredictable large-scale distur-bance may be important for both the temporal and spatial coexistence of tree species in riparian zones.

In riparian areas, tree species are well adapted to disturbances throughout their life history. The question of which is more important as a cause of species diversity in the forests, niche or chance, is not fully resolved (Brokaw & Busing 2000; Nakashizuka 2001). It is likely that the coexistence of riparian tree species is maintained through some niche partitioning in the early life-history stages, but chance events such as unpredictable large-scale disturbances are also important.

21.2.2 Alluvial fan

21.2.2.1 Disturbance regime

The major disturbance types in alluvial fans are debris flows and flooding in mountain regions (see Chapter 9). Flooding is divided into two types (see Chapter 3). One type of flooding is caused by periodical snowmelt that occurs each spring in the Sea of Japan climate. This flooding is small in magnitude and is not accom-panied by large topographical changes. The other type of flooding is caused by typhoons. This type of flooding is larger in magnitude and more unpredictable than that from snowmelt and varies in frequency and magnitude.

The typical valley floor topography in alluvial fans consists of multiple active channels with abandoned channels (Fig. 3). The unpredictable large flooding that occurs on the order of every 10–1000 years results in channel shifts in stable floodplains. Dramatic channel shifts occur over long time periods (100–500 years) across wide floodplains (see Chapter 10). Consequently, many types of riverbed geomorphic surface emerge in the floodplain, including gravel bars, lower and upper floodplains, secondary channels, and terraces (Shin & Nakamura 2005).

21.2.2.2 Life-history strategies of riparian trees

In braided rivers on alluvial fans in Japan, Salicaceae and *Ulmus* species are dominant. Each species in these groups has a different life-history strategy for reproduction and seedling development.

In riparian *Salix* species, flowering and seed dispersal occur in spring. *Salix* species have small anemochorous seeds that are capable of long-distance dispersal. The flowering and seed-dispersal phenology of *Salix* species largely overlap, but the species can be divided into early, intermediate, and late species (Niiyama 1990; see Chapter 11). *Salix cardiophylla* Trautv. & Mey. var.

Fig. 3. Braided rivers in the alluvial fan of the Kamikouchi valley in Chubu-Sangaku National Park, Japan. The typical topographical change is channel shifts after large floods

urbaniana (Seemen) Kudo (formerly *Toisusu urbaniana* (Seemen) Kimura) is distributed mainly in upstream habitats (Niiyama 1989). In contrast with other members of the Salicaceae, it flowers in spring and disperses seed from August to September (Nakamura & Shin 2001).

The seed longevity of the Salicaceae is extremely short. Niiyama (1990) showed that seeds of *Salix* species maintain a high germination percentage for 2 weeks, but completely lose their viability within 45 days after dispersal.

The life span of *Salix* species is relatively short. *Salix dolichostyla* Seemen (formerly *Salix jessoensis* Seemen) and *Salix jessoensis* subsp. *serissaefolia* (Kimura) H. Ohashi (formerly *Salix serissaefolia* Kimura) have long life spans of approximately 100 years. In contrast, *Salix miyabeana* subsp. *gymnolepis* (H.Lev. & Vaniot) H. Ohashi & Yonek. (formerly *Salix gilgiana* Seemen) has a life span of < 40 years. The former two species are large-sized trees, whereas the latter is a shrub.

The period of flowering and seed dispersal of *Ulmus davidiana* var. *japonica*, a late-successional species, is in spring. The seeds germinate immediately after dispersal under bright conditions such as at the forest edge, as well as in spring of the following year within forests. Seedlings of this species are found in various habitats and disturbance regimes. They can survive in such sites in the presence of a thin herb layer (see Chapter 10).

21.2.2.3 Dynamics and coexistence mechanisms

The distribution patterns of Salicaceae along a river can be divided into three types. *Salix cardiophylla* var. *urbaniana* and *Salix arbutifolia* are abundant in the upper reaches of the river. *Salix dolichostyla*, *Salix rorida*, and *Salix gracilistyla* are typical willows of alluvial fans. *Salix udensis*, *Salix miyabeana*, and *Salix schwerinii* are widely distributed in the middle and lower reaches of a river. The cause of this segregation may be adaptation to different fractions of substrata throughout the life history, especially in the germination or seedling establishment stages.

The coexistence mechanisms among *Salix* species at the local scale result from the heterogeneity of site substrata (see Chapter 11). Substrata texture and moisture affect the regeneration success of Salicaceae species. Seed dispersal timing also promotes coexistence along the narrow shoreline when the high water level decreases gradually in spring as snowmelt declines.

Ulmus davidiana var. *japonica* is a typical late successional species on alluvium in Japan and its only invades forest floors that are disturbed by floods (see Chapter 10). This tree depends largely on large-scale disturbances that have long return intervals (Sakai et al. 1999).

21.2.3 Lowland regions

21.2.3.1 Disturbance regime

In lowland rivers, the dominant disturbance is flooding (Fig. 4). Flooding occurs during the rainy season or typhoons and tends to be periodical. The disturbance regime changed considerably after the construction of artificial embankments and dams in upstream regions. Channel shifts became limited to the narrow area enclosed by the banks. The scale of flooding was diminished after dam construction in the headwaters. Inundation and sediment deposition by flooding are major disturbance factors in lowland regions.

21.2.3.2 Life-history strategies of riparian trees

The dominant riparian species are members of the Salicaceae such as *Salix chaenomeloides* Kimura and *Salix gracilistyla* Miq. in Shikoku and Chugoku in western Japan. In addition to the Salicaceae, *Elaeagnus umbellata* Thunb. spreads to riparian gravel bars in the middle reaches.

Salix gracilistyla is a dominant species at the waterside, where trees suffer frequent disturbance, and it has a strong ability to sprout and to form adventitious roots (see Chapter 13). After flooding, this species restores its plant organs by sprouting new shoots and forming adventitious roots from stems buried in the sand. This tendency is often evident among trees of the Salicaceae (Naiman & Décamps 1997). The stand area of *Salix gracilistyla* expands via periodic sand burial and subsequent vegetative growth. This species maintains high productivity

Fig. 4. Flooding after heavy rain from a typhoon in September 2007 in Arakawa River of Kumagaya, Saitama Prefecture, Japan

by the absorption of sufficient soil nitrogen, soil phosphorus, and hyporheic water, with contributions from mycorrhizal symbiosis.

Salicaceae species are often flood tolerant. Seedlings of *Salix triandra* L. (formerly *Salix subfragilis* Anders.), which is frequently observed in lowland regions of western Japan, have high flood tolerance (Ishikawa 1994). However, the growth of *Salix gracilistyla* seedlings is limited by periods of long inundation (Nakai & Kisanuki 2007).

Elaeagnus umbellata achieves effective seed dispersal to safe establishment sites via both endozoochory and hydrochory (see Chapter 14). Frugivorous birds carry seeds upstream and to places where floodwaters cannot reach. This species does not create a long-term seed bank. *Elaeagnus umbellata* has multiple life-history strategies, including reproduction by seed germination and vegetative regeneration by sprouting and root sucker formation.

21.2.3.3 Dynamics and regeneration mechanisms

Human-made structures such as dams and embankments have affected the distribution and dynamics of lowland vegetation. Johnson (1994) found an increase in woodlands along the Platte River, Nebraska, USA, by comparing aerial photographs. Kamada et al. (1997) noted the expansion of riparian woodlands after the completion of dams in the 1970s. The stabilization of river systems caused by civil engineering enhances the establishment of willow communities (see Chapter 12). The multiple life-history strategies of *Elaeagnus umbellata* allow it to become established in wide, elevated areas along regulated river channels.

21.2.4 Wetlands

21.2.4.1 Flooding and ecophysiology of riparian trees

The major disturbance in wetlands is flooding. Flooding events do not cause the physical destruction of riparian vegetation. Rather, such events result in ecophysiological effects on tree species.

The dominant tree species of swamp forests in the cool-temperate zone of Japan are *Alnus japonica*, *Fraxinus mandshurica* var. *japonica*, and *Ulmus davidiana* var. *japonica*. Their distribution in the floodplain is restricted by the source of water, flooding regime, soil physiochemical features, and groundwater fluctuation regime (see Chapter 15). *Alnus japonica* adapts to most wet conditions, and *Ulmus davidiana* var. *japonica* grows in mesic sites without stagnant water. *Fraxinus mandshurica* var. japonica shows intermediate tolerance to flooding.

Flood-tolerant plant species such as *Alnus japonica* and *Fraxinus mandshurica* var. *japonica* have anatomical and physiological adaptations to flooding (see Chapter 16). In these species, flooding induces the development of adventitious roots, aerenchyma tissue, and epicormic shoots. The photosynthesis and growth of riparian trees are affected by flooding. However, damage caused by flooding is more serious in non-riparian than riparian species.

21.2.4.2 Dynamics and life-history strategies

The life-history characteristics of wetland riparian trees are discussed in Chapter 15. *Alnus japonica* forests regenerate by the sprouting of stems, rather than by seedling establishment (see Chapter 15). Flooded *Alnus japonica* trees have multiple stems derived from epicormic shoots; this is advantageous for the maintenance of trees (see Chapter 16). However, the regeneration dynamics are unclear for these tree species. There is little information on the early life-history stages such as seedling establishment and survival.

The decline of seedling establishment sites is a concern. Most wetlands in Japan have been changed to farmland and industrial areas. In addition, the disturbance regime has been changed by the construction of dams and artificial riverbanks, and the frequency of flooding has decreased considerably. Establishment sites for wetland tree species have been reduced by the increased stability of the riparian zone. In the future, it is important for the conservation of wetland ecosystems to understand the life-history strategies and regeneration mechanisms of these tree species. It is also important to reproduce the natural disturbance regimes.

21.3 Conservation of species diversity

In general, riparian forests have higher biodiversity than do mountain forests (Tabacchi et al. 1990; Nilsson et al. 1991; Suzuki et al. 2002). The riparian forests of northern Honshu Island, Japan, are characterized by a high diversity of tree species (see Chapter 17). The species diversity of forest floor plants is greater in riparian forests than in forests on mountain slopes (see Chapter 18).

Some experiments have shown that the cause of the higher species richness in riparian forests is the heterogeneity of riparian microenvironments influenced by topography and water (Vivian-Smith 1997). The riparian zone has many types of topography, including active channel, abandoned channel, floodplain, terrace, and slope (e.g., Kovalchik & Chitwood 1990; Gregory et al. 1991; Kawanishi et al. 2004; Nakamura et al. 2007). These various topographies result from various disturbance regimes in riparian zones. Disturbances are critical to the maintenance of riparian species diversity (Duncan 1993).

There are many kinds of substrata, including large rocks, gravel, mineral soil, woody debris, and litter (e.g., Gregory et al. 1991; Nakamura & Swanson 1994). The germination sites of riparian plants differ among species (Masaki et al. 2007). Small seeds such as those of *Cercidiphyllum japonicum* need fine mineral soil, whereas large seeds of other species need gravel and litter (Seiwa & Kikuzawa 1996). The seeds of *Aesculus turbinata*, which are very large, can germinate by rodent seed hoarding underground (see Chapter 7). In contrast, many forest floor plants respond to small-scale landforms and differentiate among habitats at the microscale (see Chapter 18).

Canopy gaps result in variation in light conditions. The causes of small gaps and large gaps are single treefalls and large landslides with destruction of the canopy layer, respectively. The herb layer is also a source of heterogeneity in light conditions on the forest floor.

Water is an important factor determining the establishment of tree seedlings. The water conditions are saturation near the active channel and relatively dry gravel deposits in higher terraces. The responses to flooding differ greatly among riparian tree species. *Fraxinus mandshurica* var. *japonica* and *Alnus japonica* are distributed in wetlands and are adapted to soil flooding by the emergence of adventitious roots and development of aerenchyma tissue (see Chapter 16). In headwater streams, the response to flooding in cool-temperate deciduous forests differs among tree species (see Chapter 5).

Species diversity in riparian zones in Japan is promoted by the variety of disturbance regimes originating from climatic and geological effects. The conservation and restoration of rare or endangered riparian species are major issues. The habitat of *Salix hukaoana* is wide floodplains at the upper reaches of large river systems in the foothills of northern Honshu (see Chapter 19). Isolated distribution patterns between the Azusa River in Kamikouchi, Nagano Prefecture, and Hokkaido are evident for *Salix rorida* and *Salix arbutifolia* (see Chapter 10). These species have habitats similar to that of *Salix hukaoana*. However, the habitat of these *Salix* species is changing and is being destroyed by dams and artificial riverbanks.

In the warm-temperate regions of Japan, mountainous riparian forests have been heavily exploited or converted to plantations of *Cryptomeria japonica* or *Chamaecyparis obtusa* (see Chapter 20). Consequently, the habitat of rare riparian species such as *Quercus hondae* and *Lagerstroemia subcostata* var. *fauriei* has declined sharply. For the purpose of restoration of these rare species, habitat estimation at the landscape level and remote-sensing data would help to determine efficient reallocation plans for converting plantations to semi-natural forests.

21.4 Riparian forest management

Management plans for the conservation and restoration of riparian forests were developed from the findings presented in this book. The following initial proposal will be improved by feedback from future restoration projects (The Japanese Riparian Forest Research Group 2001).

21.4.1 Natural forest management

Natural riparian forests that are at primitive or climax stages must be protected by excluding human activity. These forests show the original condition of natural environments in the riparian landscape of Japan (see Chapter 1) and are important for the future of riparian forest management. Natural riparian forests are models for the future restoration and rehabilitation of riparian forests and can also supply gene resources for such projects.

These forests are valuable not only for plant biodiversity, but also for wildlife habitat. In addition, natural riparian forests play an important role as corridors for animal movement among forests.

21.4.2 Forestry management

Secondary riparian forests under natural conditions undergo succession toward potential natural vegetation. If the species composition does not have components of potential natural vegetation, dominant canopy tree species of the natural vegetation should be introduced.

In Japan, the amount of area composed of plantations is 41% of all forests. Riparian rehabilitation of these forests is a major issue. For plantations with saplings of riparian species, the thinning of canopy trees is effective to promote riparian species regeneration. If plantations have no saplings of natural vegetation, the invasion of riparian trees is expected by seed dispersal or buried seeds (Kawanishi et al. 2007). Because plantation thinning is not effective to promote the invasion of trees, natural riparian tree species that compose the potential canopy layer must be planted.

In clear-cut sites after harvest, the planting of riparian species is important for rehabilitation. Although the vegetation may invade the cut site by natural succession, these species are limited to pioneer shrubs, except for natural canopy tree species (Sakai et al. 2005).

The channel floor regions of riparian zones are highly vulnerable to invasion by exotic vascular plants (Hood & Naiman 2000). Exotic trees (e.g., *Robinia pseudoacacia* L.) that were planted in the mid-20th century for erosion control should be thinned to allow the growth of natural riparian tree species in riparian zones. If native tree species dominate the sublayer, thinning is effective for the replacement of exotic trees (Sakio 2003).

In the bare land around woodland paths or erosion control areas, the planting of riparian tree species is recommended. Saplings for planting should come from seeds near the plantation site to prevent genetic contamination.

21.4.3 Riparian ecosystem management

An important aim of riparian ecosystem management is to maintain continuity between small headwater streams and downstream rivers. Allan (1995) demonstrated the physical and biological effects of dams. Dams prevent the movement of fishes and aquatic insects from downstream rivers to headwater streams (Holmquist et al. 1998). A dam can block the return of salmon and sweetfish (ayu) from ocean to river. Multiple dams can also act as barriers to plant dispersal along rivers (Jansson et al. 2000). Additionally, it is important in riparian ecosystems that organisms are able to move between terrestrial and aquatic ecosystems. Amphibians such as salamanders and frogs lay eggs in streams or ponds in spring, and their larva live in water. Concrete riverbanks are obstacles to the movement of amphibians.

In relation to riparian forests, the continuity of headwater streams and downstream rivers is important for various ecological functions. The corridor of riparian forests preserves water quality and temperature, links the water systems in basins, and contributes to the maintenance and expansion of wildlife, as well as aquatic and terrestrial plants. Johansson et al. (1996) suggested that continuous river corridors are important to maintain regional biodiversity. In addition, the recovery of disturbance regimes in rivers and lakes is necessary for the restoration of riparian vegetation. Natural disturbance is a key factor in the regeneration and coexistence mechanisms of riparian tree species. Once a riparian ecosystem loses vegetation through human impacts, recovery can be very difficult if the disturbance regime changes considerably by the construction of dams or riverbanks. The restoration of riparian ecosystems is an expensive, long-term project. Therefore, the most important prescription for riparian management may be the preservation of existing natural riparian vegetation.

For the optimal management of riparian resources, riparian management zones should have variable widths that are delineated at ecological boundaries (Gregory & Ashkenas 1990). Riparian management zones along headwater streams in forestry regions in Japan require a horizontal width of > 30 m on both sides of active channels.

21.5 Future research

In conclusion, I propose some areas for future research based on past results. In headwater streams and alluvial fans, unpredictable large disturbance is important for the dynamics and coexistence of riparian forest species. To understand such effects, long-term studies should be conducted at large scales that involve large disturbances and the recovery of riparian forests such as the Ogawa Forest Reserve (Nakashizuka & Matsumoto 2002). The Japan Long-Term Ecological Research Network (JaLTER) plays an important role in this proposal. This network consists of many long-term and large-scale research sites in Japan that cooperate with the Monitoring Sites 1000 Project of the Ministry of the Environment, Japan. If unpredictable large disturbances occur, it may be necessary to include new research sites for the resolution of vegetation recovery systems. Moreover, it is desirable to promote ecophysiological research on riparian tree species. For example, it is important to study the life-history strategies and regeneration process of each characteristic species of lowlands and wetlands (e.g., Salicaceae, *Alnus japonica* and *Fraxinus mandshurica* var. *japonica*).

Today, the rehabilitation and restoration of riparian forests are major emphases of ecosystem management. Further studies of biodiversity using molecular techniques are important for the conservation and restoration of vegetation in riparian areas. Remote-sensing techniques are also an effective tool for the restoration of riparian landscapes. Such studies will offer new information and lead to the improved management of riparian ecosystems.

References

Allan JD (1995) Stream ecology. Chapman & Hall, London, 388p
Baker WL (1990) Species richness of Colorado riparian vegetation. J Veg Sci 1:119-124
Bellingham PJ, Sparrow AS (2000) Resprouting as a life history strategy in woody plant communities. Oikos 89(2):409-416
Brokaw N, Busing RT (2000) Niche versus chance and tree diversity in forest gaps. Trend Ecol Evol 5(5):183-188
Denslow JS (1980) Patterns of plant species diversity during succession under different disturbance regimes. Oecologia 46:18-21
Duncan RP (1993) Flood disturbance and the coexistence of species in a lowland podocarp forest, south Westland, New Zealand. J Ecol 81:403-416
Gregory S, Ashkenas L (1990) Riparian management guide (Willamette national Forest), Forest Service, United States Department of Agriculture, USA, 120p
Gregory SV, Swanson FJ, McKee WA, Cummins KW (1991) An ecosystem perspective of riparian zones: focus on links between land and water. BioScience 41(8):540-551
Holmquist JG, Schmidt-Gengenbach JM, Yoshioka BB (1998) High Dams and marine-freshwater linkages: effects on native and introduced fauna in the Caribbean. Conserv

Biol 12:621-630

Hood WG, Naiman RJ (2000) Vulnerability of riparian zones to invasion by exotic vascular plants. Plant Ecol 148:105-114

Hoshizaki K, Suzuki W, Sasaki S (1997) Impacts of secondary seed dispersal and herbivory on seedling survival in Aesculus turbinata. J Veg Sci 8:735-742

Ishikawa S (1994) Seedling growth traits of three salicaceous species under different conditions of soil and water level. Ecol Rev23:1-6

Ito S, Nakamura F (1994) Forest disturbance and regeneration in relation to earth surface movement. Jpn J For Environ 36(2):31-40 (in Japanese with English summary)

Jansson R, Nilsson C, Renöfält B (2000) Fragmentation of riparian floras in rivers with multiple dams. Ecology 81(4):899-903

Johansson ME, Nilsson C, Nilsson E (1996) Do rivers functions as corridors for plant dispersal? J Veg Sci 7:593-598

Johnson WC (1994) Woodland expansion in the Platte River, Nebraska: patterns and causes. Ecol Monogr 64(1):45-84

Kalliola R & Puhakka M (1988) River dynamics and vegetation mosaicism: a case study of the River Kamajohka, northernmost Finland. J Biogeogr 15:703-719

Kamada M, Okabe T, Kotera I (1997) Influencing factors on distribution changes in trees and land-use types in the Yoshino River, Shikoku, Japan. Environmental System Research 25:287-294 (in Japanese with English abstract)

Kawanishi M, Sakio H, Ohno K (2004) Forest floor vegetation of *Fraxinus platypoda-Pterocarya rhoifolia* forest along Ooyamazawa valley in Chichibu, Kanto District, Japan, with a special reference to ground disturbance. Veg Sci 21:15-26 (in Japanese with English summary)

Kawanishi M, Sakio H, Yonebayashi C (2007) A comparative study of buried seed assemblages in conifer plantation and secondary broad-leaved forest. Bull Geo-Env Sci Rissho Univ 9:31-41 (in Japanese with English summary)

Kovalchik BL & Chitwood LA (1990) Use of geomorphology in the classification of riparian plant associations in mountainous landscapes of central Oregon, USA. For Ecol Manage 33/34:405-418

Kubo M, Shimano K, Sakio H, Ohno K (2000) Germination sites and establishment conditions of Cercidiphyllum japonicum seedlings in the riparian forest. J Jpn For Soc 82(4):349-354 (in Japanese with English summary)

Kubo M, Kawanishi M, Shimano K, Sakio H, Ohno K (2008) The species composition of soil seed banks in the Ooyamazawa riparian forest, in the Chichibu Mountains, central Japan. J Jpn For Soc 90(2):121-124 (in Japanese with English summary)

Malanson GP (1993) Riparian landscapes. Cambridge Univ Press, Cambridge, 296p

Masaki T, Osumi K, Takahashi K, Hoshizaki K, Matsune K, Suzuki W (2007) Effects of microenvironmental heterogeneity on the seed-to-seedling process and tree coexistence in a riparian forest. Ecol Res 22:724-734

Naiman RJ, Décamps H (1997) The ecology on interfaces: riparian zones. Annu Rev Ecol Evol Syst 28:621-658

Nakai A, Kisanuki H (2007) Effects of evaluation above the waterline on the growth of current-year Salix gracilistyla seedlings on a gravel bar. J For Res 89:1-6

Nakamura F (1990) Analyses of the temporal and spatial distribution of floodplain deposits. J Jpn For Soc 72:99-108 (in Japanese with English summary)

Nakamura F, Shin N (2001) The downstream effects of dams on the the regeneration of riparian tree species in northern Japan. In: Dorava JM (ed) Geomorphic Processes and Riverine Habitat. American Geophysical Union, Washington DC, pp 173-181

Nakamura F, Shin N, Inahara S (2007) Shifting mosaic in maintaining diversity of floodplain tree species in the northern temperate zone of Japan. For Ecol Manage

241:28-38

Nakamura F, Swanson FJ (1994) Distribution of coarse woody debris in a mountain stream, western Cascade Range, Oregon. Can J For Res 24:2395-2403

Nakashizuka T (2001) Species coexistence in temperate, mixed deciduous forests. Trend Ecol Evol 16(4):205-210

Nakashizuka T, Matsumoto Y (eds) (2002) Diversity and interaction in a temperate forest community, Ogawa Forest Reserve of Japan. Springer-Verlag, Tokyo 319p

Niiyama K (1989) Distribution of *Chosenia arbutifolia* and soil texture of habitat along the Satsunai River. Jpn J Ecol 39:173-182 (in Japanese with English summary)

Niiyama K (1990) The role of seed dispersal and seedling traits in colonization and coexistence of *Salix* species in a seasonality flooded habitat. Ecol Res 5:317-331

Nilsson C, Ekblad A, Gardfjell M, Carlberg, B. (1991) Long-term effects of river regulation on river margin vegetation. J App Ecol 28:963-987

Sakai A, Ohsawa T, Ohsawa M (1995) Adaptive significance of sprouting of *Euptelea polyandra*, a deciduous tree growing on steep slopes with shallow soil. J Plant Res 108:377-386

Sakai A, Sato S, Sakai T, Kuramoto S, Tabuchi R (2005) A soil seed bank in a mature conifer plantation and establishment of seedlings after clearcutting in southwest Japan. J For Res 10:295-304

Sakai T, Tanaka H, Shibata M, Suzuki W, Nomiya H, Kanazashi T, Iida S, Nakashizuka T (1999) Riparian disturbance and community structure of a *Quercus-Ulmus* forest in central Japan. Plant Ecol 140:99-109

Sakio H (1997) Effects of natural disturbance on the regeneration of riparian forests in a Chichibu Mountains, central Japan. Plant Ecol 132:181-195

Sakio H (2002) What is a riparian forest? In: Sakio H, Yamamoto F (eds) Ecology of riparian forests. Univ Tokyo Press, Tokyo, pp 1-19 (in Japanese)

Sakio H, Kubo M, Shimano K, Ohno K (2002) Coexistence of three canopy tree species in a riparian forest in the Chichibu Mountains, central Japan. Folia Geobot 37:45-61

Sakio H (2003) Can an exotic plant, *Robinia pseudoacacia* L., be removed from riparian ecosystems in Japan? J Jpn For Soc 85(4):355-358 (in Japanese with English summary)

Sakio H (2005) Effects of flooding on growth of seedlings of woody riparian species. J For Res 10:341-346

Sato T, Isagi Y, Sakio H, Osumi K, Goto S (2006) Effect of gene flow on spatial genetic structure in the riparian canopy tree *Cercidiphyllum japonicum* revealed by microsatellite analysis. Heredity 96:79-84

Seiwa K, Kikuzawa K (1996) Importance of seed size for the establishment of seedlings of five deciduous broad-leaved tree species. Vegetatio 123:51–64

Shin N, Nakamura F (2005) Effects of fluvial geomorphology on riparian tree species in Rekifune River, northern Japan. Plant Ecol 178:15-28

Suzuki W, Osumi K, Masaki T, Takahashi K, Daimaru H, Hoshizaki K (2002) Disturbance regimes and community structures of a riparian and an adjacent terrace stand in the Kanumazawa Riparian Research Forest, northern Japan. For Ecol Manage 157:285-301

Tabacchi E, Planty-Tabacchi AM, Décamps O (1990) Continuity and discontinuity of the riparian vegetation along a fluvial corridor. Landscape Ecol 5:9-20

The Japanese riparian forest research group (2001) Guide-line for the management of riparian forests. Japan Forestry Investigation Committee, Tokyo, 214p

Vivian-Smith G (1997) Microtopographic heterogeneity and floristic diversity in experimental wetland communities. J Ecol 85:71-82

White PS (1979) Pattern, process and natural disturbance in vegetation. Bot Rev 45:229-299

Subject Index

1-aminocyclopropane-1-carboxylic
 acid (ACC) 243

abandoned channel 42, 75, 145, 162,
 255, 318
Abies homolepis 78, 155
Abies sachalinensis 64, 140, 146
Abies sachalinensis var. *mayriana* 64
aboveground biomass 194, 195
aboveground net production 194
Acer argutum 55, 60, 78, 79
Acer carpinifolium 52, 54, 56, 78, 79
Acer micranthum 257, 263
Acer mono 53, 56, 78, 140, 253, 287
Acer mono f. *marmoratum* 56
Acer mono var. *glabrum* 64
Acer mono var. *mayrii* 65
Acer shirasawanum 61, 79
Aceraceae 53
Aceri-Fraxinetum 68
Acero-Fagetea 53
active channel 75, 77, 150, 286, 308,
 316
adventitious root 41, 87, 141, 145,
 160, 168, 192, 240, 241, 322
aerenchyma 240, 322
Aesculus turbinata 51, 62, 63, 93,
 107, 252, 316
alluvial cone 38, 58, 154, 157
alluvial fan 8, 32, 49, 75, 123, 139,
 171, 226, 227, 269, 273, 318
alluvial plain 32, 41, 227, 230
alluvial soil 65, 229, 230
alluvium 141, 161, 191
Alnetea japonicae 67
Alno-Ulmion 68

Alnus hirsuta 40, 140, 232
Alnus hirsuta var. *microphylla* 155
Alnus hirsuta var. *sibirica* 262
Alnus japonica 41, 65, 225, 230,
 238, 322
Alnus japonica swamp forest 67
Alnus swamp forest 67
alpha diversity 253
Amagi Mountains 58
Amphibian 325
anatomical characteristics 242
annual plant 275
annual variation 118
antagonistic predator 117, 118
Apodemus speciosus 110, 111
Arakawa River 269
artificial bank 33, 156
Aruncus dioicus var. *tenuifolius* 65
Aster dimorphophyllus 58
autumnal floods 212, 213, 216
auxin 243, 244
azonal distribution pattern 58
azonal edaphic climax 49
azonal edaphic climax vegetation 53,
 65
azonal vegetation 49, 65–67

back swamp 33, 41, 225, 227, 228,
 230
bar 178, 191, 200
bar area 187
Betula 39, 78, 232, 257
Betula ermanii 155
biodiversity 3, 208, 300, 323, 324,
 326
biomass allocation 115

boulder   33, 214
braided channel   153, 157
braided river   8, 35, 153, 318, 319
brook-channel Ash forest   67
brook-channel forest   49, 51

Cacalia delphiniifolia   275
Cacalia farfaraefolia   60, 275
Cacalia tebakoensis   57, 58
Cacalia yatabei   60
cacalietosum delphiniifoliae   63
*Cacalio yatabei-Pterocaryetum
      rhoifoliae*   59, 63
Cacalio-Pterocaryetum   65
Camellietea   58, 66
Camellietea japonicae   58
Camellietea japonicae region   69
Camellietea region   58
canopy gap   77, 84, 92, 115, 116,
      233, 252, 323
Cardamine leucantha   55
Carici remotae-Fraxinetum   68
Carpinus cordata   53, 79
cell division   242
Cephalotaxus harringtonia var. *nana*
      62, 63
Cercidiphyllum japonicum   51, 56,
      63, 64, 76, 82, 83, 87, 252, 256,
      316, 317
Chamaecyparis obtusa   299, 324
channel   33, 41, 127, 140, 177, 186,
      205, 255, 268, 271, 301, 316
channel process   39
channel shift   9, 157, 314, 318
channel straightening   42
Chosenia arbutifolia   153, 172, 281
Chrysosplenio-Fraxinetum   60
*Chrysosplenio-Fraxinetum
      spaethianae*   56, 58, 59
Chrysosplenium fauriei   63
Chrysosplenium macrostemon   60,
      61, 274, 275
Chrysosplenium ramosum   65
Chugoku Mountain   61, 62
Cirsium effusum   58

Cirsium nipponicum var. *yoshinoi*   62
class group   51, 53, 54, 68, 69
Clerodendrum trichotomum   262
climate   8
climatic climax   49
climatic climax forest   53, 64, 68
coarse substrata   169–171
coarse woody debris (CWD)   151
coexistence   3, 75, 84, 94, 123, 165,
      169, 173, 252, 273
coexistence mechanism   9, 75, 87,
      165, 313, 317, 320
colluvial processes   144
colluvial slope   255, 256
colonization   113, 116, 119, 161, 182,
      188, 200, 205
competition   115
conservation   10, 150, 208, 264, 281,
      293, 295, 299, 300, 313, 323
conservation efficiency   300, 306
cool-temperate zone   49, 51, 53, 58,
      66–69, 76, 225, 230
coppicing   238
Cornopteris crenulato-serrulata   60
corridor   300, 325
cotyledon removal   114
crest slope   16, 23, 268, 269
Cryptomeria japonica   299, 324
current-year seedling   82, 84–86, 91,
      169, 214

dam   27, 150, 151, 187, 205, 320,
      325
dam regulation   151
dam structure   151
debris flow   9, 19, 20, 37, 38, 49, 76,
      126, 139, 271, 272, 315
debris flow terrace   58, 273, 302
deciduous broad-leaved forest   69, 77
deep-seated landslide   24, 25
deep-seated slide   25–28
Deinanthe bifida   57, 61, 274
delta   8, 32, 33, 41, 49, 171
demographic structure   91, 94
demography   75, 108, 116

density 92, 111, 116, 131, 169, 288, 303
density-dependent mortality 116
deposition 16, 19, 37, 39, 127, 139, 159, 191, 269, 316
depth 18, 36, 241
dichotomy 150
digital elevation model 300
Diphylleia grayi 63
directed dispersal 116, 117
disaster-prone country 150
dispersal distance 81, 111, 112, 292
dispersal distance curve 116
distribution 112, 169, 209, 225, 238, 283
disturbance regime 3, 8, 9, 42, 67, 75, 76, 91, 92, 96, 123–126, 130–134, 139, 157, 171, 216, 252, 267, 313, 314
driftwood 26, 41
drought 82, 119, 159, 169, 180, 191
Dryopteridenion monticolae 63
Dryopterideto-Fraxinetum comanthosphacetosum 56
Dryopteridieto-Fraxinetum commemoralis 61
Dryopteridieto-Fraxinetum spaethianae 61
Dryopterido-Fraxinetum commemoralis 59, 61
Dryopterido-Fraxinetum spaethianae 59
Dryopteris monticola 63
Dryopteris polylepis 60, 61, 275
duration of flooding 241
dwarf bamboo (Sasa spp.) 150, 270

earthflow 37
East Asian Flora 53
ecological distribution 92, 95, 285, 295
ecological function 3, 325
edaphic climax forest 55, 58, 68
edaphic climax type 67

edaphic climax vegetation 49
Elaeagnus umbellata 205–219, 320–322
elasticity analysis 96
Elatostema umbellatum var. *majus* 275
elevation 143–145, 169, 179, 180, 209, 215, 255
embankment 3, 186, 205, 320
environmental gradient 92, 94–96, 172
endangered tree 299
endozoochory 206, 214, 216, 219, 321
enlarged stem 241
Eothenomys andersoni 118
epicormic shoot 238–240, 322
Eragrostis curvula 187
erosion 4, 15–18, 27, 28, 32–34, 125–127, 139, 157, 268, 316
erosion control 77, 293, 325
escape hypothesis 116, 117
ethylene 243, 244
Euptelea pioneer scrub 66
Euptelea polyandra 52, 56, 65, 66, 276, 317, 318
Eupteleion polyandrae 66
Eutrema japonica 272
even-aged forest 51
extra-order 51, 68, 69

*F*agaceae 53
Fagetalia 68
Fagetea 66
Fagetea crenatae 49, 53, 54, 68
Fagetea crenatae region 53, 69
Fagetea region 59, 61, 63
Fagus crenata 52, 112, 118, 197, 252, 269
Fagus-Quercus class group 54
Fisher's α index 291
fitness 96, 103, 218
flood 151, 168, 187, 216, 237, 321
flood-disturbance regime 147

flooding 9, 32, 33, 41, 76, 85, 92,
    139, 171, 237, 243, 314, 316,
    318, 320, 322
flooding tolerance 231, 316
flow regime 151
fluvial process 15, 35, 76, 161, 315
fluvial terrace 32, 49, 51
food habit 118
foot slope 268, 271
forest dynamics 139, 150
forest floor plant 269, 323
forest vegetation 75, 267
Fossa Magna element 59
Fossa Magna region 58, 59, 66
Fraxinenion speathianae 61
Fraxino-Ulmetalia 50, 54, 66, 68
Fraxinus mandshurica var. *japonica*
    64, 67, 155, 225, 232, 238, 322
Fraxinus platypoda 51, 60, 61, 76,
    81, 269, 316
Fuji volcanic zone 58

Galio paradoxi-Pterocaryetum 56
gamma diversity 253
gap formation 76, 84, 233
gap properties 92
geographic race 58, 57, 62
geographic vegetation unit 51
germinating period 182, 183
germination 82, 165, 168, 169, 180,
    192, 213, 216, 232
gleyed horizon 230
grain size 272
gravel bar 162, 205, 206, 214
Grey soil 229, 230
groundsurface 15, 17, 18, 20, 21, 23–
    25, 27, 28, 127, 227, 228

habitat 91–96, 98–103, 107, 119,
    130–133, 159–162, 165, 191,
    192, 209, 226, 233, 253, 267,
    273, 282, 285, 292–295, 299–
    308, 317
habitat connectivity 306

habitat effect 99, 101
habitat heterogeneity 267
habitat quality 303, 306, 308
habitat segregation 123, 124, 171,
    292
Hakone Mountains 58
head hollow 22–24, 26–28
headmost wall 22, 24, 26, 28
headwater forest 269
headwater stream 8, 123, 314–316
heavy snowfall 9, 31, 150, 285
herb 79, 251, 267
herb layer 162, 267, 316, 323
herbaceous plant 52, 267, 268, 270,
    273
herbaceous plant diversity 270
herbivore 99, 107–109, 115, 116,
    118, 119
herbivory 114–119
heterogeneity 125, 126, 130, 131,
    133, 161, 166, 169, 252, 253,
    267, 271, 323
high terrace 43, 146, 150
hillslope process 15, 25, 315
Hosiea japonica 63
Hydrangea macrophylla var. *angusta*
    63
Hydrangea sikokiana 56
*Hydrangeo sikokiani-Pterocaryetum
    rhoifoliae* 56, 58
Hydrangeo-Eupteleetum 66
Hydrangeo-Pterocaryetum 59
hydrochory 208, 214, 217–219, 321
hydrogeomorphic processes 139, 215
hydrological condition 182, 183,
    185, 186, 188
hypertrophic stem growth 243
hypertrophied lenticel 240, 243
hypogeal cotyledon 109, 115
hypogeal germination 109
hyporheic water 199, 321

Impatiens hypophylla 57
Impatiens noli-tangere 275
indirect interaction 118

infrequent species 128, 129, 131–134
intra-order 51, 55, 68
inundated 187, 209, 293
invasive alien plant 187
Isopyro-Fraxinetum spaethianae 58
Isopyrum stoloniferm 58
Izu Peninsula 58

Japanese archipelago 8, 49, 52, 53,
    57, 64, 68, 150
Japanese horsechestnut 97, 107
Juglandaceae 51

Kanumazawa Riparian Research
    Forest 108, 251
Kii Mountains 58
Kii Peninsula 56–58
Kirengeshoma palmata 57
Köppen Cfa 64
Köppen Df 63
Korean Peninsula 53, 68, 282
Kuromatsunai lowland 64
Kushiro Mire 4, 41, 238

Lagerstroemia subcostata var. *fauriei*
    301, 324
Lamium humile 56
landform type 267, 268, 272
landscape 3
landslide 9, 35, 268, 315
landslide slope 22, 34, 271, 302
Laportea macrostachya 55, 275
larderhoard cach 111
large seed 81, 82, 107, 108, 169
large seed mass 115
large seed size 115, 116
Larix kaempferi 155
late successional species 150, 161,
    162, 319
lectotype 51, 54, 56, 58, 59, 61, 63
Leucosceptrum japonicum 272
Leucosceptrum japonicum f.
    barbinerve 59

Leucosceptrum stellipilum 58
Leucosceptrum stellipilum var.
    tosaense 57, 272
life table response experiments
    (LTRE) 96
life-history strategies 216, 218, 292,
    313, 316, 318, 320
life-history trade-off 82, 317
light environment 84, 232
long-distance dispersal 112, 119, 318
long-term monitoring 265
lower convex break of slope 16, 17
lower sideslope 16–18, 20, 21, 23,
    24, 26–28, 126, 268, 269
low-moor peat soil 229
LTRE contribution 103

Machilus thunbergii 124, 317
mammal 107, 116, 206
management objective 307
mass-movements 15
mast year 81, 100, 317
masting 119, 212, 308
matrix analyses 91
Matteuccia struthiopteris 63
meandering channels 41, 151, 153
Meehania urticifolia 52, 55, 79, 272,
    275
micro-landform 16, 123, 269–273,
    308
microtopographic scale 169
microtopography 57, 65, 80, 107,
    169, 251, 255, 304, 316
mire 41, 232, 238
Mitella pauciflora 56
mixed Maple-Ash forest community
    67
montane belt 51, 59
moorland 41
morphological changes 241–244
Mt. Fuji 58
mudflow 39
multiple dam 325
mutualism 117
mutualistic seed dispersers 108, 117

mycorrhizal association 199

N-1-naphtylphthalamic acid, (NPA)
    244
Nakatsugawa River 269
naphthaleneacetic acid (NAA) 244
natural disturbance regime 43, 91,
    92, 313
natural disturbances 9, 35, 76, 91,
    92, 293, 313
natural levee 33, 41, 66, 225–228,
    230
new landslide site 273
niche differentiation 262
niche partitioning 76, 87, 95, 317
nutrient resorption 196–198

O₂ deficiency 237
old landslide slope 273
Ooyamazawa Riparian Forest 6, 77,
    83, 268
organic matter content 142, 272
Oshima Peninsula 51, 64

Pacific climatic division 58, 59
Pacific distribution type 61
Pacific species 52
Pacific type 52, 60–62
pan mixed forest 64
particle size distribution 142, 163
peak flood 215, 216
peat soil 229, 230
Peracarpa carnosa var. *circaeoides*
    272
Persicaria debilis 61, 275
Philadelphus satsumi 56
photosynthesis 238, 241
Phyllitido-Aceretum 68
phytosociological system 53
Picea glenii 39
Picea jezoensis 41, 64, 140, 145,
    146, 155
pioneer scrub community 65, 66

pioneer species 156
pioneer tree 161, 194, 317, 318
Poisson loglinear model 301
Polysticho-Aesculetum 63
Polysticho-Aesculetum turbinatae
    61, 62
Polysticho-Pterocaryetum 61, 63
Polystichum braunii 65
Polystichum retroso-paleaceum 52,
    62, 63
Polystichum retroso-paleaceum var.
    ovato-paleaceum 52, 61, 62
population dynamics 91, 99, 116,
    294, 302
population growth rate 96–99
population growth rate λ 96
Populus maximowiczii 140, 146,
    155, 172
potential habitat 300, 303
primary succession 66
productivity 194, 195, 230, 262, 299,
    307
prospective analysis 101–103
Pterocarya forest 50, 51
Pterocarya rhoifolia 37, 51, 60, 62,
    63, 76, 79, 93, 126, 155, 252,
    269, 286, 316
Pterocaryion 66
Pterocaryion rhoifoliae 50, 56, 65,
    68
Pterostyrax hispida 61, 62, 79

*Q*uercetea mongolicae 51, 53, 54, 68
Querco monglicae-Fagetea crenatae
    53
Querco-Fagetea 68
Quercus 112, 197
Quercus crispula 64, 95, 140, 252,
    255, 258, 273, 287
Quercus gilva 124, 130, 317
Quercus hondae 303, 324

*R*abdosia shikokiana 63
ravine Ash-Maple forest 68

ravine *Zelkova* forest 65, 67, 68
ravine *Zelkova serrata* forest 65
recaching 110
reed marsh 67
regeneration 10, 41, 76, 87, 107, 114,
    161, 165, 171, 225, 292, 321
regeneration traits 165, 171
regional flora 132, 134
regional scale 166, 171
rehabilitation 3, 324, 326
relative dominance 144, 290
relative elevation 169, 209
relic species 155, 281
reproductive strategy 76, 81, 218,
    292
resorption efficiency 197, 198
resprout 115
restoration 3, 300, 313, 324, 326
retrieval 110
retrospective analysis 103
riparian ecosystem 3, 9, 186, 253,
    313
riparian ecosystem management 325
riparian forest 3, 15, 40, 49, 50, 76,
    91, 107, 123, 139, 153, 165, 194,
    251, 269, 281, 299, 313
riparian habitat 9, 91, 92, 107, 119,
    160, 166, 192, 206, 253, 281, 299
riparian management zones 325
riparian zone 9
risk spreading 115
river discharge 34, 140, 166, 167
river management 3, 295
riverbed 33, 127, 139, 153, 177, 185,
    214, 285
Robinia pseudoacacia 325
rodent 107
rodent population 119
Rubus pectinellus 63

Salicetea sachalinensis 67
Salix 165, 177, 191, 281
Salix arbutifolia 41, 153, 172, 320,
    323

Salix cardiophylla var. *urbaniana*
    40, 140, 155, 172, 283, 318, 320
Salix chaenomeloides 179, 180, 193,
    205, 320
Salix dolichostyla 167, 283, 319, 320
Salix gracilistyla 173, 179, 191, 192,
    208, 292, 320
Salix hukaoana 281, 323
Salix jessoensis 167
Salix jessoensis subsp. *serissaefolia*
    319
Salix miyabeana 167, 320
Salix miyabeana subsp. *gymnolepis*
    319
Salix pet-susu 166
Salix riparian grove 67
Salix rorida 155, 166, 282, 320
Salix sachalinensis 167
Salix schwerinii 166, 320
Salix subfragilis 167
Salix triandra 167, 321
Salix udensis 40, 166, 282, 295, 320
Sasamorpha borealis 270
Sasamorpho-Fagion 52
Saso-Fagetea 53, 55
Saso-Fagetea crenatae 50
Saso-Fagion 52
scatterhoard cach 111
Scopolia japonica 60
Scutellaria pekinensis var. *ussuriensis*
    65
Sea of Japan element 63
Sea of Japan side 52, 63, 285
Sea of Japan species 52
Sea of Japan type 52, 62
seasonality 165, 167
sediment 15, 31, 52, 127, 141, 160,
    177, 191, 226, 255, 302, 316
sediment deposition 255, 320
sediment dynamics 31
sediment transport 39
sedimentation 26, 41, 49, 76, 125,
    168, 214, 271
seed 81, 107, 165, 213, 232, 292,
    304

seed dispersal   81, 107, 111, 159, 167, 169, 177, 182, 205, 206, 281, 301, 304
seed dispersal curve   112
seed disperser   111, 117, 169
seed fate   109
seed germination   213, 321
seed hoarding   107
seed predator/disperser   118
seed reserve   107, 115
seed shadow   112, 117
seed/seedling predators   108
seedfall   109, 112
seedling   82, 109, 160, 180, 316
seedling bank   82, 119
seedling emergence   83, 109, 112
seedling establishment   82, 102, 107, 160, 169, 215
seedling growth traits   160
seedling morphology   115
seedling shadow   112
seedling survival   114, 116, 214
seed-shadow enlargement   117
semi-climatic division of the Sea of Japan   62, 63
semi-endemic woody species   52
semi-natural forest   300
shade-tolerant species   87, 150
shallow landslide   16–19, 23–25, 27, 28, 144
shallow slide   18–20, 24–28
shoot clipping   114
sink of diversity   264
site stability   143, 262
slope-failure   36
slope-foot Maple-Ash forest   68
slump   37
snow depth   254, 285
snow flood   140, 150
snowmelt   31, 94, 143, 167, 168, 281, 292, 318
Sohayaki element   57, 58
Sohayaki region   57
soil creep   17, 23
soil disturbance   270
soil flooding   237, 323

soil redox potential   238
soil texture   169, 173
Sorbus alnifolia   53
spatial distribution   39, 113, 118, 132
spatio-temporal variations   96
species diversity   124, 127, 130, 131, 149, 251, 264, 267–271, 291, 323
species pool   253
species richness   128, 148, 251, 267, 269, 270
species-area curve   131, 251
spring ephemeral   273
sprout   83, 192, 231, 232
sprouting   83, 141, 146, 232, 317
stabilization of river-system   186, 322
stable period   148
stage-classified projection matrix   97
Stellaria diversiflora   52
Stewartia monadelpha   61
stomatal conductance   238, 241
stream gradient   125, 127, 133, 134
sub-boreal   63, 67
submerged condition   145, 180
submergence tolerance   145
substratum   15, 141, 169–171, 320
survival ratio   85, 316
swamp forest   65, 225–227, 322
Symplocos coreana   56
synchronous regeneration   79
synonym   51, 56, 58, 59, 61, 63
syntaxonomic system   49, 50, 68, 69
Syringa reticulate   65

talus   20, 52, 268, 271–273
Tanzawa Mountains   58
taproot   160
temperate zone   49, 67, 69, 230
temporal change of bar area   187
terrace of debris flow   271
terrace scarp   269, 273
Tertiary Period   69
thinning   324
Tilia japonica   64
Tilio-Acerion   68
toeslope   141–144

Toisusu urbaniana 40, 140, 155, 172, 283
tolerance 86, 115, 161, 180, 192, 226
topographical change of riverbed 185
Trigonotido brevipedis-Fraxinetum platypodae 61
Trigonotis brevipes 62
two-storied forests 149
typhoon 9, 31, 49, 76, 85, 96, 101, 167, 178, 316
typhoon-induced storm 150

Ulmion davidianae 54, 65, 66, 68
Ulmo laciniatae-Cercidiphylletum japonici 65
Ulmus davidiana var. *japonica* 64, 124, 155, 225, 233, 286, 319
Ulmus forest 51
Ulmus laciniata 41, 51, 55, 64
upper sideslope 16–18, 20–24, 27, 268, 269

valley 15, 32, 76, 92, 126, 144, 153, 171
valley bottom 15, 49, 51, 126, 229, 268, 269, 285
valley bottom plain 227, 229, 230
valley bottom *Pterocarya* forest 51, 54, 55, 65–68
valley-bottom forest 49–53, 56, 61, 63, 64, 67
valley-bottom *Fraxinus-Pterocarya* forest 60
valley-bottom *Ulmus-Cercidiphyllum* forest 64
vegetation class group 51
vegetation-geographic unit 53, 54, 69
vegetative growth 194
vegetative reproduction 41, 273–275
Veronica cana var. *miqueliana* 58

Viburnum plicatum var. *glabrum* 63
Viola shikokiana 59, 272
Viola vaginata 62
volcanic eruption 9, 38
vole 118

warm-temperate region 124, 177, 299, 300
warm-temperate zone 66, 67, 69, 133
water content 142
water stress 84, 85, 170, 253
water table fluctuation 227
water-dispersed species 145
Weigela-Alnus pioneer scrub community 66
Weigelo-Alnetalia firmae 66
wetland 5, 10, 225, 237, 322
wetland forest 230, 238
wetland forest element 65
wetland *Ulmus davidiana* var. *japonica* forests 65
wetland *Ulmus* forest 51, 66, 67
willow community 177
wind-dispersed species 145
within-quadrat diversity 269
within-unit diversity 269
wood production 242

Yakushima Island 301
year effects 100
Yoshino River 166, 178, 205

Zelkova 58, 65
Zelkova serrata 58, 62, 65
Zelkovenion 66, 68
Zelkovetalia 66
Zelkovion 66, 68
zonal vegetation 49
zonation 169

DATE DUE

TN: 4531184

Pieces: **1**

ILL: 180933973

SFR 09/18/17